ALSO BY MICHAEL GRUNWALD

The Swamp

The New New Deal

WE ARE EATING THE EARTH

The Race to FIX OUR FOOD SYSTEM and SAVE OUR CLIMATE

MICHAEL GRUNWALD

SIMON & SCHUSTER

NEW YORK AMSTERDAM/ANTWERP LONDON
TORONTO SYDNEY/MELBOURNE NEW DELHI

Simon & Schuster
1230 Avenue of the Americas
New York, NY 10020

For more than 100 years, Simon & Schuster has championed authors and the stories they create. By respecting the copyright of an author's intellectual property, you enable Simon & Schuster and the author to continue publishing exceptional books for years to come. We thank you for supporting the author's copyright by purchasing an authorized edition of this book.

No amount of this book may be reproduced or stored in any format, nor may it be uploaded to any website, database, language-learning model, or other repository, retrieval, or artificial intelligence system without express permission. All rights reserved. Inquiries may be directed to Simon & Schuster, 1230 Avenue of the Americas, New York, NY 10020 or permissions@simonandschuster.com.

Copyright © 2025 by Michael Grunwald

All rights reserved, including the right to reproduce this book or portions thereof in any form whatsoever. For information, address Simon & Schuster Subsidiary Rights Department, 1230 Avenue of the Americas, New York, NY 10020.

First Simon & Schuster hardcover edition July 2025

SIMON & SCHUSTER and colophon are registered trademarks of Simon & Schuster, LLC

Simon & Schuster strongly believes in freedom of expression and stands against censorship in all its forms. For more information, visit BooksBelong.com.

For information about special discounts for bulk purchases, please contact Simon & Schuster Special Sales at 1-866-506-1949 or business@simonandschuster.com.

The Simon & Schuster Speakers Bureau can bring authors to your live event. For more information or to book an event, contact the Simon & Schuster Speakers Bureau at 1-866-248-3049 or visit our website at www.simonspeakers.com.

Interior design by Ruth Lee-Mui

Manufactured in the United States of America

1 3 5 7 9 10 8 6 4 2

Library of Congress Cataloging-in-Publication Data has been applied for.

ISBN 978-1-9821-6007-4
ISBN 978-1-9821-6011-1 (ebook)

For Max and Lina,

the best reasons to save the earth

CONTENTS

	Introduction: The Land Problem	1
One	The Guy Who Figures Stuff Out	21
Two	Land Is Not Free	49
Three	Weird Science	75
Four	Destroying the Climate in Order to Save It	95
Five	The Menu	119
Six	It's the Food That Needs to Change	145
Seven	The Fake Meat Hype Cycle	178
Eight	The Soil Fantasy	211
Nine	More Beef, Less Land	239
Ten	More Crops, Less Land	263
Eleven	How to Save the World	289
	Epilogue: Can We Do This?	323
	Acknowledgments	335
	Notes	339
	Image Credits	353
	Index	355

WE ARE EATING THE EARTH

INTRODUCTION
THE LAND PROBLEM

FANTASIES AND REALITIES

Imagine we did it. Imagine we quit fossil fuels.

We'd have to retrofit our entire Ozymandian blob of a global economy to run on clean energy instead of ancient hydrocarbons—one billion vehicles, two billion homes, every school, mall, skyscraper, data server, airport, seaport, factory, and cryptocurrency on Earth. No more gas stoves, gas stations, or gas-fired power plants. No more petrochemicals, petrostates, or petroleum jelly. Somehow, we'd have to shut down international energy conglomerates that rake in trillions of dollars every year extracting the fossilized remains of the Carboniferous Period.

Even as a thought experiment, it's hard to imagine. Fossil energy is so ubiquitous, so useful, so entrenched. Donald Trump has been elected twice to lead the free world while ridiculing the idea that we should even consider producing or consuming less of it. Going fossil-free is one of those idle fantasies that seem absurd when you think about them for a second, like cartwheeling across the country or building a city out of taffy.

The thing is, it's not really an optional fantasy. The science is clear:

Climate change is real, it's us, it's here, it's very bad, and fossil fuels are responsible for about two-thirds of it.

There hasn't been this much carbon dioxide in the atmosphere in three million years. The earth hasn't heated up this quickly in 485 million years. It has warmed by 1.2 degrees Celsius (2.2 degrees Fahrenheit) since the Industrial Revolution, and that blip in temperature is already weirding the weather, accelerating vicious wildfires in California, Australia, and even Siberia; unprecedented ice melts in Greenland and Antarctica; and biblical swarms of locusts in India and Kenya. The Weather Channel airs apocalypse on loop: a freakish heat wave in Phoenix, two feet of rain in Fort Lauderdale, a mega-drought in the Horn of Africa, a third of Pakistan underwater.

This isn't a thought experiment. It's a real experiment, a test of what happens when we swaddle our planet in the same heat-trapping greenhouse gases that make Venus uninhabitable. The locusts are just one hint it's not going so well.

As I write these words at the end of 2024, the hottest year ever recorded, the last 10 years were the 10 hottest years ever recorded. Climate change is already contributing to civil wars and mass migrations, the degradation of most of the world's coral reefs and the disappearance of most of Glacier National Park's glaciers. We've emitted more heat-trapping greenhouse gases since the Kyoto Protocol first aimed to limit them in 1997 than we emitted in all human history before Kyoto. And the atmosphere is like a bathtub: If we manage to reduce our flow of carbon, it won't fill as fast, but the level will keep rising until we stop the flow and start draining the tub.

Scientists give us almost no chance to limit warming to 1.5 degrees Celsius, the official goal of the Paris climate accord, and little chance to stay within 2 degrees, unofficial shorthand for climate chaos. We're on course for nearly 3 degrees. The UN's Intergovernmental Panel on Climate Change, the global authority on global warming, warns that we're careening toward a variety of irreversible tipping points, like the melting of the Arctic permafrost and the collapse of the Gulf Stream. Activists fear that President Trump's fossil-fueled return to power could create a political tipping point.

The world is getting better in so many ways—less poverty, more literacy, amazing technologies—but our self-interested, now-focused species often seems ill-equipped to deal with a self-inflicted, long-term threat like climate change. The whole subject can get depressing, like we're confirming the Terminator's cinematic assessment of humanity: "It is in your nature to destroy yourselves." Not only do three-fourths of young people report "climate anxiety" about their futures in a carbon-whacked world, there's an epidemic of climate fatalism, the belief we're so inescapably screwed that nothing we do will matter. And there's still plenty of climate denial—not just the Trump kind that calls global warming a hoax, but the softer kind that prefers not to dwell on its implications for our planetary home.

After all, Earth is still a nice home for most of us, with breathable air, pizza, and reliable Wi-Fi. Screechy Armageddon rhetoric can be a buzzkill for Earthlings who aren't on #ClimateStrike with Greta Thunberg. Alarmist warnings about future devastation, and even alarming news about today, can become background noise. The existential dangers can seem remote, the necessary changes hopelessly unrealistic. I'm writing this book at home in low-lying Miami, where the heat index sweltered over 100 degrees for a record 46 straight days last summer, and even I tune out gloom and doom about its possible future as Atlantis.

But as you've probably heard, not all the news has been bad. Some of those hopelessly unrealistic changes have started to happen. The fossil-free fantasy has gotten a bit less absurd. We're not inescapably screwed.

We made almost no progress toward quitting fossil fuels before 2010, which is why so many climate books have read like soliloquies by Debbie Downer. But we're now in the midst of a clean energy revolution. Two-thirds of America's coal plants have shut down since 2010, while wind power quintupled, solar power expanded 60-fold, and our electric vehicle fleet grew 5,000-fold. Europe is shifting even faster, while China is adding more renewable energy than the rest of the world combined.

Fossil fuels still energize most of the planet, so the transition away from them will take years—and it really matters how many years. But the global transition has begun, and Trump can't stop it. Most nations and corporations have made net-zero commitments. Global transportation is trending electric, and most new electricity is zero-emissions electricity, because clean now routinely outcompetes dirty. There's even progress in "hard-to-abate sectors" like steel and cement. A decade ago, the IPCC climate panel's business-as-usual scenario envisioned 4.3 degrees Celsius of warming by 2100, a recipe for 30 feet of sea-level rise, but that's not business as usual today. Thunberg and her anxious generation of fossil fighters are winning, even if they think they're losing, even if they're not winning fast enough. The rise of carbon-free energy is yet another way the world is getting better.

Radical change always seems unimaginable before it happens. It wouldn't be radical if it didn't. But we can now see how the fossil fuel story ends, even if we don't know when it ends.

As you've probably guessed, there's more bad news, too. And while this isn't another pure Debbie Downer book, it is about that additional bad news. Remember, the fossil fuel story is only two-thirds of the climate story. Even if the fantasy becomes reality, even if we wean ourselves off fossil fuels, we'll still find ourselves slouching toward disaster unless we deal with the rest of the story.

The rest of the story is just as consequential and connected to our daily lives, but the world has been much slower to recognize it and grapple with it. And we have no idea how or when it ends.

THE LAND PROBLEM

The rest of the climate story, and the rest of this book, is about the food we eat, the farms that produce it, and the fight to feed the world without frying the world. It's also about the forests and other natural landscapes that reduce climate change by absorbing carbon dioxide from the atmosphere, when they don't get cleared and converted into agricultural landscapes.

Which is to say this is a book about land. The one-third of our climate problem that isn't about fossil fuels is mostly about land, and so are many of our other nastiest environmental problems. We need to start treating the limited land we've got on the only habitable planet we've found as our most precious resource, because it needs to produce much more food to sustain us and absorb much more carbon to save us. And as Mark Twain said, they're not making more of it.

To avoid the worst climate outcomes, we won't just need to quit fossil fuels that emit carbon. We'll also need to preserve ecosystems that store carbon—and they're endangered by the relentless spread of farms and pastures that already cover two of every five acres of land on Earth. Every farm, even the scenic ones with red barns and rolling hills that artists paint and writers sentimentalize, is a kind of environmental crime scene, a pastoral echo of whatever carbon-absorbing wilderness it once replaced. We need agriculture, because we need food, but it can't keep overrunning the planet, because we can't keep losing a soccer field's worth of tropical forest every six seconds.

Our land problem, it turns out, is mostly an agriculture problem. Our agricultural footprint is already the size of all of Asia plus all of Europe, and the more it expands, the more nature's footprint shrinks, expelling the carbon stored in its soils and vegetation into our overheated atmosphere. The conversion of natural land into farmland and rangeland is also wiping out biodiverse habitats, driving the worst mass extinction since an asteroid wiped out the dinosaurs 66 million years ago. We're clear-cutting and broiling the planet to stuff our faces.

But the march of crops and livestock into the wilderness won't stop just because it needs to stop. We'll have to stop it. And producing enough calories to nourish our growing population without chewing up the carbon sinks that can stabilize our warming climate will be as monumental a challenge as ending oil. If current trends hold, the world's farmers will need to clear at least a dozen more Californias' worth of land to fill nearly 10 billion human bellies by 2050. That would wipe out the Amazon rainforest and other natural carbon storehouses that are not only refuges for wildlife but our best defenses against climate change.

So current trends better not hold.

The IPCC says that to meet the Paris goals, we must end all deforestation by 2030, and also start a global campaign of reforestation—because the best technology we've discovered to drain the carbon from our atmospheric tub is the miraculous 3.5-billion-year-old process of photosynthesis. Not only does it produce the oxygen that keeps us alive, photosynthesis transforms sunlight, water, and carbon dioxide into vegetation that offsets our greenhouse gas emissions as it grows. Thanks to photosynthesis, forests now soak up one-fourth of the carbon we emit, preventing the earth from broiling even faster.

That's why deforestation is such a disaster for the climate as well as the woodlands and woodland creatures it wipes out. It creates a double whammy of more emissions now and fewer emissions avoided in the future. We simply can't decarbonize the atmosphere if we keep vaporizing trees. It's like trying to clean the house while smashing the vacuum cleaner to bits in the living room.

Agriculture is by far the leading driver of deforestation. It's also the leading driver of biodiversity loss, wetland destruction, water shortages, and water pollution. It's a terrible driver. But if we want to stop releasing carbon by displacing nature and start removing carbon by restoring nature, we'll need to overcome some terrible math. We'll need to expand agricultural production about 50 percent by 2050 to feed our hot and hungry planet—the caloric equivalent of a dozen extra Olive Garden breadsticks every day for everyone alive—while shrinking our agricultural footprint.

In other words, we'll need to stop eating the earth, or at least so much of the earth.

We'll need to make more food with less land, so that we don't devour the natural sanctuaries that help prevent climate chaos. We'll need to consume and produce food more sustainably, stop wasting so much of it, and take better care of the ground beneath our feet.

Ultimately, we'll need to leave more of the earth alone, which, historically, is not how we've rolled. We started eating the earth long before we started burning fossil fuels, and our land hunger has been as damaging as our oil addiction.

Scientists like to say we live in the Anthropocene, the age of humans, a new epoch marked by the domination and discombobulation of the earth by *Homo sapiens*. Climate scientists especially like to say that, to highlight how our big-brained species is reconfiguring our planet with climate-warping fossil fuels, altering the nature of nature in a way no smaller-brained species ever has. Our self-immolation is usually associated with modernity—with private jets, Bitcoin mining, and SUVs that weigh more than elephants. A scientific committee set the Anthropocene's official start date around 1950, when oil consumption and atmospheric carbon began to soar. The famous "hockey stick" graph of global temperatures got its nickname from its sharp spike upward after World War II.

In reality, though, humanity has been reshaping the earth ever since the invention of agriculture around 12,000 years ago. The planet only had five million people then—fewer than the Miami area today—but farming sparked the first population boom, up to a billion before the fossil fuel era even began. And early farmers didn't need diesel tractors or toxic agrochemicals to transform their environment. They subdued nature with fire and the axe, converting forests and savannas into farms and pastures that sustained permanent settlements. By 1800, they had already cleared a landscape the size of South America.

In recent years, scientists have documented that early agriculture's liberation of the terrestrial carbon in nature also changed the climate, long before the combustion of the primordial carbon in fossil fuels turbocharged those changes. They've used ice-core data and ancient pollen samples to argue that the clearing of forests for preindustrial farms created enough warming to avert another ice age.

"The ice cores show something unnatural happened," says the father of this Early Anthropocene hypothesis, University of Virginia paleoclimatologist William Ruddiman. "What happened was agriculture and widespread deforestation."

Indigenous people deforested so much of the Americas for agriculture that one study of their tragic depopulation after European contact reached a macabre conclusion: The natural regrowth of forests on their abandoned farmland after their "Great Dying" reabsorbed enough carbon to create global *cooling*. Precolonial Americans are often stereotyped as primitives who lived in harmony with nature, but many were sophisticated farmers who manipulated nature, so much so that their disappearance helped nature reclaim some territory, however briefly.

"There's a modern hubris that says 'Oh, we're so powerful,' so it's sobering that much smaller populations of early humans had so much influence on the environment and climate," said Steve Vavrus, a University of Wisconsin scientist who models the Early Anthropocene. "When you think about the impact of agriculture then, it raises questions about agriculture now."

More expansive agriculture has always meant less expansive nature, which has meant more carbon in the atmosphere. And agriculture has expanded even faster in the fossil fuel era, commandeering 20 million square miles of land, an area nearly 400 times the size of Iowa.

We tend to think of humanity's dominion over the earth as a building story about the spread of cities and towns, highways and driveways, industry and commerce. But it's mostly a farming story. Of all the planet's land that isn't ice or desert, barely 1 percent is developed. Half is cropped or grazed. Urban sprawl is a rounding error compared with agricultural sprawl. Look out the window on a cross-country flight: The land we use to live, learn, work, and play is dwarfed by the land we use to make food.

It's worth remembering that the farmers and ranchers who work all that land empower the rest of us to make different choices.

Because they alchemize sunshine, rain, and soil into nutrition, we can be teachers, healers, influencers, and authors of books exploring agriculture's downsides. Our food doesn't magically materialize on grocery shelves; farmers do a great job of growing it, and we owe them, as they often remind us, a great debt. They do get paid, and often generously subsidized, for what they grow, but in rich countries, they get a much smaller share of the profits than powerful seed, fertilizer, grain-trading, meatpacking, food-processing, retail, and restaurant conglomerates. In developing countries, they're often among the poorest of the poor.

Since one theme of this book is the need to make agriculture more productive so that it can feed us with less land, it's also worth remembering that agriculture in much of the world is already far more productive than it used to be. Modern farmers with GPS-enabled, 500-horsepower tractors can do a lot more than their ancestors did with one-horsepower horses.

They can also use herbicides and pesticides to protect their crops; fertilizers and irrigation systems to supersize their harvests; and cutting-edge breeding, feeding, and veterinary care to get more meat, milk, and eggs out of their livestock. Old MacDonald would marvel at their efficiency.

This is the legacy of the Green Revolution, which began in the 1960s when the legendary American agronomist Norman Borlaug bred a hardy new variety of higher-yielding wheat, and now refers to a long era of innovations ranging from the artificial insemination of cattle to the genetic modification of corn. It was an efficiency revolution that tripled crop and livestock production yields per acre, saving billions of people from malnutrition. Those amped-up yields have also slowed the expansion of farmland, sparing billions of acres of nature from destruction. This idea that making more food per acre reduces the pressure to clear more acres to make food—originally known as "the Borlaug hypothesis," now usually called "sustainable intensification"—is central to this book.

But the Green Revolution didn't end agricultural sprawl. Since 1960, fields and pastures have replaced one-sixth of the Amazon. And while the Green Revolution's intensification has produced incalculable benefits for humanity, supplying most of our daily bread, not all of it has been sustainable. Agriculture now uses 70 percent of our fresh water, depleting aquifers and sucking rivers dry. Its weed killers and fertilizers poison lakes and estuaries, as do the 10 million tons of manure its livestock unload each day. Individually, most farmers and ranchers are proud of their stewardship of their land, but collectively, they're stewarding a mess.

Agriculture also generates one-fourth of all climate-warming emissions, as much as electricity and heating combined. Tractors, crop dusters, and other fossil-fueled farm equipment emit carbon dioxide. Fertilizers and manure release nitrous oxide, a greenhouse gas 300 times more potent than CO_2. Cattle and sheep burp and fart methane that's 80 times as potent as CO_2, and the climate doesn't care how goofy that sounds. And the conversion of natural land into agricultural land does even more damage than all those emissions from agricultural production, releasing as much carbon every year as the entire United States.

Our overall food system now generates about one-third of global emissions, so it's not yet as humongous a climate problem as fossil fuels. But to meet the Paris targets, the world will need to eliminate three-fourths of those food-related emissions by 2050, the equivalent of decarbonizing transportation and shipping. So it's a pretty humongous problem. And we're directing less than 3 percent of the world's climate finance at solving it.

The bottom line is that what we eat and how we grow it will shape our future. We'll have to get more out of the earth without using more of the earth; this book focuses on food, as well as our misguided use of land to grow energy, but farms and forests also provide lumber, paper, cotton, leather, rubber, and more. We'll have to start thinking as urgently about the fresh carbon in trees, grasses, and soils as we think about the prehistoric carbon in coal, oil, and gas.

That means we'll have to make changes as unimaginable as quitting fossil fuels—and we can't quit food. Feeding the world without frying it is another absurd fantasy that eventually needs to become reality.

This carbohydrate problem will be even trickier to solve than the hydrocarbon problem.

As hard as it will be to ditch fossil fuels, we now basically know how: electrify the economy and run it on clean electricity. Some activists question whether we need zero-emissions nuclear plants (yes) or solar farms in tortoise habitat (probably also yes) or ugly transmission lines for renewable power (yes again), but the general path is clear.

It's much less clear how to fix food.

Until recently, there was only dim public awareness of any connection between food and the environment—a vague sense that pesticides and GMOs were bad, while eating local, natural, and organic was good. A backlash was brewing against the excesses of industrial agriculture: factory-farm assembly lines that treat sentient animals like widgets and worse; row-crop monocultures that rely on perennial chemical warfare; fertilizers that choke the Great Lakes with algae and create a dead zone the size of Connecticut in the Gulf of Mexico; intensive production of ultra-processed crap that makes us fat and sick; overuse of antibiotics; an agribusiness sector that spends even more than the fossil fuel sector lobbying politicians; lavish government subsidies; negligible environmental regulation; and the danger that confined livestock operations could spawn avian flus and other deadly pandemics.

But none of that was about the climate.

In the last few years, food and farming has entered the climate chat. President Joe Biden's 2022 climate bill blasted an astounding $20 billion into a new budget item called "climate-smart agriculture," and the 28th annual global climate summit in Dubai in 2023 was the first to devote a day to food. Voices as diverse as Thunberg, Bill Gates, Billie Eilish, Pope Francis, King Charles, and the Indian mystic Sadhguru have all called for transforming the food system to save the climate, while angrier voices keep accusing climate activists of trying to ban cheeseburgers and force everyone to eat bugs. Outright denial of the eating-the-earth problem is becoming more common—"Farming has no material effect on climate change," Elon Musk tweeted, preposterously, in February 2024—because nobody bothers to deny problems that nobody talks about.

Now people are talking.

It's become a cliché that if food waste were a country, it would rank third in the world in emissions. There's been incessant buzz over climate-friendly meat and dairy alternatives—initially focused on plant-based unicorns like Impossible Foods, Beyond Meat, and Oatly, more recently on the future of "cultivated meat" grown from animal cells. There's even more hype over "regenerative agriculture" that aims to stabilize the climate and rejuvenate soils by sequestering carbon in farmland, with support from the UN, environmentalists, farm-to-table chefs, leading philanthropies, documentaries like *Kiss the Ground* and *Carbon Cowboys*, Big Food, Big Ag, and politicians of all stripes. And as food and ag policy has moved to the center of the climate plate, climate investors have moved into food tech and ag tech—not only animal-free animal foods but Uber for tractors, Tinder for cattle, Fitbit for pigs, gene-edited crops, vertical farms that use hardly any land, invisible biotech peels that prevent fruit and vegetables from spoiling, biofertilizers that replace chemicals with microbes, and biopesticides that use the RNA technology behind the coronavirus vaccines to constipate beetles to death.

Still, the world's approach to food and land is at least two decades behind its approach to fossil fuels. The first scientific blueprint for solving the eating-the-earth problem, a 564-page doorstop of a report that hardly anyone read cover to cover, was only published in 2019. Two of the world's most knowledgeable climate analysts, journalist David Roberts and entrepreneur Michael Liebreich, admitted during a 2024 podcast about the heroes and villains of decarbonization that they were both haunted by and uninformed about farming.

"I'm just going to put my hand up and say I have not done enough work on agriculture," Liebreich said.

"This has been my approach, too," Roberts replied.

The fossil fuel story has obvious villains—Big Oil, King Coal, the politicians who do their bidding—but food and land is less clear-cut. It's trendy to demonize "industrial agriculture," and "ruthless efficiency" has become an epithet associated with rapacious corporations squeezing dollars out of the dirt. But when the fate of the climate depends on manufacturing quadrillions more calories, Big Ag's factory farms can help, while the low-yield inefficiency of ancient farmers explains how so few of them did so much damage. It's also trendy to romanticize small family farms where soil is nurtured with love and animals have names rather than numbers, but kinder and gentler agriculture that makes less food per acre can induce more deforestation and emissions.

The inconvenient truth is that it's complicated. Michael Pollan writes beautifully about rustic farmsteads that honor the rhythms of nature, but organic, local, and grass-fed are often worse for the climate than conventional, imported, and feedlot-finished. Fertilizer is a climate killer, because it's made of natural gas and generates twice as many emissions as Germany, but also a climate savior, because it helps farmers grow more food per acre. The efficiency of hated agribusinesses like Cargill, Tyson, and Archer Daniels Midland cuts emissions, while their recent embrace of beloved regenerative practices may increase emissions. Forest protections can be pointless if they shift deforestation to unprotected areas, while boycotts of deforestation-linked commodities like soy and palm oil can backfire if they induce farmers to plant less efficient crops. Even "Paper or plastic?" is a complicated climate question, because paper uses land.

While the fossil fuel problem is now a mostly political story, because we know what we need to do, the eating-the-earth problem is still an analytical story, because we're still figuring out what to do. Fortunately, its most important analyst has a knack for figuring things out.

GRILLING THE PLANET

This book owes its existence to a fact check. I didn't know something I should've known, so I called someone I knew would know.

It was 2018, and I was finishing a magazine story about my new green life with solar panels and an all-electric Chevy Bolt. My thesis was that clean energy was ready to go mainstream, because it was already saving me money. My electric bill was down 90 percent. My gas budget was zero. To emphasize that I wasn't an eco-saint, just an eco-mercenary exploiting the new economics of decarbonization, I threw in a line listing other climate-friendly things I was still too lazy or selfish to do, like line dry my laundry or stop eating meat.

When I reread my draft, though, I tripped over that last bit. People talked about quitting meat as if it were self-evidently climate-friendly, but I realized I had no idea if that was actually true. I had heard that farting cows, a common denialist punch line, were a real problem. I knew that *100 Things You Can Do to Save the Planet*–type lists pushed Meatless Mondays. But I knew nothing about meat beyond its deliciousness. Maybe left-wingers had unfairly tarnished its eco-reputation in solidarity with innocent animals? I didn't want to perpetuate a virtue-signaling myth, even in a throwaway line.

I decided to call Tim Searchinger, a senior research scholar at Princeton and senior fellow at the World Resources Institute think tank in Washington, D.C. I had known Searchinger for years as a brilliant thinker happy to express unpopular thoughts, a relentlessly logical, brutally candid, earnestly persuasive policy wonk with a flair for debunking conventional wisdom. He was passionate about what he believed but obsessive about backing it up, which helped explain his unique résumé: After nearly two decades as an environmental litigator and advocate, he had stopped lawyering and started science-ing; the top-flight journals *Science and Nature* have now published nine of his scientific papers, even though he still has no scientific degree. His critics saw him as an egotistical, motormouthed, Manhattan-raised, Yale Law School–trained mansplainer, and he did have a lot of confidence in his judgment. But he mansplained to men as well as women, and his mansplanations were always evidence-based. I had a lot of confidence in his judgment, too.

It wasn't only because he once gave me a tip about the troubled restoration of the Florida Everglades that led me to move to Miami, write my first book, and meet my wife. I liked that he always told the truth, always did his homework, and often zigged when the herd was zagging. He was a committed enviro who adored trees and drove a Prius, but he was even more committed to facts and science, which often led him into spats with less rigorous advocates—and eventually drove him into academia. He was an unapologetic yucker of consensus yums. The welcome mat on his porch read: "No Dogmas Allowed."

I had last worked closely with Searchinger a decade earlier, when his first scientific paper had exposed farm-grown fuels like corn ethanol as climate disasters masquerading as climate solutions, and I had mined his work for a man-bites-dog *Time* cover story on "The Clean Energy Myth." We hadn't talked in a while, but I figured he'd have a similarly data-driven take on meat, and it occurred to me that an eco-defense of carnivores could be an even juicier man-bites-dog story.

So I called to ask: Is meat really so bad for the climate?

He replied: Yes.

He added: *Duh.*

In public, Searchinger usually tries, with varying degrees of success, to disguise his disdain for the uninformed and illogical. In private, he has an enjoyably raucous tendency to express that disdain with withering mock-fury. He says "duh" a lot. Also "durr," a synonym; "duuuuhhyee," the sound of a schmuck walking into a lamppost; and "un-fucking-believable," to express similar incredulity about laziness and stupidity. He's an intellectual savant with a compulsion to get every answer right, and he's never understood why everyone else isn't. He's an idealistic mensch who yearns to repair the world, but he's perpetually frustrated by all the sloppy thinking and motivated reasoning he encounters on his repair jobs.

He sounded a bit miffed that I had expected an ignorant contrarian take on meat, so I mock-apologized for failing to keep up with the cow-fart literature. Actually, he explained—this time, the "duh" was merely implied—bovine burps emitted much more methane than farts. But what made animal agriculture truly devastating didn't come out of either end of a cow.

"It's land," he said. "Meat uses too much land. Just like ethanol."

Huh. Searchinger's original insight about ethanol, the hidden-in-plain-sight epiphany that rerouted his career into climate research, was that using farmland to grow fuel instead of food would induce the clearing of forests to grow more food. I hadn't stayed in close enough touch with him to know that his work exposing biofuels as a deforestation bomb had led him into a decade-long investigation of the entire food and agriculture sector. I had inadvertently cold-called the leading authority on the eating-the-earth problem.

Meat was bad for the climate, he explained, because we devoted a huge portion of our land to livestock, especially cattle, that produced a small portion of our calories. Our appetite for bacon and wings, and especially burgers and steaks, was obliterating jungles and boiling the planet. Nearly 80 percent of the world's agricultural land was used to feed livestock. Pound for pound, beef generated 50 times more emissions than coal. A nation formed by the world's cattle would rival the Food Waste Republic for number three on the global emissions list.

Searchinger was still fighting to stop biofuels, because farmers were using a Texas-sized land mass to grow them. But livestock already used enough land to cover 50 Texases, and he was convinced the future of humanity depended on stopping its spread. When I called, he was finishing the most important work of his career to date, a comprehensive food and agriculture report intended to alert humanity that we are eating the earth and show us how to stop—the aforementioned 564-page doorstop, *Creating a Sustainable Food Future*.

"If we don't get serious about land, we're FUCKED!" he shrieked.

When Searchinger gets agitated, usually about the folly of people who ought to know better, he's like a pedantic Yosemite Sam, a whirlwind of furious verbiage. His nasal baritone rises a couple of octaves as he reaches his crescendo of denunciation. His rants are always educational, though, and his riff about land hit me like an anvil falling on Wile E. Coyote. I had written a fair amount about climate change, including the most prominent article linking it to inefficient land use. How could I be so ignorant about meat? I knew agriculture used pesticides and fertilizers that polluted the air and the Everglades. Why didn't I know its worst impacts came from using half the planet's habitable land?

Compared with the fossil fuel problem, the eating-the-earth problem was clearly underpublicized and under-understood. It was unfolding in the working lands and wild lands that most of us only flew over or hiked through—although we contributed to it whenever we ate, or used wood, toilet paper, and other products of the land. And unlike the fossil fuel problem, which was getting better much too slowly, the eating-the-earth problem was getting worse.

I know, I know, that's more bad news. Sorry to be the bearer. But every day, the number of people on Earth increases, while the amount of land on Earth does not. Those people are eating more meat, which means more emissions from burps and manure, and more land to grow grass and grain for animals to eat before we eat them. And there's one more complication making the problem even more problematic.

That complication, ironically, infuriatingly, is climate change.

This book isn't about dying reefs, endangered penguins, and other distressing climate impacts. It's about the eating-the-earth problem, and our fledgling efforts to fix it in order to limit those impacts. But some climate impacts are already making the problem harder to fix, by damaging the earth's ability to produce food and store carbon. It's an unnerving doom loop.

Agriculture is riskier than ever in a warmer world, as more intense droughts and floods stress and kill crops and livestock, while pests and diseases invade new regions. The media often cover this with cute trend stories about climate change coming for our beer, wine, coffee, and chocolate, but it's coming for our agriculture, posing a threat, as the 2024 Global Report on Food Crises put it, of "global-scale loss of capacity to grow major staple crops."

Climate change is also a threat to the forest sinks we need to help mitigate it. California's climate-worsened wildfires are creating more emissions than its two million solar rooftops are preventing, while sapping the ability of its forests to suck carbon out of the air in the future. Agriculture has annihilated so much of the Amazon that it's no longer an overall carbon sink, and scientists fear it's approaching a tipping point that could transform it into an arid savanna.

Lower farm yields, higher forest emissions, and disappearing carbon sinks are the exact opposite of what's needed to solve the eating-the-earth problem. While he was researching his report, Searchinger began to consider the problem so enormous, and the trajectory so troubling, that he joked to colleagues they ought to euthanize their children.

But all enormous problems are hard to solve. Otherwise, they wouldn't have gotten enormous. And the Terminator was too harsh about our self-destructive tendencies. We do have some advantages over smaller-brained beasts, especially our capacity to learn and change.

What sets human beings apart is our ability to grasp what doing the same thing and expecting different results is the definition of. And we are capable, as rafting guides and self-help gurus remind us, of participating in our own rescue.

The really good news is that remarkable people are working on the eating-the-earth problem, and their work can be antidotes to climate fatalism.

They're reengineering soybean plants to grow dairy proteins, developing feed additives to help cattle burp less methane, restoring carbon-rich peatlands, upcycling food waste into snacks, and even improving the miracle of photosynthesis. They're growing superefficient salmon in indoor tanks, superefficient trees on barren land, and superefficient crops genetically manipulated to survive droughts and floods. And they're making all kinds of meat and dairy substitutes—not only plant-based burgers and milks but deli turkey slices fermented from fungi, cream cheese brewed from bacteria from the hot springs under Yellowstone National Park, pet food cultivated from cells from chicken eggs, and bacon grown from cells from living pigs.

While many of those solutions can address pieces of the problem, Searchinger thinks about the entire problem, and he'll be the rafting guide for this book's float through the eating-the-earth mess. The next four chapters will trace his intellectual journey, guided by his North Star of better land use. He initially figures out that using land to grow bioenergy—first the brewing of crops into fuel, then the burning of wood for electricity—is a dire threat to nature and the climate, an ineffectual fix to the fossil fuel problem that could make food and land unfixable. From 2007 until 2019, as policymakers start taking climate change seriously and embracing bioenergy as a solution, he helps lead the global fight to prevent it from overwhelming the earth, trying to stop bad science and bad politics from creating bad incentives and bad outcomes.

While he's fighting to stop the use of land to grow energy, he's also learning that food production is an even worse climate threat for similar land-use reasons. Chapter 5 is all about *Creating a Sustainable Food Future*, his dense blueprint for how better approaches to food and farming could defuse that threat. And the rest of this book follows how others have tried to follow that blueprint since 2019, focusing on the three most prominent kinds of solutions: meat and dairy substitutes designed to rein in animal agriculture, regenerative farming and grazing designed to sequester carbon in farm soils, and efforts to boost crop and livestock yields in order to grow more food with less land. It also follows Searchinger's own efforts to figure out which solutions are making headway, and how to make policy change as well as technological change happen faster.

I spent those five years learning about these issues as a journalist, and this is a book of nonfiction. Every fact and quote comes from my research and reporting. But there's also plenty of analysis. If you think climate change is a hoax, you're wrong, so you'll hate my analysis. You also might get annoyed if you think the world simply needs to end capitalism, factory farming, or food waste to fix the land sector, or we all simply need to go vegan. The world isn't that simple. I've quit beef to reduce my own climate hypocrisy, but another theme of this book is that changing the consumption patterns of eight billion people and the production patterns of 600 million farms will be extremely complex. Food is the ultimate hard-to-abate sector.

While fossil fuels are now mostly a political test, the eating-the-earth problem is still an intellectual test. It's a puzzle. This book is about the struggle to fix it, or shrink it, or at least understand it. It's about what's been learned, what's being done, and what lies ahead. Don't euthanize your children. The fossil fuel story has proved that change is possible, and this is another story about change, about thinking hard and working hard to try to make things better.

It begins, though, with a realization that things can change for the worse.

ONE
THE GUY WHO FIGURES STUFF OUT

CLEAR, SIMPLE, AND WRONG

Something felt off.

Tim Searchinger lacked the proper credentials to say exactly what was off that day in the spring of 2003. He was a lawyer, not a scientist or economist. He was reading a complex technical paper on an unfamiliar topic, produced by well-respected researchers at the world-renowned Argonne National Laboratory. Sitting at his cluttered desk in the Environmental Defense Fund's sixth-floor offices in Washington, D.C., overlooking the famous back entrance to the Hilton where President Ronald Reagan was shot, he just had a sense the paper didn't add up.

Searchinger tended to distrust new information until he could study it to a pulp. He never assumed consensus views were correct, conventional wisdom was wise, or sophisticated-looking scientific analyses reflected reality. He questioned everything, so his unease that day didn't feel particularly unusual. He had no inkling it would eventually lead him to a new profession—and the world to a new way of thinking about food, farming, land use, and climate change.

The Argonne study analyzed whether fueling cars with corn ethanol rather than gasoline reduced greenhouse gas emissions, which did not seem like a particularly urgent question in 2003. And Searchinger was a wetlands guy fighting to save the streams and swamps that provide kitchens and nurseries for fish and wildlife, not an energy-and-climate guy trying to keep carbon out of the atmosphere. So it was a bit odd that he would slog through such an obscure report.

But not too odd.

He was also an agriculture guy, because farms were the main threat to the wetlands he wanted to protect. And he was above all a details guy, a data sponge willing to soak up minutiae far too technical for less obsessive laymen. The revelatory stuff usually seemed to be hidden in arcane modeling assumptions and other fine print. He was a compulsive reader of boring papers, all the way through the footnotes, and he had learned from his uphill legal and political battles that knowledge could be a powerful weapon against money. He always did the reading, and his burden in life was that others didn't.

Ethanol was just his latest uphill battle.

It was the most common form of alcohol, the fermented magic in beer, wine, and liquor. It was also a functional automotive fuel; it had powered the first internal combustion engine, and Henry Ford once called it the future of transportation. Gasoline turned out to be more efficient and better for engines, so ethanol mostly ended up in solvents and booze. But in the 1970s, ethanol distilled from corn—the "field corn," or maize, grown by grain farmers, not the "sweet corn" you eat off the cob—had carved out a small role as an additive in US fuel markets.

That was the start of a twisted political love story. Farm interests, whose outsized political influence dated back to America's origins as an agrarian nation, seized on ethanol as a new government gravy train. The U.S. Department of Agriculture, founded under President Abraham Lincoln for the express purpose of supporting farmers, backed ethanol as enthusiastically as it backed farm subsidies, farm loans, and other federal farm aid. And presidential candidates sucked up to farm interests so reliably that a *West Wing* episode lampooned the quadrennial tradition of ethanol pandering before the Iowa caucus, as the fictional future president Matt Santos considered denouncing subsidies he considered stupid and wasteful.

"You come out against ethanol, you're dead meat," an aide warned Santos. "Bambi would have a better shot at getting elected president of the NRA than you'll have of getting a single vote in this caucus."

The Midwestern grain interests behind ethanol did have serious political swat. The top ethanol producer was agribusiness giant Archer Daniels Midland, whose former CEO helped finance the Watergate burglary, and whose reputation as an all-powerful force of corporate darkness would soon be satirized in *The Informant!* The U.S. industry owed its existence to a lavish tax break for domestic ethanol and a punitive tariff on foreign ethanol, both of which owed their existence to Big Ag lobbyists. The corn the industry distilled into fuel was also subsidized through "loan deficiency payments," "counter-cyclical payments," and a slew of other bureaucratically differentiated programs that all diverted taxpayer dollars into farmer wallets. The farm lobby usually got what it wanted out of Washington—not only subsidies and tax breaks, but exemptions from wetlands protections, pollution limits, and other regulations. Even the federal rule limiting the hours truckers could drive had a carve-out for agricultural deliveries.

Still, barely 1 percent of America's fuel was ethanol, and barely 1 percent of America's corn became ethanol. The issue wasn't on Searchinger's radar until Big Ag began pushing an ethanol mandate, and he began worrying it could become the corn industry's new growth engine.

His concern had nothing to do with climate change, because that wasn't on his radar, either. It wasn't yet a front-burner issue in Washington, and he knew no more about it than the average newspaper reader. He was focused on preserving what was left of nature in farm country, and preventing polluted farm runoff from fouling rivers and streams. More ethanol would mean more cornfields, more pollution, and more drainage of the Midwest's few remaining wetlands.

Most Americans seemed to think the middle of the country was somehow ordained to be amber waves of grain—he used to think so, too—but he always kept in mind that it had once been a vibrant landscape of tallgrass prairies and forested swamps, a temperate-zone Serengeti with spectacularly diverse plant communities and birds that darkened the sky. Washington had accelerated the near-total obliteration of that ecosystem, with incentives as well as rhetoric encouraging farmers to grow crops from "fence row to fence row," and ethanol seemed like the latest excuse to complete Middle America's metamorphosis into an uninterrupted cornfield. Searchinger was on the prowl for science he could use to prevent that, so when he heard about the Argonne paper, in those days before studies were routinely posted online, he called the lead author, a Chinese-born environmental scientist named Michael Wang, and asked him to FedEx it.

Unfortunately, Wang's team had calculated that ethanol generated 20 percent fewer greenhouse gases than gasoline, a modest but measurable improvement. Wang had helped pioneer the "life-cycle analyses" that were becoming standard in the field, and the emissions model known as GREET that he developed at Argonne was considered state-of-the-art, while Searchinger had never even read a climate study. So he didn't really have standing to object.

But he did know models could mislead, because one of his professional obsessions was exposing how the U.S. Army Corps of Engineers cooked the books of cost-benefit analyses to justify its own ridiculously destructive water projects. He had learned from Army Corps documents how economic and scientific models could be structured and twisted to reach convenient conclusions, how garbage in plus garbage assumptions could produce garbage out. And when he started thumbing through the ethanol study, he had familiar bad vibes.

Wang had found that drilling oil and refining it into gasoline emitted much *fewer* greenhouse gases than planting, fertilizing, and harvesting corn and refining it into ethanol. Initially, Searchinger was confused: If the agro-industrial complex was twice as carbon-intensive as the petro-industrial complex, why would ethanol have a *smaller* carbon footprint?

The study's answer was that cornfields, unlike oil wells, were carbon sinks. The Argonne team assumed that growing corn on a farm offset the tailpipe emissions from burning corn in an engine, because cornstalks sucked carbon out of the atmosphere through photosynthesis. The climate case for farm-grown fuels was that ethanol merely recycled carbon, while gasoline liberated carbon that had been buried for eons. It made sense that ethanol, a renewable fuel, would be climate-friendlier than gasoline, a fossil fuel. "Renewable" sounded clean and green, while "fossil" evoked zombies coming back from the dead to destroy the earth.

Searchinger's spidey-sense kept tingling, though. His father, another question-everything guy, liked to quote H. L. Mencken: "For every complex human problem, there's a solution that's clear, simple and wrong." That's what ethanol felt like. And the more he thought about the study, the less he understood its conclusions.

Yes, corn soaked up carbon as it grew. But it soaked up just as much carbon whether it was grown for fuel or food! Why would growing corn for ethanol and burning it in an engine be any climate-friendlier than growing that same corn for food and burning an equivalent amount of gasoline in an engine? The carbon absorbed in the field wouldn't change; neither would the carbon emitted from the car. If the only difference was that producing ethanol emitted much more carbon than producing gasoline, where were ethanol's benefits?

That led back to his original concern: If more corn was diverted from food to fuel, how would the lost food be replaced? Presumably, Midwest farmers would plant more corn, converting more wetlands into farmland that would get blasted with more chemicals. Again, he wasn't focused on the climate impact, just the environmental impact of losing habitat and increasing pollution. But he had a hunch the Argonne researchers and their spiffy analytical tools were also understating the climate costs of using grain to fuel our cars instead of ourselves.

Searchinger loved figuring things out, and he was on the verge of figuring something out that would transform climate analysis.

Uncharacteristically, though, he lost interest.

For one thing, it became clear that climate would be irrelevant to the debate over the proposed "Renewable Fuels Standard." With America at war in Iraq, ethanol's boosters were touting the mandate as a win-win that would reduce reliance on Middle Eastern oil while propping up demand for Midwestern corn. They weren't touting it as a climate solution, because

Washington wasn't looking for climate solutions. The Senate had unanimously rejected the Kyoto Protocol a few years earlier, and Congress had ignored the issue ever since.

It also became clear the biofuels debate would be another charade controlled by farm interests and farm-friendly politicians. President George W. Bush had genuflected to ethanol in Iowa, as future presidents always do. (Even *The West Wing*'s Santos caved.) Senate Democratic Leader Tom Daschle of South Dakota, whose top aide later became an ethanol lobbyist, and Republican House Speaker Dennis Hastert of Illinois—who also became an ethanol lobbyist, before going to jail in a child molestation scandal—were both farm-state biofuels boosters.

Searchinger did try to lobby some non-Midwestern politicians to oppose the mandate, arguing it would punish their constituents at the pump to subsidize out-of-state agribusinesses. But even an aide to Democratic Senator Jon Corzine, a former Wall Street titan from corn-free New Jersey, sheepishly admitted his boss couldn't buck the ethanol lobby, because he might need Iowans someday.

Come on, Searchinger pleaded, the guy who ran Goldman Sachs thinks he's running for president?

"Tim, they're *all* running for president," the aide replied.

Searchinger sometimes joked that he was the patron saint of almost-lost causes, because he spent his days failing to save wetlands, failing to stop farms from degrading the environment, and failing to reform the Army Corps. He didn't go looking for uphill battles—he's a generally friendly guy with no particular lust for conflict—but he didn't shy away from them, and as an enviro in ag world, he ended up in a lot of them. Even his victories felt temporary, because defenders of nature have to win again and again to keep wild places wild, while despoilers of nature only have to win once. And unlike campaigns to save the whales or the Grand Canyon, causes that inspired public outrage and sympathetic press, his fine-print fights to limit the damage from American agriculture went mostly unnoticed.

Usually, he was fine with that. He was a relatively happy warrior who believed knowledge could at least sometimes be power. But sometimes, power was power, and the anti-ethanol cause felt unusually lost. ADM, which owned half of America's ethanol plants, seemed to own half of Congress, too. The proposed mandate wasn't big enough to transform the Midwest, anyway, so he moved on to issues where victory was at least conceivable.

In retrospect, he's embarrassed by how much he failed to grasp in 2003. At the time, he was totally unaware of the climate benefits of the wetlands he was fighting to save. He also knew almost nothing about international agriculture and its intrusions into tropical rainforests, so he overlooked how mandating farm-grown fuel in America could trigger deforestation and food shortages abroad. It certainly hadn't dawned on him that biofuels represented a larger land-use problem that threatened humanity's future on a planet with limited land to use.

Then again, it hadn't dawned on anyone else, either.

Searchinger would later return to ethanol and climate, making scientific and economic connections the field's scientists and economists had missed. He would then figure out how agriculture was eating the earth, and create the first serious plan for preventing that. It was an odd plot twist for an urban lawyer whose closest encounter with farm life growing up had been the petting zoo in Central Park.

But not too odd.

Taking on biofuels, and then the broader food and climate problem, required a wonk-crusader smart and stubborn enough to master the intricacies of esoteric models in unfamiliar disciplines, intellectually arrogant enough to believe he could parachute into the new fields and prove the experts wrong, and foolishly romantic enough to believe his impertinent crusades could help save the world. That's always been who he is.

BORN ELITE

His fourth-grade report card left no doubt that even by the elite standards of Manhattan's Dalton School, Timmy Searchinger was precocious. His science teacher, Mrs. Raskoff, marveled with twee prep-school prose that "the quality of his frequent class participation is so fine that he usually leads the discussion to advanced levels heretofore never attempted."

He was an undersized kid with a severe lisp, but he had supreme confidence in his ideas, and he was never afraid to call out thtupid ones. To his older brother Brian, a shy boy with a similar lisp, Tim seemed remarkably comfortable in his skin, whether belting out "The Impossible Dream" as an off-key Don Quixote in their family room, announcing his plans to become president of the United States, or shredding the logic of less rigorous classmates. Mrs. Raskoff did hint at some personality challenges that accompanied his brainpower: "Timmy is now making a conscientious effort to tolerate those less able than himself."

More than half a century later, Searchinger is still supremely confident, and now speaks in fully spooled paragraphs with no lisp, high speed, and high volume. He's five foot ten with chestnut-brown hair, a slight pudge that testifies to a career spent indoors, and the bespectacled, disheveled, distracted aura of the absent-minded Ivy League academic he's become; during a few days in Denmark, I saw him get lost walking to a meeting, board a wrong train, and leave his phone in an Uber. When he's absorbed in thought, his face squinches into a real-life hmm emoji with a hint of suspicion, as if he can't find his keys and is trying to deduce who stole them.

But even in his mid-sixties, he's got some boyish charm; he describes himself, accurately if not entirely self-deprecatingly, as "an aged cherub." When he figures something out, he still exudes the ooh-ooh-I-know enthusiasm of a kid with his hand up in the front row. And while that fourth-grade report card now hangs in his bathroom, proof he's learned to laugh at his foibles, tolerating the less able still isn't his forte. He's learned to be deferential in formal settings, but his ability to mask his frustrations about the ignorance of others remains a work in progress.

His brother says he hasn't changed much since fourth grade.

"I think his mind worked that way in the womb," Brian said.

Timothy Darrow Searchinger left the womb on June 17, 1960, a true child of Manhattan in the '60s.

He grew up in the Upper West Side when it was already home base for New York's liberal intelligentsia, caricatured by the famous Steinberg poster of their view of the irrelevant world beyond the Hudson River. But it was grittier then, and the city-sculptor Robert Moses was using slum clearance money to raze a chunk of it for the Lincoln Center arts complex. Tim was mugged for his bus pass in elementary school. His family's rent-stabilized apartment overlooking Central Park would be worth millions today, but back then some of his Dalton friends from the tonier Upper East Side weren't allowed across town to visit.

Still, he was a privileged kid, born into New York's cultural elite. His father, Gene, was a high-concept filmmaker who made documentaries for high-culture institutions like the Metropolitan Museum of Art and the Metropolitan Opera. His mother, Marian, was a theatrical agent who represented stars like the actor Jane Alexander and playwright Langston Hughes. They didn't think they could conceive a child, so they adopted two before Tim proved them wrong, a pastime he'd come to enjoy. They chose his middle name in honor of Clarence Darrow—prophetically, a bulldog lawyer who defended science and challenged orthodoxy.

Tim spent long hours shooting hoops in the park, and his ferocious competitiveness in darts and Monopoly became a family joke. But he enjoyed a pretty rarefied childhood, visiting museums with his art-world dad, watching musicals with his theater-world mom, dissecting *The New York Times* every morning over breakfast. And he was a pretty rarefied kid. After another client of his mom's, an accomplished director, took him to a James Bond flick, Tim launched into an in-depth critique of the plot. The director suggested he should try having fun sometime.

Dalton was rarefied, too: Tim's schoolmates included *Dirty Dancing* star Jennifer Grey, whose father Joel starred in *Cabaret*; *Commentary* editor John Podhoretz, whose father Norman edited the same journal; and *Fast Food Nation* author Eric Schlosser, whose father ran NBC. The Searchingers weren't rich by Dalton standards—Rupert Murdoch's daughter was also in Tim's class—but they were comfortable enough to afford a second home on a lake upstate in the Hudson Valley.

Tim got his intellectual and moral intensity from his charismatic father, an unreconstructed leftist who marched to his own creative beat. Tim's leftism was later reconstructed, but kibitzing with his dad shaped his worldview.

Gene Searchinger grew up in London, where his own father, Cesar, became a radio pioneer as the first overseas correspondent for CBS, before getting replaced by a young up-and-comer named Edward R. Murrow. Gene enjoyed a highbrow expat childhood—he once asked his father's pal George Bernard Shaw to write a play about him—before serving as a U.S. spy during World War II. He kept his service revolver and the cyanide pill he was supposed to swallow if he ever got captured, and on his 50th birthday, he canoed into that Hudson Valley lake and dumped them.

Gene was an extrovert, a left-wing chatterbox with a passion for ideas. He built a satisfying career pursuing nerdy obsessions, an example Tim would follow. *The Human Language,* his deep-think documentary about how we communicate and how that defines our species, was one of the highest-rated programs ever on PBS, introducing the lefty linguist Noam Chomsky to wider audiences, using comedians like George Carlin to make complex ideas accessible.

Gene was amused by the world's circus of human folly. He tacked Mencken's "clear, simple, and wrong" quotation above the lake house stairs. He claimed credit for Searchinger's Law: "Everything takes longer than you think." But he was also appalled by the world's injustices. He sued a golf course for discriminating against an African American friend of Brian's. He joined some blacklisted filmmakers in a lawsuit against the Directors Guild's loyalty oath that went all the way to the Supreme Court. He also joined a dissident group of Dalton parents battling the school's conservative headmaster, Donald Barr, who reminded him of the European fascists appeased for too long during his childhood. Barr mocked the rebellion as the outgrowth of a permissive counterculture where "a child evinces an interest in photography at breakfast and by dinner he has a Nikon"—his similarly reactionary son William later served as President Trump's attorney general—but Gene attended PTA meetings with near-religious fervor until he was ousted.

Tim is his father's son, an intellectual explorer, a rebel against conventional thinking. But he got his practical problem-solving side from his mother, the family's main breadwinner in an era when that was still rare for a woman.

Marian grew up in a working-class Jewish home in Springfield, Massachusetts; her father, a cobbler, was a former amateur boxer. She headed off to New York to pursue acting, married a State Department diplomat, then followed him to the University of Cambridge after he was fired during Joe McCarthy's anticommunist purge. She felt cloistered as a don's wife, so she left him and returned to New York, working as a secretary to a show-business agent who was excessively fond of ethanol in its nonautomotive form. She went into business herself after realizing she was already doing the drunk's job for him.

She then met Gene, who was similarly smart and outgoing. He wasn't Jewish, but a hyperverbal left-wing Upper West Side filmmaker seemed Jewish-ish, and they both cared more about morality than religion. Their main difference was that Gene was a child of affluence who didn't fret about money, while Marian was a Depression kid who darned holes in Tim's socks. She had an aphorism of her own: "Celebrate today, because tomorrow everything will fall apart."

The Searchingers were the opposite of modern helicopter parents, loving but not doting, treating their kids like mini-adults. They both smoked four packs a day, and neither freaked out about their little darlings inhaling the fumes. Tim was six and Brian seven when they first rode the crosstown bus alone to school. Tim's mom put him in ballet class when he was eight, so he had to walk home alone through a neighborhood Saul Bellow described as "a gathering place for drunks, narcotics addicts, and sexual perverts." When Tim was twelve, he and Brian jetted off on a two-week ski vacation in Austria, using their own cash from babysitting and tutoring gigs.

"I scratch my head when I think how much freedom we had," said Brian, who ran a marijuana dispensary and a bike business in Boulder. "There was definitely some lost childhood, but it prepared us for life."

This premature adulthood was not so unusual in that turbulent New York era, especially at Dalton, which earned notoriety as the school Mariel Hemingway's teenage character attended while dating middle-aged Woody Allen in *Manhattan*, and later with the news that the nonfictional sex trafficker Jeffrey Epstein attended student parties while teaching math there in the '70s. Tim's lifelong friend Roger Platt said the stereotypes of unsupervised and unrestrained Dalton students had some basis in reality.

"The school encouraged an overwhelming sense of confidence and independence, with some New York chutzpah mixed in," said Platt, a retired real estate lobbyist.

Tim's Dalton years certainly did not diminish his confidence or chutzpah. In high school, the administration asked him to teach three history courses to his fellow students. He also launched a newspaper called *The Awful Truth*, landed a larger role than Jennifer Grey in a school play, and wooed his first girlfriend while persuading her to join a Model UN resolution about transporting radioactive materials. He also continued his frequent class participation, often to call out teachers for their factual and logical errors. He was like an intellectual Batman ridding Gotham of bad ideas, a predigital version of the comic strip character who can't come to bed yet because someone is wrong on the Internet. He could be pretty intense.

"Tim cared—about society, what was right—more than anyone I knew," said the novelist Allen Kurzweil, his partner on *The Awful Truth*. "I was interested in storytelling, but for Tim it only mattered in the service of some greater purpose."

He led his first moral crusade his senior year, organizing a school-wide protest after two students he barely knew were expelled for smoking pot. What he cared about was the principle, which he denounced in a righteous speech at an assembly comparing Dalton's selective rule enforcement to Britain's before the American Revolution. To his shock, the administration caved. He was shocked again when he tried to draft a thank-you note, and his fellow revolutionaries accused him of selling out to The Man. He still thought of himself as a leftist like his dad, but unlike his fellow travelers, he cared more about progress than purity.

At Amherst College, he continued to exude the self-assurance of a prep-school smart aleck, and to care perhaps a bit too deeply about what was correct. History professor Robert Gross remembers Searchinger interrupting one of his lectures to dismantle his analysis of 19th-century working-class cultures. "A lot of students were intimidated by Tim's intellect," Gross recalled, "while Tim of course wasn't intimidated by anyone." After he attended his first lecture in microeconomic theory, his roommate, Laurence Ball, had to endure a rant about everything the professor had butchered.

"I thought it was amusing he had such strong feelings on the first day," said Ball, who is now a Johns Hopkins economics professor. "He was right, though."

He had a lot of strong feelings. One night while he was walking to a house party, another friend wanted some of the Jack Daniel's they were bringing as a gift. He was aghast: That would be RUDE! The friend was dismissive: Why would bringing a 95 percent–full bottle of whiskey be rude? He insisted the proper distinction wasn't quantitative between 95 and 100 percent full, but qualitative between open and not—and bringing an open bottle would be flat-out WRONG!

"That was a rare debate Tim lost," Ball said. "The guy just took a swig."

Searchinger channeled his argumentative talents into Amherst's Oxford-style debate team; he placed third at nationals his senior year, and it still eats at him that he didn't win. Chris Coons, a teammate from Delaware who is now a U.S. senator, thought he had the ideal qualities for that arena: clear thinking, quick wit, and cocky unflappability in the face of raucous heckling.

"Tim was so confident, so *New York*, kind of rumpled and tragically hip, like he should've been smoking clove cigarettes and hanging out with people who played jazz in the Village," Coons said. "I was wowed."

He was always interested in politics, especially left-wing politics that aimed to empower ordinary people; he wrote his thesis, summa cum laude of course, about populist movements. But he also broadened his horizons beyond his Steinberg-poster view of the world. One summer, he got a hands-on education in proletarian drudgery working in a salmon-packing factory in an Inuit village in Alaska, slicing open fish and pulling out entrails while standing ankle deep in blood and guts. He got a break every two hours, and worked in a near-constant state of disbelief that it hadn't been two hours yet. After a month of hard labor, he drove a New York City cab for the rest of the summer, another real-world education. He later spent an adventurous semester at the University of Zimbabwe, then hitchhiked around Africa, where he chatted about the moon landing with Pygmies in Zaire and fended off the amorous advances of a silverback gorilla in Rwanda.

"Tim was learning how people who weren't Dalton and Amherst kids lived, and it made him even more determined to do big things," Brian said.

The more he learned, the less ideological he became. When he finally read Marx, he found it mostly gibberish. He arrived in Zimbabwe during the joyous early days of majority Black rule, but quickly saw Robert Mugabe's leftist revolution was headed in ugly directions. After college, he won a fellowship to study worker-owned businesses in Europe, where the evidence of his eyes again conflicted with the dogmas of his childhood. He could see the co-ops he was visiting could never scale, and he was too empirical to let a nice empowerment story trump the facts. He was drifting away from his dad's utopian leftism, becoming a pragmatic progressive Democrat who respected the power of markets even though he disliked the excesses of capitalism.

The best thing he did that year was visit Dijon, where he had a long chat in a friend's apartment with a tall, vivacious Amherst teaching fellow from Burgundy named Brigitte Szymanek. They argued about a Truffaut love-triangle film; he thought the woman in the triangle was selfish for tormenting the two men, while she liked the idea of an empowered woman with the freedom to choose. She really liked meeting an idealistic guy who knew Truffaut and asked her dorky questions about nuclear power.

Searchinger felt like he had been transported into a French film with a captivating heroine, stylish without makeup, witty without pretension. He later wrote an impulsive letter inviting her to a getaway in the Black Forest, and after months of awkward message tag in that time before mobile phones or email, they agreed to meet at the Freiberg train station at 10 p.m. on May 7, 1983. They both showed up, and they've been a couple ever since.

LAWYER FOR NATURE

His next stop on the liberal-elite track was Yale Law, where he was again notorious for trying to educate his professors, though widely recognized as a brainiac among brainiacs. He wrote one journal article chastising the Supreme Court for botching the concept of due process. He didn't find the law as interesting as history or economics, but he liked its rules, its logic, the way it could be weaponized to right wrongs. He volunteered at Yale's prison clinic, where he shepherded one petty drug dealer's run-of-the-mill parole case all the way to a federal appeals court. Jay Pottenger, the professor who oversaw the clinic, said he sculpted the case like pottery clay; Pottenger met plenty of brainiacs, but couldn't remember another with "such an unusual combination of creativity, grit, and sense of justice."

After graduation, he clerked for a moderate Republican federal appeals judge, alongside a right-wing Republican co-clerk, but the political differences didn't produce many legal ones. The correct answers didn't seem to depend on politics. He took the conservative side of one rare disagreement, ribbing his judge as a "bleeding-heart Republican" for awarding damages to a plaintiff who stuck his hand in a lawnmower. While his instinct was to support victims, even idiots, against corporations, the facts and the law came first.

"Tim loved to argue, but he was never on an ideological bender," said his co-clerk, Jerome Marcus, who later played a key role in President Bill Clinton's impeachment, then argued one of Trump's challenges to the 2020 election. "The world was a friendlier place then, before we were constantly assaulted by politics on the Internet, and Tim was a rational guy."

He enjoyed drafting legal opinions about whether chocolate fudge soda could be trademarked, but he wanted to start doing stuff that mattered. He wasn't interested in getting rich shuffling paper for corporate clients at a big firm, the default path for Ivy League law graduates in the Reagan era. He was a born litigator, and even though he had never taken an environmental law class or learned much about nature, he thought he could be most useful as a litigator for the earth. He doubted he could sue people out of poverty, or cross-examine dictators into respecting democracy, but he thought a good lawyer could use America's strong green laws to make a real difference. He had also discovered in Africa that he was an instinctive tree-hugger. Nights in the Congo rainforest, with a natural soundtrack like a Tarzan movie, stirred his urbanite soul.

So he took a job as a deputy counsel to Pennsylvania's Democratic governor, with a broad portfolio that included the environment. Thick piles of proposed regulations and other action items constantly landed on his desk, and he quickly realized he couldn't effectively do stuff until he knew stuff. He made a point of trying to learn stuff, from the proper liner width for a landfill to the governor's legal authority to withhold tax refunds during a budget crisis.

"He'd be totally unfamiliar with an issue, so he'd read everything there was to read," recalled his boss, Morey Myers, a prominent civil rights lawyer. "I was working until I was blurry-eyed, and Tim worked even harder."

Myers was 93 when we spoke, and still remembered that after the Nuclear Regulatory Commission shut down a Pennsylvania nuclear plant where operators were caught sleeping on the job, an NRC official called to ask if he was aware a pushy lawyer from his office kept butting into federal jurisdiction. Sounds like Tim, Myers said. "Well, he's doing such a good job, we're going to cede jurisdiction!" the official replied. Searchinger could see the melted-down Three Mile Island reactor out his office window, and to help prevent a sequel, he was willing to ramp up his New York pushiness in the pursuit of stricter safety rules.

He was too empirical a liberal to assume the answer to policy questions was always more government. When Myers asked him to investigate state elevator inspections after a high-profile fatality, he came back with the politically inconvenient conclusion that private insurance inspectors were already doing a fine job. But he did enjoy using government to fight injustices—especially injustices to nature, which couldn't defend itself in court. In his first wetlands litigation, he had to defend some regulators who let a businessman fill in marshes along the Susquehanna River. He soon concluded that what his clients had done was flat-out wrong, like bringing an open whiskey bottle as a gift, so he scrapped protocol and settled the case.

Harrisburg whetted his appetite for environmental law. After he left government in 1989 to join a boutique firm that agreed to let him do pro bono work, his first pro bono case was representing the Environmental Defense Fund in a high-profile federal lawsuit over water quality in the Everglades. He arrived in Miami ignorant about the world's best-known wetland, but plowed through scientific reports until he became an Everglades expert.

"My job was to be the guy who knows shit," he says. "My dream job!"

He soon accepted a much lower-paying job with EDF in New York. On his first day as a full-time environmental lawyer, he went to a hearing about pollution from his hometown's sewage plants, and realized he literally needed to learn about shit. His dream job.

In the Everglades, trying to protect nature from influential sugar barons, Searchinger began to learn how farm politics could clash with science and the law.

He once got a tip that a Florida environmental lobbyist, Charles Lee, was cutting a deal with Big Sugar to weaken the state's water-quality laws, in exchange for some farmland the state could use to build artificial marshes to filter polluted runoff from sugar farms. Searchinger was all for filter marshes, but he realized the deal would neuter the federal lawsuit, so he rushed to the airport without a suitcase to spend a week in Tallahassee warning legislators and reporters. He wore the same tan suit to walk the halls every day, and the same borrowed T-shirt festooned with smiley-faced condoms to work the phones every night. He got into one screaming fight with Lee, who mocked him as a clueless interloper swooping into a town he didn't understand to save laws nobody followed. But he had enough of a clue to help scuttle Lee's deal, preserving those state laws and a federal case that is still driving progress in the Everglades three decades later.

Lee recalled Searchinger as a know-it-all who, if you asked him the time, would explain how to build a watch, "basking in his role as the perfectionist." Still, Lee conceded that "even if Tim wasn't always fun to deal with, he always had valuable knowledge," perhaps because events vindicated his belief the Everglades would fare better in court than in Florida's sugar-dominated political arena. He wasn't opposed to all deals, just bad deals.

The proof arrived at a 1991 hearing, when Florida's folksy new Democratic governor, Lawton Chiles, after initially arguing the federal government's lawsuit should be dismissed because he had a plan to clean up the Everglades, reversed course and announced he wanted to surrender his sword. It was a pivotal moment in Florida history, leading to a consent decree that would help convert 80,000 acres of farmland into filter marshes, far more than the governor's plan or Lee's deal.

Searchinger noticed the governor's top environmental aide, Carol Browner, glaring at him in court as he detailed the weaknesses of the Chiles plan, and he still suspects that's why he didn't land a Clinton administration job after Browner took over the Environmental Protection Agency (EPA). Really, though, he was better suited outside government, where he faced less political pressure to suppress his opinions and tolerate the less able. EDF's former chief counsel, Jim Tripp, recalled that he had an uncanny ability to spot key technicalities and nuances in fine print, but was less adept at disguising his disgust for nuance-ignorers and fine-print-skippers.

"Tim definitely rubs some people the wrong way," Tripp said.

When Florida's enviros, farmers, and business leaders later set aside their conflicts over water quality to support a bipartisan $8 billion Army Corps plan to replumb the entire Everglades ecosystem, Searchinger became a lonely and unpopular critic of the plan, channeling the dissident views of muzzled Everglades National Park scientists who saw the largest environmental restoration effort in history as a Rube Goldberg mess. He wasn't looking to pick fights with other enviros by barging into Florida and attacking their pet project, but the fact that almost every stakeholder supported it did not temper his conviction that it was political garbage.

"Tim was never super-worried about anyone's feelings," said Shannon Estenoz, an Everglades activist who later became an assistant Interior secretary under Biden. "But even folks who didn't like him couldn't dismiss him. He was too frigging smart, and he knew the details."

Searchinger fought another uphill battle after Chiles surrendered in the Everglades, another fine-print science fight on a political battlefield that honed his unusual set of skills.

The battle began in August 1991 after he noticed a *Times* story on page A7 with the soporific headline "Bush Announces Proposal for Wetlands." President George H. W. Bush had pledged "no net loss" of wetlands on the campaign trail, but his White House was proposing to redefine what counted as a wetland. And while the story made that sound like a routine tweak, Searchinger read the actual proposal and realized Bush's aides were trying to gut protections for millions of seasonally flooded acres. He wrote his first-ever *Times* op-ed savaging the plan as "probably the largest weakening of an environmental regulation in U.S. history."

He and Brigitte got married the next week, and he had planned to telecommute from Vermont while she taught French at Middlebury College. Instead, he stayed in New York and barely saw her for six months. He spent the time picking the brains of scientists, binge-reading studies, and writing a 175-page report called *How Wet Is a Wetland?* that explained in painstaking detail why the plan's seemingly innocuous approach to "hydric soils" and "oxidized rhizospheres" would break Bush's no-net-loss promises. It also displayed damning photographs of wetlands that would no longer be considered wetlands, including a quarter of the Everglades. He again tried to be a marketing department for obscure scientists, making a complex and confusing issue accessible and comprehensible, just like his dad had done with linguistics.

"Tim was better than anyone I'd ever seen at mobilizing scientific evidence to move policy," said former EDF chief scientist Michael Oppenheimer, an acclaimed physicist who is now his colleague at Princeton.

How Wet Is a Wetland? shook up the debate in a way interest-group reports rarely do anymore, emboldening Bush's own EPA head to bash the plan publicly, stirring up opposition from Republican-aligned hunting and fishing groups. The White House soon scrapped its plan. The president had already broken his no-new-taxes pledge, so abandoning his no-net-loss pledge would have reinforced a read-his-lips narrative in an election year.

For Searchinger, again, the lesson was that facts at least sometimes mattered. Many policy wars were so information-deprived that evidence could actually move the political needle. A more personal lesson was that synthesis was his superpower. He didn't know as much about wetlands as real scientists, but he knew how to highlight what was important about what they knew. He could translate science into English, getting into the weeds without dragging policymakers or the public there. And after devouring hundreds of papers and interrogating dozens of experts, he thought he understood the sweep of the science as well as any of them.

"Tim was always a few steps ahead," said ecologist David Wilcove, another EDF and Princeton colleague. "He has that terrier quality that makes him a pain in the ass, and it can feel like walking into the Spanish Inquisition if you disagree. But the odds are overwhelming that he's right, he's done his homework, and your soul really is at risk if you don't do it his way."

The baby-faced Manhattan lawyer's reputation as an unlikely wetlands expert was rising fast enough that after the devastating Mississippi River flood of 1993, he was invited to brief Vice President Al Gore and other senior White House officials on how wetland losses made the disaster worse. There was just one problem: He had only seen the Mississippi out of airplane windows. It was one time when the populist conspiracy theories about Acela Corridor elites making policy for flyover country rang true.

But he did the reading, consulted the experts, and synthesized their views: Drainage and development in the floodplain had put more water in the river and more people in harm's way. And flood politics had always emphasized short-term relief and rebuilding, ignoring long-term sustainability and resilience. His briefing helped persuade the White House to try to do more than just restore the preflood status quo—and while most of the aid still went to relief and rebuilding, some of it helped create a new wildlife refuge and relocate an Illinois town to higher ground. That was modest but tangible progress, against the odds, informed by science.

As usual, he approached the issue with borderline monomania. One night he shared a hotel room in Omaha with another young enviro, Scott Faber, who watched him pace around for hours in his tighty-whities, reading technical reports and providing running commentary.

"Tim's pouring out insight after insight, oblivious to the fact that this was incredibly weird," Faber said. "He's just offended by irrational policy."

Searchinger had found his calling, trying to make policy somewhat less irrational. He liked playing a real-life Don Quixote, dreaming impossible dreams and tilting at special-interest windmills. It felt right, and occasionally, it worked.

BELTWAY CRUSADER

After the flood, he started working on imperiled Midwestern ecosystems. That meant working on agriculture, because Midwestern ecosystems were surrounded by farmland.

He sometimes quipped that he was indisputably one of the environmental movement's top few agriculture experts, because there were only a few. Most enviros preferred fighting to save charismatic megafauna or shut down dirty factories. Nobody was writing fat checks or running glitzy media campaigns to get farmers to clean up their runoff or stop draining wetlands. One green group CEO told him that following ag policy—a tangle of overlapping subsidy programs with inscrutable acronyms—was like listening to hieroglyphics.

But Searchinger figured sexy causes didn't need his help. Agriculture was an underscrutinized, oversubsidized environmental scandal that covered half of America, degrading its air, water, and land under the protection of Washington's most influential lobbying groups. Farmers were basically unregulated manufacturers who happened to manufacture food. Somebody on nature's side had to learn about "nonpoint source pollution," "pesticide tolerance levels," and other ag trivia, so why not the bookworm from the Upper West Side?

He moved to Washington for EDF in 1994, joining the permanent class of Beltway wonks who track the federal leviathan, laboring in obscurity except for occasional newspaper quotes and C-SPAN hits. Brigitte taught high school French. They bought a Sears-catalog bungalow in the left-wing bubble of Takoma Park—a "nuclear-free zone" whose name, like Cambridge or Berkeley, was often preceded by "the People's Republic of"—just as Newt Gingrich's angry band of right-wing Republican insurgents was reclaiming Congress for the first time in a generation.

It was a stressful time to fight for the earth in Washington, with all the federal safeguards that attracted him to environmental law under assault. Green groups had to play defense against simultaneous crusades to abolish the EPA, rewrite the Clean Water Act, and gut green protections through "regulatory reform." He led a successful behind-the-scenes fight to block a lower-profile but equally radical effort to define any rule harming any property owner as an unconstitutional "taking," which would have forced governments to compensate any farmer or developer restricted from draining wetlands. But he also used his fine-print skills to spot an ingenious opportunity to play offense after working on the 1996 farm bill, a $60 billion behemoth that continued the bipartisan tradition of shoveling cash to farmers.

One of the shovels was the Conservation Reserve Program, which had begun paying farmers not to farm some of their land in the 1980s, when crop surpluses created by "fence row to fence row" policies crushed prices, drove thousands of farmers into bankruptcy, and inspired Willie Nelson's iconic Farm Aid concerts. Although the agriculture committees touted CRP as an environmental program, its real purpose, in addition to shoveling cash to farmers directly, was to curb overproduction in order to raise crop prices and shovel cash to farmers indirectly. But Searchinger found language, designed to shovel cash to farmers even faster, that allowed the agriculture secretary to make payments directly to states. He realized an environmentally friendly administration could use that loophole to finance large-scale wetlands restoration, diverting farm aid that Congress had disguised as conservation aid into actual conservation.

Soon he was writing plans for Maryland's Democratic governor and Minnesota's Republican governor, seeking hundreds of millions of dollars to restore wetlands and create natural buffers around the Chesapeake Bay and Minnesota River. David Folkenflik, then a *Baltimore Sun* political reporter, was amazed to watch a public-interest lawyer working the levers of power, orchestrating politicians and bureaucrats like a corporate lobbyist, undermining congressional intent on behalf of Mother Nature.

"What Tim did was genius: He forced Washington to live its stated values!" said Folkenflik, now National Public Radio's media reporter. "I had never seen anything like it."

Searchinger's policy scheme got its moment in the sun after manure pollution from chicken farms fueled a nasty parasite outbreak in the Chesapeake Bay. When the White House wanted something to announce to show it cared about fishermen with rashes and fish with oozing sores, he served up his plan to preserve buffer zones along the bay. And when Gore visited to announce it, Searchinger helped arrange for him to declare the Minnesota River next in line. Afterward, he noticed one USDA farm traditionalist who had tried to defend the status quo giving him the same death glare Carol Browner had given him during that Everglades hearing.

"The Minnesota Nice folks needed a pushy New York lawyer to make sure they didn't get left out," he says. "For once, everything fell into place."

For once. Most of his fights were losing fights, over everything from subsidized crop insurance, which encouraged farmers to cultivate vulnerable floodplains, to the manual the Army Corps used to manage the Missouri River, which prioritized farms over endangered fish. But he was learning about agriculture, nature, and Washington. And I got a chance to learn from him.

When I first met Searchinger at a Union Station coffee shop in 1999, he was already a 39-year-old senior attorney at EDF. But he looked like a skinny college kid dressed up for his first job interview, with that cherubic face and bouncy energy. He quickly began barraging me with meticulously documented information, peppered with outraged commentary. Every few minutes, after explaining some un-fucking-believable idiocy, he unleashed a "HA! HA! HA!" like an unaccented Count from *Sesame Street*. He was an exhausting but bizarrely honest source, insistent about what he knew yet transparent about what he didn't, providing counterpoints to his own arguments and even acknowledging the few he didn't consider complete nonsense.

I was a reporter for *The Washington Post*, and he wanted me to expose the Army Corps as a battalion of concrete-addicted marauders, throttling wild rivers into barge channels that barges didn't even use. I thought he must be exaggerating, but he had piles of proof that most federally channelized rivers had virtually no barge traffic, and that the Corps cherry-picked and manipulated data to justify preposterous make-work. He had unraveled the lunacies of overcomplicated Corps computer models—like a wetlands model for a Missouri flood-control project where he caught a junior-high-level math error, and a hydrological model of the $8 billion

Everglades restoration project that produced physically impossible results. The turgid models churned out conclusions that sounded authoritative, but when he delved into their assumptions, it felt like pulling back the curtain on the Wizard of Oz.

Soon I was almost as obsessed as he was. I spent the next year investigating how the Corps cooked its books to approve nutty projects that kept its employees busy and its congressional patrons happy. I wasn't a good investigator, but I didn't have to be, because he had already uncovered much of the nuttiness, and the rest was hidden in plain sight.

For example, the Corps was promoting that Missouri flood-control project as a lifeline for a low-income town, but he had found Corps documents acknowledging it wouldn't reduce the town's flooding at all. It was really a drainage project for well-connected farmers in the floodplain, and it was poised to destroy tens of thousands of acres of ecologically sensitive wetlands along the Mississippi. He even got a Corps official to admit under oath that the agency's math error—the wetlands equivalent of mixing up part-time and full-time employees—had nudged the project's benefit-cost ratio to 1.01, barely above the threshold for construction.

Searchinger also gifted me a classic D.C. scandal, introducing me to the Corps economist who led the analysis of a billion-dollar navigation project on the Mississippi—until he concluded the costs outweighed the benefits. Wrong answer. Corps leaders yanked him off the study, rigged their economic model to justify construction, and, in violation of the Washington rule not to write down anything you don't want on the *Post*'s front page, copied him on emails ordering his team to "get creative" in order to "get to yes as fast as possible." *Duuuhhyee.* The scandal deepened when Searchinger leaked me a secret "Program Growth Initiative" that Corps generals had launched to get Congress to double their budget.

My last article about the Corps was supposed to be a redemption story—until he gave me more documents exposing the agency's green poster child, Everglades restoration, as another make-work mess. That was the tip that led me to Miami and my wife. It was another classic example of Searchinger at work—finding the best scientists and publicizing their ideas; dissecting complex models and dismantling their assumptions; connecting technical dots like the L-67 levee and Water Conservation Area 3B into a narrative of how the restoration plan wouldn't restore the flow of the River of Grass. He had a gift for ferreting out the logical fallacies in quantitative models, then explaining them in ways judges or even journalists could understand.

But all his ferreting only helped produce a few tweaks to the Everglades project—and it's now a $25 billion project. The other Corps scandals didn't change much, either, even though watchdogs issued scathing reports and the agency's reputation was tarnished in Washington. Politicians still loved using water projects to steer jobs and money to constituents and donors, and the Corps loved to help. After one congressional hearing dominated by its defenders, I asked Searchinger if it would have to kill someone for Congress to reform it. Not even then, he groused. Sure enough, after Corps levee failures killed thousands in Louisiana during Hurricane Katrina, little changed.

Nevertheless, Searchinger would carry the lessons of the Corps into his climate work: Nonsense can be hidden in plain view, especially when only insiders pay attention. Authoritative-sounding analyses backed by sophisticated-looking models can be super-wrong, especially when analysts have incentives to be super-wrong. Groupthink is a potent force, so consensus policies can be dumb policies. And policy is a long game, so it's worth getting the truth out even if the politics aren't ready for action. He updated his dad's droll observation with Searchinger's Second Law: "Even adjusting for Searchinger's First Law, everything still takes longer than you think."

One lesson I learned was that if Searchinger thought something mattered, it probably did.

THE NEXT CHAPTER

"Sorry, Tim, gotta run."

Searchinger ignored the brush-off and kept strategizing at warp speed. The new farm bill was stuffed with more subsidies than ever for big farmers to plant endless rows of commodity crops, and he was explaining to a reform-minded congressional aide named John Mimikakis how they could pass an audacious amendment to divert $20 billion into conservation.

"Tim, I'm serious! Stop talking and turn on the news."

Two planes had crashed into the World Trade Center.

"We'll have to finish another time," Mimikakis deadpanned.

Why? Manhattan was a thousand miles away! Searchinger kept game-planning, until Mimikakis interrupted again. The Pentagon had been hit. The Capitol was being evacuated.

"When Tim is full throttle, he's not easily deterred," Mimikakis said. "And it was an important amendment, although, you know, not that day."

When Congress returned, traditional farm groups and their friends in Congress threatened to scuttle the entire Farm Security and Rural Investment Act if the green amendment passed. The House finally rejected it in a surprisingly close vote; Bill O'Conner, a traditionalist who led the Agriculture Committee staff, admitted the reformers almost won: "I don't think that ever came close to happening again." Still, when a reformer told him the aggies had been smart after 9/11 to insert "Security" into their bill title, O'Conner shot back that it was enough to insert "Farm."

The politics of agriculture was just incredibly difficult to disrupt. America was no longer a nation of self-sufficient yeomen; 0.03 percent of its citizens produced half its food. But it remained an article of political faith in Washington that the rural heartland was a font of virtue that justified showering farmers with tax dollars. To argue otherwise was to betray elitist disdain for rural America. Anyway, farm interests paid closer attention to farm policies than anyone else, and the frogs and birds that relied on Midwestern wetlands lacked Big Ag's lobbying budget.

Searchinger encountered similar frustrations in his war on that Army Corps farm boondoggle in Missouri. He persuaded a judge to block the project, so the Corps reworked its get-to-yes analysis, so he had to find its new gimmicks to get the judge to re-block the project; the Corps rejiggered its analysis four more times before the project was finally scrapped. The ordeal reminded him of the Monty Python sketch where a pet shop owner keeps insisting a dead parrot is merely resting, or stunned, or perhaps pining for the fjords. He felt like the irate customer who keeps pointing out the parrot is bereft of life, no more, an ex-parrot.

Of course, this was the life he had chosen. Fighting powerful interests wasn't supposed to be easy. He had helped stop some bad stuff, like the

Bush assault on wetlands and the radical Republican takings bill, and advance some good stuff, like judicial oversight of the Everglades and conservation of the Chesapeake Bay. He hadn't reformed the Corps, but he had helped embarrass the agency enough to derail its empire-building Program Growth Initiative.

He was also a happy husband and father. Like his parents, he and Brigitte adopted their son Noah when they thought they couldn't conceive, then were proved wrong by their daughter Chloe. She was born with a severe heart defect, but modern medicine fixed it, giving Searchinger even more appreciation for science and life. And life was good. He loved walking his rescued golden retriever, cooking dinner for Brigitte and friends, and hanging around nature, especially big trees. Big trees always made him happy.

Professionally, though, he thought it was time for a new chapter. He summarized his angst in Searchinger's Third Law: "For anything worth doing, by the time you get it done, you no longer feel satisfaction, only relief." How satisfied could he feel about protecting a few acres of nature here and there when millions of acres were being lost worldwide?

He had spent nearly two decades failing to save the American environment. He thought it might be time to start failing to save the global environment.

His next chapter began on January 23, 2007, while watching George W. Bush deliver the State of the Union address. Halfway through, the president announced that to reduce reliance on foreign oil, he wanted to mandate 35 billion annual gallons of alternative fuels.

What?

Searchinger thought he must have misheard. The current mandate was only seven billion gallons.

"That is five times the current target," Bush declared.

Are you fucking kidding me? Searchinger had not misheard.

He did some quick math: 35 billion gallons of ethanol would require the entire U.S. corn harvest. That would mean no corn for feeding cattle, hogs, or chickens, no corn for corn syrup or cornflakes, no corn for the world's leading corn exporter to export. Just about every ear would have to be brewed into fuel.

He didn't know what impact that would have on the Midwest landscape, but he doubted it would be small. He had no idea what impact it would have on the climate, either, but he remembered his queasy feeling about the Argonne study. And while he still wasn't a climate guy, global warming was starting to feel like an existential threat that made other environmental issues irrelevant. Atmospheric carbon levels had soared from 317 parts per million when he was born to 383, well above the threshold of 350 that scientists considered necessary for a stable climate. The world was catching fire. Maybe it was time to join the firefighters.

That week, during a networking chat with David Sandalow, a former Clinton White House energy expert, he happened to mention that a pumped-up biofuels mandate could be disastrous. Sandalow replied that big money was already lining up behind it.

"I'd hate to be the guy who tried to jump in front of that train," Sandalow mused.

That banal observation staggered Searchinger. He hadn't jumped in front of the ethanol train in 2003, and it was probably still unstoppable. But it would definitely be unstoppable if nobody tried to stop it. This sounded like a job for the patron saint of almost-lost causes.

Dammit, he thought. *I guess I gotta be that guy.*

TWO
LAND IS NOT FREE

WHAT THE EXPERTS MISSED

The prevailing science still considered corn ethanol climate-friendly. The most authoritative study, published in *Science* in 2006 by a team from the University of California, Berkeley, had confirmed ethanol's emissions were 20 percent less than gasoline's. The IPCC, the gold standard on global warming, was still promoting biofuels as low-carbon alternatives to fossil fuels. And Searchinger was still laughably undercredentialed to challenge the consensus. He was not a professional researcher or even an unbiased amateur. He still knew almost nothing about climate change.

But he had learned one thing he hadn't known when he first got suspicious about ethanol: Wetlands, forests, and natural lands in general were better carbon sinks than cornfields.

That may sound obvious. Of course wild vegetation that can photosynthesize for decades absorbs more carbon than crops that get harvested after growing for a few months. Of course there's more carbon in a forest of tall, thick trees than a field of short, spindly crops. But he didn't assume true-sounding things were true until he did the reading, and only now did he know it actually was true: An acre of nature almost always stored vastly more carbon than an acre of farmland. That meant the conversion of nature into cornfields increased greenhouse gas emissions.

This no-duh notion that protecting nature helped the climate had dire implications for ethanol. The problem was the chain reaction he had worried about in 2003: If cornfields produced more fuel instead of food, new cornfields would be needed to replace the lost food—and common sense suggested they wouldn't be planted in parking lots. Natural land would have to be converted into farmland, which would release carbon the experts weren't counting.

Searchinger had no idea how much carbon. He couldn't say if ethanol was bad, very bad, or horrifically bad for the climate. But since the original Argonne study had established that ethanol's production emissions were much worse than gasoline's, the argument that it was better for the climate assumed that growing corn for ethanol dramatically reduced emissions from land use. If growing corn for ethanol actually increased emissions from land use, by inducing the clearing of wilderness, then gasoline had to be better for the climate. Even though it wasn't renewable. Even though it was a fossil fuel.

It took chutzpah for a layman to second-guess an entire scientific discipline, but Searchinger found it quite plausible that emissions analysis, like Everglades restoration, could be widely accepted garbage, and that an entire cottage industry of climate analysts could be wrong. He had a lot in common with the short sellers who were defying conventional wisdom that year by betting against the housing market, the brash contrarians lionized in *The Big Short* after the market imploded. The fact that his theory fell outside the mainstream made him question the mainstream, not his theory.

The core of his insight was that ethanol only looked good for the climate if you assumed that using land to grow ethanol had no cost to the climate. But it did have a cost, what economists call "opportunity cost," the cost of losing the opportunity to use the land in other ways. Even though opportunity cost is a staple of introductory economics, it hadn't infiltrated life-cycle analyses. And Searchinger's friend Steven Berry, a distinguished economics professor at Yale, liked to call him the world's greatest introductory economics student.

Berry thought opportunity cost was so vital an economic concept that he taught it the first day of his own introductory class. He always read aloud the Robert Frost poem "The Road Not Taken," to show that opportunity cost is about the what-if value of the not-taken road. For example, a student considering graduate school should consider not only the direct cost of tuition, but the opportunity cost of several years without income when she could otherwise earn money in the workforce. There are opportunity costs to getting married, buying a home, spending tax dollars on Army Corps pork, and almost any other choice in life.

Searchinger's epiphany was that using land has opportunity costs, too. An acre used to grow ethanol can't help feed the world or store carbon in native vegetation. Yet the models only counted the carbon benefits.

"When Tim first explained the problem, I said, 'That's classic opportunity cost. They couldn't have missed it,'" Berry said. "But they did miss it."

Searchinger summarized his epiphany with a four-word mantra: *Land is not free.*

The earth has lots of land, but not an infinite amount, and while ethanol is renewable, the land used to produce it is not. Land used to grow fuel can't grow trees or grow food that reduces the pressure to cut down trees elsewhere—and that has a climate cost as real as the emissions from diesel tractors or burping cows. When he did some back-of-the-envelope math, he estimated the average carbon benefit of reforesting an acre of cornfields would be several times larger than the benefit of using an acre of corn to replace gasoline, which suggested ethanol had a huge "carbon opportunity cost," a concept he would introduce to the scientific community a decade later in a paper in *Nature*. He couldn't pinpoint the specific cost of using a specific piece of land to grow fuel, but the cost could never be zero, because land on a finite planet is never free.

Yet life-cycle analyses of biofuels assumed land was always free. Searchinger asked an agricultural economist he trusted, Ralph Heimlich, to review the literature, and it mostly ignored land conversion or merely suggested the issue should be studied. The Berkeley team buried a typical caveat in *Science*, acknowledging biofuels could induce damaging land-use changes, especially "the conversion of rain forest into plantations for fuel production." But they punted, warning that "estimating the magnitude of such effects would be very difficult," requiring complex analyses of hard-to-predict variables: "For these reasons, we ignore emissions due to potential changes in land use."

In retrospect, ignoring land-use emissions because they were hard to predict was a dubious strategy. That *Science* paper has been cited thousands of times, and Berkeley professor Michael O'Hare, one of its six authors, says all six came to regret its pro-ethanol conclusions—although all six were relieved their caveat about the dangers of land use covered their posteriors.

"I'm sure I inserted that sentence right before it went to the publisher," O'Hare quipped, "and I'm sure all the others would say the same thing."

Searchinger knew that predicting emissions from land-use change would be hard. It just seemed clear that draining carbon-rich wetlands or clearing carbon-rich forests to grow fuel would create a lot of them. It didn't seem scientific to pretend the emissions didn't exist just because their magnitude was uncertain.

In fact, the Berkeley paper's nightmare scenario of rainforest conversion was already happening in Southeast Asia, where peatlands and woodlands were being drained and destroyed to grow palm oil, some of it for European biodiesel. This rush to convert nature into palm plantations had turned Indonesia, the world's 20th-largest economy, into the third-largest carbon emitter. A U.S. ethanol mandate seemed likely to create similar incentives, encouraging farmers to convert the Midwest's remaining natural lands into cornfields, releasing the carbon in their vegetation and soils.

Nevertheless, when land was considered free, using land to grow fuel appeared to have no downsides. Even an early IPCC report suggested there were four billion acres of "potential croplands" on Earth—an area twice the size of the continental U.S.—that could one day grow enough bioenergy to replace all fossil energy. That sounded like a great way to end oil, until Searchinger realized those supposedly surplus lands encompassed just about anything potentially arable that didn't already support crops—including the Amazon and Congo rainforests, the Everglades, the Serengeti, and much of the world's best grazing land.

Yes, the official arbiter of climate science suggested that wiping out the planet's best carbon sinks would *help* the climate. The IPCC simply assumed all that uncropped land was free, as if it were doing nothing rather than storing carbon and nourishing livestock. Tufts University ecologist William Moomaw, a lead author of the offending IPCC chapter, told me its grandiose speculation about biofuels was a reminder that scientists can screw up.

"That was pure madness, a perfect example of scientists ignoring the actual biology of the world," he said. "Unfortunately, I was one of those scientists."

The simple insight that bulldozing nature to grow fuel has a climate cost wasn't Searchinger's only epiphany. He had a related insight that would be even more disruptive: Farm-grown fuels also create *indirect* land-use change.

A biofuels mandate wouldn't only encourage Iowa farmers to clear natural land into cropland to grow ethanol. It would also encourage Iowa farmers to use existing cropland to grow ethanol instead of food. Since the world would still need food, another farmer elsewhere on the planet would probably convert natural land to cropland to replace the diverted Iowa grain. And if that elsewhere had lower yields than Iowa's super-fertile breadbasket, that farmer might have to convert several acres of natural land to grow enough grain to replace one acre of Iowa farmland.

The point was that using land had impacts beyond that patch of land. Mandating ethanol could create a chain of unintended consequences: Higher corn prices might induce Midwestern farmers to plant corn instead of soybeans, which might induce foreign farmers to convert cattle pastures to soybeans to replace the lost U.S. soybeans, which might induce foreign cattlemen to convert wilderness into pastures to replace the lost beef. More ethanol would mean more crops and less nature worldwide, whether or not the ethanol crops directly replaced nature.

The sheer size of Bush's proposal made this clearer. How could diverting America's 10 billion bushels of corn into fuel and forcing its overseas customers to look elsewhere for food have zero impact on the world's agricultural footprint? Indirect land-use change was another intuitive concept, and it would become so central to climate analysis that Searchinger's clunky acronym ILUC (pronounced "eye-luck") would become standard academic jargon. A literature review a decade after he coined the phrase found 191 studies with quantified ILUC estimates.

At the time, though, only one study even contemplated ILUC—ironically, the Argonne paper by Michael Wang, whose friend Mark Delucchi, a UC Davis scientist, had urged him to consider ethanol's effects on land use. Wang had concluded the impact was trivial, and even teased Delucchi that they could offset the minuscule land-use emissions themselves by collecting cornstalks from local farms and brewing them into fuel.

When Searchinger dug into Wang's model to investigate why it spat out such tiny numbers, he found a bunch of questionable assumptions minimizing ethanol's impact. It felt like the get-to-yes studies the Army Corps used to rubber-stamp its own projects, where every assumption made the projects look cheaper. Searchinger called this Reverse Murphy's Law, when everything that can go right suspiciously does go right. Published science has an aura of authority, but pulling back the curtain can be a reminder that it's produced by fallible people.

The Argonne model assumed that most corn for ethanol would be grown on existing farmland, which made sense, but that none of the farmland that previously grew food would be replaced elsewhere, which made no sense. It also assumed any new farmland created to grow ethanol would never replace forests, which seemed arbitrary. And it predicted the ethanol mandate would boost grain prices so high that lower food consumption in developing countries would sharply reduce demand for new farmland, which seemed terrible; it meant ethanol would derive many of its climate benefits by making poor people hungrier.

One footnote justifying those assumptions cited "personal communication" with Wang's friend Delucchi, but when Searchinger tracked him down, Delucchi disagreed with Wang's take. He not only believed ethanol was much worse for the climate than gasoline, he had argued in an unpublished manuscript that a key culprit would be the conversion of wild land into cropland. It turned out that Delucchi, another unconventional thinker who regularly informed wrong people of their wrongness, had been worried about the land-use impacts of farm-grown fuels since Searchinger was in law school. He had raised sporadic alarms for years without waking anyone up, and was convinced biofuels were protected by a force field of agro-political influence—not only in government, but in Big Ag–funded academic agriculture departments.

"I recognize as a scientist that it sounds like a crazy conspiracy theory, but I can assure you I've never gotten pushback as hostile as when I started questioning whether biofuels were good for the climate," Delucchi told me.

Knowing he wasn't the first heretic to question ethanol's land-use impacts made Searchinger even more confident that the high priests of lifecycle analysis were the ones missing something. And he had spent enough time around farm politics to know there was nothing crazy about suspecting it could skew science.

This time, he intended to fight back. Fulfilling Bush's entire renewable fuels mandate with ethanol would require as much land as America's national parks. It could undo all his work protecting nature in the Midwest, just as palm oil was wiping out orangutan habitat in Indonesia. It could strain food supplies as well, since the grain it took to fill a Ford Explorer with ethanol could feed an adult for a year. It felt like a return to the horse-and-buggy era, when farmers had to grow millions of acres of oats and hay for transportation fuel.

The more he thought about it, the more gutting nature to grow moonshine for fuel sounded wacky, like setting houses on fire to produce heat for electricity. He decided to do his own dive into the science. Maybe he could discredit Bush's push for biofuels, just as he had done with Bush's dad's assault on wetlands.

"WE HAD DIFFERENT PRIORITIES."

Renewable fuels were exploding onto the national agenda at a time of dizzying disruption. A digital news startup that went live the day of Bush's January 2007 State of the Union—then called *The Politico*, though the "The" wouldn't last—called it a "moment of anxiety and upheaval."

In the previous six months, Twitter had also gone live, Facebook opened its social network to the world, Amazon launched the cloud, and Apple unveiled the iPhone. It was simultaneously a time of financial revolution, fueled by mortgage derivatives that seemed to make risk disappear, and biological revolution, featuring genetic manipulations of the building blocks of life. Politics was in flux, too. Democrats had just reclaimed Congress, making Nancy Pelosi the first woman Speaker of the House, and Barack Obama was about to launch his historic campaign. *Time*'s cover the week of Bush's speech, "Dawn of a New Dynasty," announced that China was ending America's reign as the world's only superpower.

It was during this time of transformation that climate change burst into the mainstream. The big consciousness-raiser was the 2006 documentary *An Inconvenient Truth*, which elevated Gore's nerdy slideshow into a cultural event. It also helped inspire California to enact a Global Warming Solutions Act requiring sharp cuts in carbon emissions, America's first major climate legislation. In 2007, the Supreme Court classified greenhouse gases as pollutants, while Gore and the IPCC—which had just called warming "unequivocal"—shared the Nobel Peace Prize.

Climate was having a moment on Capitol Hill, too. The Republican front-runner to succeed Bush, John McCain, sponsored a "cap-and-trade" bill combining a national cap on emissions with tradable credits rewarding companies for reducing emissions. Pelosi created a climate committee that signaled the Democratic majority's commitment to even more aggressive action. The issue had such bipartisan buy-in that Pelosi sat on a love seat in front of the Capitol with her combative Republican predecessor Newt Gingrich to film an ad for Gore's advocacy group.

"We do agree our country must take action to address climate change," Gingrich said into the camera.

Searchinger had picked an opportune moment to start working on climate—and a terrible moment to take on ethanol. Biofuels had more political juice than ever, now as a win-win-win climate fix as well as national and economic security fix. Thanks to the Renewable Fuels Standard, ethanol plants were sprouting throughout farm country, creating jobs in rural towns and windfalls for their owners, building momentum for

Bush's effort to expand the mandate as well as new subsidies for "flex-fuel" cars running on high-ethanol blends.

"The spigot of public money is open and the pigs are rushing to the trough," the farm philosopher Michael Pollan wrote in an ethanol-bashing *Times* piece.

Pollan was at the vanguard of a left-leaning "slow food" movement that saw Big Ag and especially King Corn (the title of a 2007 documentary) as destroyers of diets and landscapes, bloating Americans with corn-fed Big Macs and corn-syruped Big Gulps. The foodie critics saw ethanol as a new dumping ground for surplus corn that agribusinesses couldn't pump into confined livestock or diabetes-inducing sweeteners, requiring so much fossil fuel to produce it was practically a fossil fuel itself. But their movement didn't have much sway outside the Bay Area's organic-kale precincts, and even California was crafting a "Low Carbon Fuel Standard" to promote farm-grown alternatives to gasoline. The state's Republican governor, the former Terminator Arnold Schwarzenegger, retrofitted his Hummer to run on soybeans.

If climate action had fledgling bipartisan support, biofuels had rabid bipartisan enthusiasm. McCain, who used to call ethanol a scam, hailed it in Iowa as a "vital alternative energy source," while Bush's deputy agriculture secretary came from the Corn Refiners Association. Democrats also sounded like lobbyists for biofuels, with Obama calling them "the future of the auto industry," and Hillary Clinton outflanking him by proposing a 60-billion-gallon mega-mandate. Even Gore had backed ethanol, a stance he later attributed to "a certain fondness for the farmers in the state of Iowa."

It was never surprising when politicians pandered to farmers. But Searchinger was irritated to see environmentalists boarding the bandwagon, touting ethanol as climate-saving methadone that could wean America off the heroin of climate-killing gasoline. He groused that many enviros who had never paid attention to agriculture were suddenly convinced farm-grown fuels could cure cancer, tooth decay, and whatever problems America had with interest rates.

The venerable Natural Resources Defense Council was driving the bandwagon, pumping out exuberant reports depicting biofuels as miracle elixirs that could cut U.S. oil use in half and send $5 billion a year to farmers. NRDC even trial-ballooned the idea of paying farmers through the Conservation Reserve Program—the agricultural cash machine Searchinger once raided for large-scale wetlands restoration—to grow biofuels on their land instead of leaving it to nature.

Searchinger couldn't believe it. Since when was enriching farmers a green priority? Weren't conservationists supposed to conserve conservation land?

In fairness, most enviros were less exuberant about corn ethanol than the future of "advanced biofuels" extracted from crop residues and other inedible feedstocks, especially cellulosic ethanol from grasses or woody biomass. They weren't commercially available yet, but they seemed like the next green thing, an ideal solution to the food-versus-fuel dilemma that forced bellies to compete with vehicles for grain. BP, temporarily rebranding as "Beyond Petroleum," was pouring $500 million into advanced biofuels research. Silicon Valley icon Vinod Khosla, an investor in several alt-fuel startups, told Congress that cellulosic ethanol could fulfill Bush's entire mandate within a decade—and replace most U.S. gasoline within two decades. Even Searchinger thought converting croplands back to grasslands to grow cellulosic feedstocks sounded promising.

Still, he winced when environmentalists rushed to defend corn ethanol as a vital "bridge fuel." One NRDC report titled "Growing Energy" warned that anything detrimental to today's ethanol industry could cripple tomorrow's better biofuels, which sounded like a threat to keep quiet about the downsides of corn or else the nice cellulose gets whacked.

Nathanael Greene, the leader of NRDC's biofuels campaign, knew corn ethanol wasn't perfect, but he believed advanced biofuels would eventually free transportation from the tyranny of gasoline. "We thought corn was OK for the moment, then once we get the better feedstocks, boy, that's a real solution," he said. The world needed cleaner substitutes for gasoline, and electric vehicles didn't look viable; another new documentary called *Who Killed the Electric Car?* chronicled how General Motors literally scrapped its first plug-ins.

In 2007, there just wasn't much green technology on the market, except those squiggly energy-efficient fluorescent light bulbs that glowed the color of jaundice. Wind generated 0.3 percent of U.S. energy, solar 0.1 percent. With almost no emission-cutting policies in place, even the modest cuts predicted by ethanol studies seemed like a nice start. In another rah-rah report, NRDC claimed corn ethanol was not only a gateway drug for the biofuels of tomorrow, it provided "important fossil fuel savings and greenhouse gas emissions reductions today."

Searchinger thought that was flat-out wrong. But even his own organization was flirting with farm-grown fuels.

The Environmental Defense Fund's motto was "Finding the Ways That Work," and its boardroom-friendly reputation defied stereotypes of tree-hugging extremists in Birkenstocks. After developing successful market-based trading solutions to acid rain and overfishing, it was now leading the cap-and-trade effort, working with industry groups on another market-oriented compromise that would provide incentives for companies to shrink their carbon footprints.

At first, Searchinger thought his new ethanol obsession would fit nicely with EDF's new climate focus. But he underestimated its leadership's eagerness to seek common ground with farm groups that had the power to crush cap-and-trade.

"We were trying to build bridges to sectors of the economy and parts of the country whose support we needed to pass a climate bill," the group's former senior vice president, Marcia Aronoff, told me later. "That had to be the priority."

EDF signaled that priority by hiring Sara Hessenflow Harper, a brash young aide to a conservative Kansas senator, to lead its outreach to farmers and Republicans. Harper, who grew up on a Kansas farm, had only recently accepted the reality of climate change, but EDF's leaders hoped her heartland roots could help them build a coalition for climate action. She was working with aggies on policies that could shovel even more cash to farmers who sequestered more carbon in their soils. She was also lecturing her new colleagues that climate activists had no credibility outside their liberal bubbles.

The tensions erupted on February 12, 2007, when Aronoff arranged an internal meeting to discuss Searchinger's concerns about ethanol. He gathered with a dozen colleagues in a conference room to make his case that land is not free. He explained how the studies that blessed ethanol as a carbon sink overlooked indirect land-use change, and warned that Bush's mandate would have a devastating ripple effect on forests and wetlands worldwide. Wasn't the Environmental Defense Fund supposed to defend the environment?

It was clear, from the body language in the room as well as responses over speakerphone, that his colleagues weren't with him. EDF's chief economist, Zach Willey, mocked the notion that every life-cycle analyst had overlooked the economic concept of "leakage," until one heroic Washington lawyer had recognized that a biofuels mandate in the U.S. would cause food production to leak around the world.

"You're saying no one else thought of this?" Willey scoffed.

Searchinger didn't like how the word "leakage" implied that only tiny droplets of farmland diverted to ethanol would leak elsewhere to replace the lost food. This felt more like spillage. Otherwise, sure, it was surprising that no one else had flagged the issue, but Willey's argument evoked the proverbial economist who wouldn't pick up a $10 bill lying on the sidewalk because if it were real, someone would've picked it up already. The world wasn't always so rational. Willey's insistence that the experts couldn't have missed a concept as obvious as indirect land-use change was severely undercut by the fact that they had.

Harper brought her old Senate briefing book to the conference room, and she tried to challenge Searchinger's substantive arguments with old talking points about the glories of ethanol. She didn't get far against a college debate champion who knew all the latest science. But she did score points on politics, arguing that a futile side battle over ethanol would alienate the farm interests EDF needed for the larger cap-and-trade fight.

"This is so stupid, we shouldn't even be talking about it," she scoffed.

The political team agreed. Its priority was cap-and-trade. Why should EDF jeopardize its credibility as a consensus builder on the most important environmental problem to indulge Searchinger's latest lost cause?

"Because ethanol is part of the problem!" he protested.

It was the first time in his career his bosses didn't have his back.

"Everyone was just like 'No, no, no, la, la, la, we don't want to believe you,'" he recalled.

Aronoff acknowledges that EDF's leaders sacrificed Searchinger's ethanol work on the altar of their climate ambitions. They wanted to assure farmers that greens could be their allies. "Tim felt very strongly, as he always does, but we had different priorities," she said. She saw him as a self-righteous moralist who viewed issues in black and white, when politics required shades of gray: "Tim is a fabulous litigator. But litigators aren't always so good at seeing where other people are coming from. He had trouble accepting that his strategy impinged on our strategy."

Searchinger did not abandon his own strategy. He tried a D.C. power move, getting a lobbyist friend to arrange a secret meeting with the National Pork Producers Council, so he could make the case that a more expansive ethanol mandate would make the grain its members fed their pigs more expensive. He hoped to stoke divisions inside Big Ag, pitting livestock interests who bought grain against corn interests who sold grain.

His scheme quickly unraveled. Many hog farmers also grew corn, so the pork council wasn't as anti-ethanol as he hoped. Anyway, farm groups understood the political power of sticking together even when their economic interests diverged. His confidential memo to the pork producers soon made its way to the ethanol boosters at the National Corn Growers Association, and one of their staffers forwarded it to Harper, advising her to check the author signature in the document.

"Author: Tim Searchinger. I couldn't believe it," Harper said. "I was trying to build bridges to these people!"

Searchinger was busted. Harper told their bosses he was sabotaging her outreach, and Aronoff ordered him to stop working on biofuels. She described the beatdown delicately: "Tim valued his freedom to do advocacy wherever he felt it critical, and that had become less of an option at EDF." Clearly, he had to quit.

The problem was, he had already quit before the ethanol conflict, giving notice that he was leaving in the spring. Now he couldn't even resign in pique, which was a shame. He had a flair for pique.

He wasn't a pure black-and-white moralist, though. He understood the value of compromise. He even understood that his bosses had the right to enforce their political strategy. He just thought giving farmers what they wanted on biofuels now and hoping they'd help on cap-and-trade later was a dumb political strategy. He also thought staying neutral on ethanol was a dumb climate strategy, because ethanol was dumb for the climate—and he intended to prove it.

He arranged to spend the next year affiliated with Princeton and a think tank, to continue the research his bosses had shut down. Before moving on to the next phase of his career, he wanted to finish analyzing ethanol, and show that all the previous analysts had done it wrong.

Incidentally, Harper now concedes she was wrong on the substance.

"I had that bias that ethanol was clean and good," she says. "Tim was right about that. He just ran into a perfect storm."

NIGHTMARE FUEL

Now Searchinger had to figure out those hard-to-predict variables that had scared actual scientists away from analyzing land-use emissions. How much would biofuel mandates boost grain prices? Where would new fields and pastures emerge, and which ecosystems would be sacrificed? Clearing more land would release a lot more carbon up front, but growing more corn would absorb a little more carbon every year, so his key task would be modeling how long it would take that modest annual carbon uptake to "pay back" the massive up-front carbon debt.

He knew nobody would give the benefit of the doubt to a biased lawyer, so he chose his models carefully. The farm sector loved the Argonne lab's GREET emissions model, which had already blessed ethanol; his only problem with it involved land use, so he used it to calculate emissions from everything except land use. He then persuaded modelers from farm-friendly Iowa State University to gauge how ethanol would affect land clearing, and the prestigious Woods Hole Oceanographic Institution to calculate the carbon impact of clearing different types of land in different parts of the world. It's said that all models are wrong, but some are useful, and together, Iowa State's agricultural model and Woods Hole's carbon inventory provided useful portraits of how U.S. ethanol might scramble the planet. They were also politically bulletproof, and scientists from both institutions later signed on to his analysis as coauthors, taking some sting out of efforts to discredit it as the dabbling of a dilettante.

After a few months wrestling with the models, he reached a preliminary conclusion: Replacing gasoline with ethanol would increase emissions for 51 years. Only then would the annual carbon uptake from growing corn and carbon savings from displacing gasoline repay the up-front carbon debt from clearing land. That was a damning result for ethanol, because the world couldn't wait half a century to start reducing transportation emissions. It needed to start yesterday.

Then he realized he had made a rookie mistake. He had forgotten that every carbon atom released by clearing land would bond with two oxygen atoms to form a carbon dioxide molecule. Carbon dioxide weighs 3.67 times more than carbon, so ethanol's impact would be 3.67 times higher than he'd thought, so emissions would keep getting worse for 187 years.

Whoa. That meant Bush's mandate would keep accelerating global warming until his grandchildren's grandchildren had grandchildren. Searchinger worried that nobody would believe such cataclysmic numbers, but tweaking every debatable assumption in more conservative directions only reduced the payback to 167 years.

Corn ethanol was clearly a new emissions bomb. The earlier papers said it was 20 percent better for the climate than gasoline. He found it 93 percent worse. Argonne had projected that one-fourth of the corn diverted to ethanol would be replaced with new farmland. His estimate was five-sixths, which meant much less hunger but much more deforestation.

If "leakage" evoked a cracked glass, this evoked a bottomless glass. Argonne's Michael Wang had estimated ethanol's land-use emissions at

2 grams of carbon per megajoule of energy, prompting his joke about offsetting them with local cornstalks. Searchinger's estimate was 104 grams. The GREET model pegged gasoline's entire well-to-wheel carbon impact at 95 grams, which suggested that even if there was some way to grow corn, brew it into ethanol, and burn it in engines without emitting any carbon, the climate would still prefer gasoline.

Things got even uglier when he analyzed the soy-based biodiesel that Schwarzenegger was using in his Hummer. Searchinger thought it would at least be less damaging than corn ethanol, because soybeans are legumes that "fix" nitrogen from the air, so they require less nitrogen fertilizer than corn and create fewer nitrous oxide emissions. But soybeans have lower yields than corn, so they use more land, and that overrode everything. When he ran the numbers, biodiesel emitted more than twice as many greenhouse gases as regular diesel.

Those were wow findings. They also had wow implications for advanced biofuels.

If land-use change alone made ethanol and biodiesel worse than gasoline, then corn and soy weren't the problem. Any biofuel that used good farmland would have the same problem, even the miracle elixirs of tomorrow that were supposed to cure tooth decay. Sure enough, he found cellulosic ethanol from switchgrass would increase emissions for 52 years, because using farmland to grow switchgrass would require new farmland to grow food. It wouldn't be much of a miracle elixir if it made climate change worse until his 99th birthday.

Clearly, land-use change wasn't just an overlooked source of biofuel emissions. It was the dominant source. Replacing fossil fuels with waste products like recycled cooking oil might help the climate, but any biofuels that used arable land would not, because the carbon losses from clearing nature overwhelmed any carbon gains from displacing fossil fuels. Searchinger told his economist colleague Ralph Heimlich he thought those findings should be published in *Science*.

Good luck, Heimlich said. His impression of *Science* was that it published scientists.

"Tim was confident, though," Heimlich said. "He's a confident guy."

"*OHHHH.* THAT'S PROBLEMATIC. THAT'S DISTURBING."

Publishing a scientific paper was an agonizing example of Searchinger's First and Second Laws about everything taking too long. But he still saw himself as a change-the-world advocate, not a chin-stroking academic, so he didn't wait until publication to start using his findings to fight biofuels. Congress was expanding the Renewable Fuels Standard. California was drafting its Low Carbon Fuel Standard. The European Union was preparing a mandate, too. Maybe if policymakers knew the truth about farm-grown fuels, they'd be less anxious to launch a global boom.

But policymakers wouldn't take a climate issue seriously if climate activists didn't, so he decided his first task would be reeducating the cheerleaders at the Natural Resources Defense Council. He wasn't sure they'd be open to reeducation. His career had crashed into the no-no-la-las of his EDF bosses, who didn't have strong feelings about biofuels beyond an unwillingness to offend farmers who did. NRDC officials had very strong feelings. Nathanael Greene had spent the first decade of his career promoting renewable fuels. NRDC climate director Dan Lashof was just as bullish. They were energy and climate experts who saw biofuels as energy and climate solutions, and they didn't see land use as a problem.

In April, Searchinger took some PowerPoint slides to NRDC's Washington office to explain why that kind of wrongness could charbroil the planet. Greene and Lashof doubted a lawyer with no climate background would shake their commitment to the most politically viable climate solution—until he explained the opportunity cost of land. They already knew corn ethanol was inefficient, but they had thought cornfields absorbed enough carbon to offset that inefficiency, and they had expected advanced biofuels to avoid that inefficiency. They were floored by Searchinger's argument that even highly efficient biofuels would flunk emissions tests if they occupied decent farmland.

"That notion was completely absent from the discussion until Tim raised it," Lashof said.

They did find his dismissive attitude toward climate experts uncomfortably reminiscent of climate deniers. Greene couldn't believe an amateur who hadn't even published anything yet seemed to expect him to renounce all existing biofuels science on the spot. But there was a stony silence after Searchinger explained that NRDC's plan to limit the U.S. mandate to "sustainable biofuels" that weren't grown on newly deforested land would be ineffectual, since biofuels grown on existing cropland would just induce new deforestation to replace the lost food crops. Land was fungible that way. He repeated his main point three times: Indirect Effects Are the Key.

"We were like: '*Ohhhh*. That's problematic. That's disturbing,'" Greene said. "Tim's a lawyer, he's a character, we weren't going to throw out everything we believed that day. But he got our attention."

NRDC had been lobbying for a maximalist Renewable Fuels Standard, but after some painful internal discussions, the group decided to start pushing to restrict the mandate to biofuels that significantly reduced emissions, and to require the emissions analysis to incorporate ILUC. Greene suspected that would only end up excluding corn ethanol, while Searchinger believed an honest analysis would exclude most biofuels, but they both agreed a scientific process should decide.

NRDC lobbyist Franz Matzner was given the awkward assignment of informing his colleagues in the eco-lobbying community that they needed to jam something called ILUC into the bill, because some unpublished science by a Washington lawyer suggested green fuels might not be green. They were rather bewildered, but environmentalists are supposed to respect science, even science that conflicts with their priors, and many of them did.

"It wasn't like everyone immediately said: 'Whoops! We were wrong!'" Matzner said. "But it was impressive how people came around."

Scientists are supposed to respect science, too. They're supposed to be data-driven, evidence-based. But they're people, and people don't like to be told they're wrong. So Searchinger didn't know what to expect when he sent a draft of his paper to Berkeley professor Alex Farrell, the lead author of the pro-ethanol *Science* article.

Farrell was a latecomer to academia, a former navy lieutenant who had worked as an engineer on nuclear submarines. The *Science* paper had made him a star. The Schwarzenegger administration recruited him to help draft California's Low Carbon Fuel Standard, and other states considering mandates began seeking his advice. At the time, BP was pouring half its $500 million biofuels investment into a six-story institute at Berkeley, and money seemed to be gravitating toward biofuels researchers. It seemed unlikely that Farrell would be supportive of a paper that directly challenged his own paper, not to mention the biofuels euphoria on campus.

But he was an honest scholar who believed science was about seeking the truth whether or not it benefited his career. He not only embraced Searchinger's findings, he invited Searchinger to brief the California agency overseeing his work on the Low Carbon Fuel Standard. Afterward, he told Searchinger he had never realized how much of ethanol's purported climate benefits came from making food unaffordable to the poor.

"But it seems like a straightforward result," Farrell wrote.

His coauthors were just as gracious. Dan Kammen, a renewable energy rock star who ran Berkeley's environmental policy center, had publicly declared that their paper proved ethanol critics were "just plain wrong." He wasn't shy about expressing views; he later resigned a federal job in a letter to Trump that spelled "IMPEACH" with the first letter of each paragraph. But he wasn't too proud to adopt new views in light of new evidence, even when the evidence came from a nonscientist.

"At first, we were like, 'Who is that random guy?'" Kammen said. "Then we were like, 'Oh, man, the random guy is right.'"

The *Science* article was the first publication for Richard Plevin, a software engineer pursuing a master's at Berkeley, and the response had been so exhilarating he decided to pursue a doctorate: "For a new academic, that paper was a grand slam, a juggernaut. It hooked me on publishing." But after reading Searchinger's draft, he recognized the juggernaut was flawed.

"Our reaction wasn't: 'How dare you rain on our parade?'" Plevin said. "It was: 'Why didn't we think of that?'"

Farrell warned his California bosses that new science was scrambling the math of farm-grown fuels, and that a mandate might create "enormous amounts of greenhouse gases." It was personally embarrassing and politically inconvenient, but he believed in facts and logic.

Searchinger refused to publish until every detail was airtight, and a funky modeling result for Argentine cropland took him months to fix. So Congress ended up passing the 35-billion-gallon mandate before his paper could help change more minds about biofuels.

However, Matzner and his fellow green lobbyists did negotiate a 15-billion-gallon limit for corn ethanol, so only two of every five U.S. acres of corn would be diverted to fuel, rather than five of every five. The rest of the mandate would be filled by advanced biofuels, with performance standards requiring all biofuels to meet emissions reduction thresholds. The enviros even secured language requiring the emissions analysis to incorporate ILUC, because biofuels lobbyists didn't yet recognize the threat that would pose to their businesses.

The corn lobby did secure one huge loophole: Existing ethanol plants were exempted from the performance standards. Still, Searchinger was delighted that Congress was mandating a scientific review of biofuels, since his paper would show that any honest emissions test would disqualify most of them. And *Science*, the most prestigious peer-reviewed journal, accepted it.

"What was surprising," Heimlich said, "was Tim wasn't surprised at all."

One reviewer was the ever-gracious Alex Farrell, who probably could have buried the paper, but instead offered generous comments. He had lost faith in renewable fuels, and he now envisioned a future of electric cars running on renewable power. He was encouraging his brother Mark, who sold carpets in Portland, to sell solar panels instead.

"Alex was all about the science," said Mark Farrell, who took his brother's advice, and would later chair Oregon's solar industry association. "He always asked himself: What could I be missing? The land-use thing was what he was missing."

In January 2008, Farrell came full circle, warning California regulators in bold italics that land-use change was a ***"very large contributor"*** to emissions, and ethanol was almost certainly worse than gasoline. A *Salon* column noted that Internet trolls responded by bashing him as a shill for Big Oil: "Pity the scientist who actually strives for truth."

What was most unusual about that column was that someone bothered to write it. It was hard to convince the public it mattered that some obscure scientists had made flawed assumptions in some obscure analyses about ethanol. It did matter, though.

Chief Kotok wore a beaded belt, flip-flops, and nothing else. The leader of the Kamayura tribe had a bowl haircut and a toothy smile that made him look like an underdressed indigenous cousin of the Three Stooges. He had three wives, 24 children, and a sly sense of humor; when I visited his grass hut in Brazil's Xingu National Park in 2008, my wife was pregnant with our first child, and he teased me about how much work I had to do to match the stress he faced every day. But Kotok spoke with solemn gravity about the decline of the Amazon.

"We are people of the forest," he said, "and the whites are destroying our home."

The Kamayura had virtually no contact with whites until the 1960s, living in isolation off the bounty of the forest. Now their refuge was surrounded by soybean fields and beef ranches, modernity had pierced their solitude—they had a diesel generator, which they mostly used to watch telenovelas—and the Amazon was collapsing around them. They didn't need to read academic studies warning of altered Amazonian microclimates and the savannization of the rainforest; they already knew the native plants they used for medicine and rituals were vanishing. Drought forced them to replant their own cassava crops five times that year, and wildfires left them with hacking coughs. Kotok had no idea biofuels were driving slash-and-burn agriculture deeper into the forest, but he sensed the root of the problem: "It's all because of money."

Searchinger had persuaded me that documenting how U.S. ethanol was ravaging the Amazon could be a compelling story for *Time*, with his upcoming *Science* paper as a hook. It was striking to see reality confirm his thesis that when biofuel mandates send grain prices up, forests come down. Our guide, John Carter, a Texas cowboy who had led a reconnaissance unit in Iraq before becoming an unlikely rainforest advocate, was not easily spooked—I saw him grab an anaconda with his bare hands—but he was downright panicked about the Amazon.

"You can't protect it," he said. "There's too much money to be made tearing it down."

In just six months, as exploding biofuel demand had driven grain prices to record highs, Brazil had lost an area of forest the size of Rhode Island. Wild land that was tillable or grazable had never been more valuable or vulnerable.

I got a sense of the frenzy from the Soybean King of Brazil, Blairo Maggi, a rough-hewn crew-cut billionaire who looked like an action-movie henchman. He was the world's largest soybean farmer, with half a million acres in the frontier state of Mato Grosso, and had a convenient second job as Mato Grosso's governor. Brazil had strict laws limiting how much farmers could deforest their land, but when I asked about enforcement during our interview in his plantation's mess hall, he laughed: "It's like your Wild West!" He asked if I had posed similar questions to the 19th-century pioneers who cleared America's forests for farms.

"You make us sound like bandits, but we just want to achieve what you achieved," he said.

Maggi assured me he wanted to save trees as well as feed the world, but when I asked if the biofuels boom made that harder, he grinned: "Ah, you've hit the nail on the head!" Clearing nature was more profitable than ever, and judging from the gleaming new tractors and combines scattered around his seemingly endless fields, he didn't lack capital to expand.

Maggi and other Brazilian soybean farmers did say they were no longer expanding into the Amazon, just pastures and a vast woody savanna called the Cerrado. But cattlemen were tearing down Amazon forest to replace pastures the farmers converted to soy, the domino effect of indirect land-use change. And the Cerrado was a jewel in its own right, a region larger than Alaska with 10,000 plant varieties and more mammal species than the African bush. It was vanishing so fast that we had to fly over hundreds of miles of farms to visit a sliver of Cerrado that one nature-loving shopkeeper had spent his life savings to preserve. There we saw toucans, a carnivorous flower that lured bugs by smelling like manure, and a chunky tapir that darted out of our path with confounding speed. We also saw a dazzling profusion of fish while snorkeling a protected stretch of crystal-clear river; the other local waterways were chocolate brown from farm erosion.

"The land prices are going up, up, up," the shopkeeper said. "My friends say I'm a fool not to sell. My wife almost divorced me. But I wanted to save something before it's all gone."

The ecological carnage in the Cerrado was a problem for the climate as well as biodiversity. Farmers, ranchers, and agricultural economists often dismissed it as "marginal land," and its trees weren't as tall or dense as the Amazon's. But they stored carbon that was being vaporized every day, and not many financially irrational landowners were risking divorce to save them. Land conversion was too lucrative, which was why Brazil ranked fourth in the world in emissions, just behind Indonesia in the race to transform jungle into cash.

Kotok and Maggi were the human faces of indirect land-use change, illustrating the incentives biofuels were creating to chew up nature. At the same time, by inflating grain and vegetable oil prices, biofuel mandates were contributing to a raging global food crisis. There were tortilla riots in Mexico, flour riots in Pakistan, and similar unrest over soaring prices in countries like Burkina Faso, Haiti, Indonesia, and Yemen that hadn't been too restful before. This wasn't Searchinger's focus, but the pain of the food crisis made a more powerful argument than life-cycle analyses ever could against using farmland to grow things that weren't food.

Humanity was just starting to take climate change seriously, so it was jarring to see the first solution with any traction making the problem worse—and also deepening the suffering of the poor. It was, as Searchinger promised, a compelling story, though not an uplifting one.

The temperature in Washington hit a daily record of 64 degrees on February 7, 2008, an appropriate backdrop for the release of Searchinger's paper about biofuels warming the earth. (That record has since been broken by *nine* degrees.) There was no school that day, so he brought nine-year-old

Noah to his interviews with the *Post*, the *Times*—which misidentified him as "Dr. Searchinger" in its front-page story—and *NBC Nightly News*. When they got home, before he could tell Brigitte about the most exciting day of his career, Noah blurted: "I can't believe what a boring job Dad has. All he does is sit around and talk to people!"

Adults showed more interest. The article that awarded Searchinger an erroneous doctorate was the most emailed on the *Times*' website. A *Post* cartoon showed a man at a gas station holding his study and yelling: "Wait! It turns out biofuels may be worse for global warming!"—while Uncle Sam kept pumping subsidies into the industry's tank. *The Onion*'s satirists elicited fake man-on-the-street reactions, such as: "Just once, can't one of our poorly considered quick fixes work?" Mark Delucchi, who first identified the land-use problem but declined Searchinger's offer to collaborate, sent along wry congratulations, musing that he should have hitched his wagon to the *Science* star's: "Wow, your paper has really made quite a stir!" A trade journal ranked Searchinger the number four enemy of ethanol, though in a tribute to the power of media, it ranked me number one.

Tesla delivered its first all-electric Roadster that week, but biofuels were still seen as the future of green transportation, and Searchinger was tarnishing their reputation, just as he had done to the Army Corps. Alex Farrell sent him a *Post* story featuring a feeble industry argument that even if ethanol did take 167 years to get climate-friendlier than gasoline, it would be worth the wait. "Got 'em on the ropes!" Farrell cheered. Drew Kodjak, director of the International Council on Clean Transportation, had set up a global working group of regulators to break down barriers to biofuels, before Searchinger exploded the idea that barriers were bad.

"We had this dreamy political situation, common ground between farmers and environmentalists, and it all evaporated after Tim's paper," Kodjak said.

NRDC soon disbanded its biofuels campaign, and Greene still sounds amazed that a mouthy lawyer convinced him to drop the cause of his career. In retrospect, he sees himself as one of the blind men in the parable who only feel a part of the elephant. Even though he knew nothing about agriculture, he had wanted to believe that "sustainably farmed" biofuels, whatever that meant, could help solve the energy problems he did know about.

"My life lesson was that most of us had blind spots because we came at the issue from just one angle," Greene says. "Tim had a sense of the elephant."

Still, Greene saw Searchinger as a political liability, alienating allies who still believed in the cellulosic dream. After Searchinger argued during a conference of Washington environmentalists that promoting even advanced biofuels had become a form of climate denial, Greene unloaded on him in a hallway: "You're hurting your own cause!"

"His gist was the math is so bad, if you still support any of this stuff, you're evil," Greene says. "It was always going to be tough to bring people together, because he was right, the math was bad! But he really pissed people off."

Searchinger knew his self-assured truth bombs pissed people off. But he didn't want to sugarcoat the science with caveats that would help critics claim the issue was complex and the answers unknowable. He wasn't sure how to make his case to biofuels fans in a way that was (a) straightforward enough to make it clear they were wrong, not just possibly wrong in certain situations, yet (b) polite enough to avoid implying they were being ridiculous if they didn't admit they were wrong: "I'm sure focusing on B would be more tactful, but would it achieve A?"

Searchinger thought many enviros, after years of being stereotyped as shrill, were now too eager to look reasonable. They didn't seem to understand that applying maximum pressure led to better deals than telegraphing desperation for a deal. He joked that if a powerful industry proposed incinerating all babies for renewable fuel, some green groups would counter that only some babies should be incinerated.

"Every goddamn group accepts the 'reality,' so the 'reality' becomes automatic," he emailed a friend. "Congress is where it is in large part because biofuels have no opponents!"

He believed a big part of the problem was that most enviros were as ignorant about agriculture as Greene, who at least admitted his ignorance. They complained about pesticides and genetically modified crops, when the real problem with agriculture was that it covered two-fifths of the planet's land. Now they wanted to use agriculture to solve their climate problems, while ignoring agriculture's climate problems. They seemed determined not to see the elephant.

Searchinger was trying to do a better job tolerating people who hadn't done the reading, but he wasn't chasing congeniality awards. His family loved him, even if his son found his job dull. He wasn't going to stop telling hard truths to avoid pissing people off. Seeing honest biofuels defenders like Greene and Farrell admit they were wrong gave him hope that others might listen, too.

Greene moved on to other clean energy issues, and he's still a well-respected advocate at NRDC. Farrell did not move on. Two months after Searchinger's paper replaced his own as the definitive analysis of biofuels, Farrell shot himself in his Berkeley apartment. He had been scheduled to testify about biofuels that day in the Minnesota legislature. He was 46 years old. Everyone was stunned.

Mark Farrell remembers his brother as a workaholic who retreated to his computer after Thanksgiving dinners, a meticulous researcher determined to get every fact precisely correct. He thinks Alex may have struggled to come to grips with his role in promoting green fuels that didn't turn out to be green.

"He had such high expectations for himself," Mark says. "I don't know. It's speculation. I just know he did amazing work, and it stopped in 2008."

THREE
WEIRD SCIENCE

"YEAH, SO WHAT."

A week before his *Science* paper rocked the biofuels world, Searchinger previewed it at a Farm Foundation workshop at a Miami Beach hotel. The workshop was supposed to explore how farm-grown fuels could help the climate, so his talk was bracingly off-message. He sensed an uh-oh hush as he told a few dozen life-cycle-analysis nerds, in bland prose that almost disguised the force of his smackdown, that their rosy conclusions were flat-out wrong. They were assuming a fantasy world where using farmland for fuel wouldn't reduce the amount of farmland available for food. They didn't understand that land wasn't free.

At least one listener wasn't shaken by his bombshells: Argonne's Michael Wang, author of the only previous biofuels analysis to consider indirect land-use change. Wang could imagine the phenomenon might be worse than his estimate, but not 52 times worse. "I felt, 'Yeah, so what, I stand by my numbers,'" he said. He bantered awkwardly with Searchinger during a break, making it clear he wasn't convinced.

Imagine I'm pointing a gun at your head, Wang said. Are you certain you can precisely predict the magnitude of the land-use change?

Of course not, Searchinger replied. *Duh*, he managed not to reply. Obviously, corn ethanol's domino effect on nature might not release precisely 104 grams of carbon per megajoule. But it would release a lot of carbon! Wang had concluded it would release virtually no carbon, because he had assumed diverting tens of millions of acres of corn from food to fuel would somehow induce no forest losses and virtually no farmland expansion. How certain was Wang about that?

The fact that the answer is uncertain, Searchinger thought, *doesn't mean your answer isn't wrong.*

He noticed that in this setting, Wang was a geek-VIP, holding court in a three-piece suit. Everyone wanted to chat up the GREET Man who had done biofuels life-cycle analyses before biofuels or life-cycle analyses were trendy. Wang's admirers—agricultural economists at land-grant schools, government biofuels experts—would have much cooler jobs if he was right that land-use change was no big deal. Upton Sinclair's line that it's hard to get a man to understand something his salary depends on not understanding was true for self-image as well as salary.

These weren't Searchinger's people, and he didn't plan to spend much time in their subculture. He still saw himself as the guy who pressured the government to follow the science, or sued the government for ignoring the science, not the guy who did the science. And he no longer enjoyed telling wrong people they were wrong as much as he had when he was young; he felt more like the Holly Hunter character in *Broadcast News* whose boss tells her it must be nice to always know better than everyone else in the room. "No," she replies. "It's *awful!*"

In any case, he had hoped his paper exposing land-use change as a very big deal would shatter illusions in the research community, especially now that biofuels were driving starvation as well as deforestation. But seeing all these number-crunchers paid to crunch numbers about biofuels, he realized they might not want their illusions shattered. And as long as he had to admit there was uncertainty, they might not have to admit they were wrong.

Searchinger wasn't surprised when Wang published the first scientific critique of his paper. He was surprised how much of it was demonstrably false.

For example, Wang claimed Searchinger's study assumed no increases in crop yields, when in fact it assumed steady increases in crop yields. Wang had to retract that claim, but he didn't retract another claim that Brazil's declining deforestation rate proved Searchinger's land-use warnings were overblown, when in fact Brazil's deforestation rate was soaring.

Searchinger was perplexed that a taxpayer-funded scientist could publish such nonsense. He liked to say he only played a scientist on TV, but he earnestly believed science should be a quest for truth. That's why he had delayed his paper to sort out that Argentine data hiccup, trying to compensate for his thin credentials with over-the-top rigor. To err was human, but Wang and other critic sseemed to be flinging spaghetti against Searchinger's wall to see what might stick.

"They make a combination of bad misrepresentations about our paper, some quite funny if they weren't being offered seriously, and a host of other factual and logical errors," Searchinger complained to an ally.

Many of the errors involved another economic concept known as "additionality," the idea that not everything that happens after a biofuel mandate matters, only *additional* things that happen *because* of the biofuel mandate. For example, Wang argued that since U.S. corn exports increased a bit during the ethanol boom, Searchinger was wrong to suggest biofuels depress exports. But Searchinger never predicted an absolute decline in exports, only fewer exports than there would've been if some grain hadn't been diverted to ethanol. (In fact, U.S. corn *harvests* increased a *lot*, so the smaller increases in exports suggested ethanol probably did drag them down.) If you ignore additionality, the fact that one child got a C after studying for a test while another got a B without studying could imply studying is counterproductive, when in reality the first child might have flunked if he hadn't studied, while the second might have gotten an A if she had.

He was appalled how frequently his critics botched additionality. Some argued biofuels shouldn't be blamed for deforestation because there were other causes—like demand for animal feed or timber, or, to quote one chunk of academic gobbledygook, "interactions among cultural, technological, biophysical, political, economic and demographic forces within a spatial and temporal context." Yeah, so what? The fact that forests faced other threats was irrelevant to the impact of biofuels. It was like arguing that cigarettes don't cause cancer because asbestos causes cancer.

Other critics floated scenarios that could free up farmland to grow biofuels without deforestation, like a global shift away from meat. Searchinger called them "anyway errors," because they gave biofuels credit for unrelated shifts that might happen anyway. It was like arguing that cigarettes don't increase the risk of heart disease because smokers might eat healthier to reduce their risk of heart disease. Yeah, so what? The extra land that would become available if the world ate less meat could solve a lot of problems—much more on that later—but it had nothing to do with the question of whether growing biofuels plowed up more land.

"Someone, please, take introductory economics!" Searchinger vented to another ally.

The critics portrayed him as an amateur with an agenda, which was true enough, but some had their own agendas. Wang criticized him for failing to use a "general equilibrium model" of the whole economy—a debatable technical point, except, as Searchinger noted in a prickly response in *Science*, Wang's GREET model wasn't general equilibrium, either. The glass-house stone-throwing suggested the dispute might not be purely technical. Even Wang's friend Mark Delucchi believed he was never likely to embrace findings that challenged his department's support for biofuels or his GREET Man stature.

"He had a lot of institutional and psychological incentives to dismiss land-use change," Delucchi said.

It quickly dawned on Searchinger that from a public relations standpoint, the merits of the debate barely mattered. The mere existence of the debate ensured his paper would be just one side of a he-said-he-said story. (There weren't many female life-cycle analysts to offer she-said.) There couldn't be a scientific consensus, especially one challenging a political consensus, when there was scientific dissent.

That's why a biofuels trade group trumpeted Wang's critique in a press release: "Scientists Raise Doubts About Recent Studies on Biofuels and Global Warming." The doubt-raisers weren't objective, either; one was a biofuels entrepreneur who ran Berkeley's biofuels center. But there couldn't be scientific certainty if scientists were raising doubts. The doubters all argued the issue was complex, which was true, and complexity was the enemy of clarity.

Searchinger felt like he was back playing Whac-A-Mole with the Army Corps, knocking down sloppy arguments only to see new ones pop up. He had to do hours of research every time, and his critics rarely admitted their errors. Why would they? He was learning what's known online as Brandolini's Law, or the "bullshit asymmetry principle": The energy required to refute misinformation is an order of magnitude larger than the energy required to produce it.

Science's editor-in-chief told Searchinger his paper attracted so many rebuttals the staff worried he must have overstated his case, until they began reading his lawyerly point-by-point rebuttals to the rebuttals. The environmental publication *Grist* declared that his response to one critique in *Science* by the cellulosic ethanol investor Vinod Khosla "left a glowing crater where Khosla's argument once stood." Most of his critics seemed to be bringing rhetoric and vibes to a fact fight. And it definitely bolstered his credibility that Brazil's deforestation rate had doubled in a year; he compiled a long list of studies by institutions like the World Bank and the Food and Agriculture Organization that all concluded biofuels had played a major role.

Still, the odor of controversy lingered. A month before his death, Alex Farrell forwarded an email from a California regulator, who warned that the indirect land-use change issue had gotten so contentious that state officials wanted to ditch ILUC entirely: "Right now what we're getting is that land use is very much in dispute and Searchinger is getting trashed."

It was discouraging to see how irrelevant the accuracy of the trashing was to its impact. Whether or not there was any signal in the noise, the noise became the signal. The issue was in dispute, who could say who was right? While the contrarian investors who shorted the housing market would soon make fortunes on their bets against the experts, he was still getting dismissed and no-no-la-la'ed.

In finance, there were consequences for being wrong and rewards for exposing the wrongness of the herd. In science, not so much.

• • •

There was one critique he didn't whack right away.

It came from Paul Hodson, a well-spoken Brit who was the top civil servant pushing biofuels at the European Commission's energy ministry. Hodson was a tall, courtly bureaucratic knife fighter who, like Searchinger, spoke fluent science and economics without scientific or economic training. And ILUC was complicating his efforts to ram a 10 percent biofuels mandate into European law.

So he sent Searchinger an email floating a tentative alternative theory: Perhaps when crop prices rise, farmers produce more crops per acre instead of clearing more acres for crops. If biofuel-driven price hikes encouraged farmers to try to boost their yields rather than expand their fields, by investing in fertilizer, seeds, and tractors rather than clearing wilderness, then ILUC might not be such a problem. And Searchinger's paper had acknowledged that rising prices could trigger small yield gains, so it made sense to measure them. "This is the first really good challenging question I have had," he replied. He promised a longer response soon.

Hodson was too savvy to wait. He immediately circulated a much less tentative memo confidently declaring Searchinger's findings were way off, because higher crop prices would induce huge yield gains and almost no land conversion. He even quoted Searchinger's email about his "really good challenging question," which almost gave the impression they agreed.

Hardly. Searchinger believed the literature was clear that the main drivers of yields were weather and technology, not prices. Farmers who got paid by the bushel obsessed about yields even when prices were low; the cliché was that their top priorities were yield, yield, and yield. They seemed to overuse fertilizer, plant the best seeds available, and splurge on John Deere's latest toys regardless of market conditions. And it seemed absurd to claim that shifting millions of farm acres from food to fuel wouldn't affect farm acreage or food production. Hodson's theory that farmers would automatically boost yields enough to replace the lost food on their existing land implied that taking land out of food production could *never* reduce the amount of available land or food.

"If Paul were right," Searchinger wrote, "we should willy-nilly convert agricultural lands to forest, and the food will magically be made up!"

It was a textbook example of the delusion that land was free. In Hodson's world, the benefits of the road not taken were apparently guaranteed even when you took the other road. And when Searchinger and his Yale economist friend Steve Berry read the papers Hodson had cited, they found no evidence supporting his conclusion that crop prices drove yields. Berry's notes were withering: "makes no sense," "exactly backwards," "a freshman economics point." Hodson even mixed up "total factor productivity" with plain old productivity, a flub Searchinger mocked as an "Emily Litella error," in honor of a hard-of-hearing *Saturday Night Live* character who delivered blistering commentaries based on basic misunderstandings.

In the real world, corn yields had doubled over three decades while corn prices had fallen 85 percent, so Hodson's insistence that prices were the main driver of yields smelled like more bullshit. He was a vegetarian who took trains to faraway conferences to limit his own climate harm, but he refused to entertain the possibility that biofuels could cause climate harm.

Hodson's memo was also a reminder that for many biofuels advocates, higher crop prices that made groceries more expensive and food crises worse were a feature of mandates, not a bug, shoveling more cash to First World farmers. To Searchinger, this was a moral outrage. Biofuels already used 5 percent of all grain and 8 percent of vegetable oil while 16 percent of kids were malnourished, which was why anti-hunger groups like Oxfam and food conglomerates like Nestlé were echoing his concerns about farm-grown fuels. The Jamaican reggae singer Livebroadkast, in his 2008 protest anthem "Biofuel Song," expressed similar concerns about this "crazy idea" with a better beat:

> *Biofuel use is gonna burn up all my food*
> *Deforestation can only mash up our nation.*
> *Evil men with that wicked intention*
> *What is your plan? Is it life or destruction?*

Searchinger was sure that if E.U., U.S., and California policymakers stuck to the science, they'd agree biofuels were a crazy idea. Humanity would need to use 30 percent of its crops to provide 2 percent of its energy, a wildly inefficient use of land.

The intrinsic problem was that photosynthesis itself was wildly inefficient. Searchinger liked to say it was invented 3.5 billion years ago by blue-green algae that weren't that smart. Corn plants converted less than 0.2 percent of sunlight into usable energy, while solar panels were 20 percent efficient, so it could take 100 acres of corn grown for ethanol to generate the energy of a one-acre solar array.

From a climate perspective, it would make far more sense to devote that one acre to solar, then use the other 99 for carbon sinks, or for food crops that would reduce the need to clear other carbon sinks. It would make even more sense to use rooftops or deserts for solar while devoting all 100 acres to doing what only land could do—growing food and forests.

But Searchinger wasn't sure policymakers would stick to the science. He was struggling to persuade scientists to stick to the science.

EUROPEAN THEATER

The food crisis of 2008 looked like it might unite the science and politics of biofuels in Europe, sparking widespread opposition to the E.U.'s draft 10 percent mandate. Siphoning grain from people to vehicles swiftly lost appeal as bread prices doubled and food inflation became a kitchen-table issue. Europe's biofuels industry launched a desperate ad blitz denying responsibility, which one pundit compared to *The Naked Gun*'s Frank Drebin shouting "Nothing to see here!" in front of an exploding fireworks factory.

But Hodson knew European politicians still wanted to appease farmers, and, more nobly, to advance climate action before the 2009 climate summit in Copenhagen. If you didn't read *Science*, alternative fuels still sounded like climate action; their green reputation had such cultural resonance that *Cars 2* was about to feature Lightning McQueen rescuing an alternative fuel from an oil-baron plot. So Hodson launched his own campaign to rescue alternative fuels in Brussels, spreading word that an American ideologue was peddling unsettled science. When Searchinger flew into town to brief legislators, Hodson finagled his way into the briefing to argue the other side. An Oxford dissertation on the mandate later quoted insiders griping that he kept battering politicians into submission with out-of-context facts and figures.

"He just bamboozles them with all the technicalities, and they just come out a bit dizzy with the impression the guy clearly knows what he's talking about, and that's it, he gets carte blanche," one critic said.

The leader of the resistance to biofuels in the European Parliament was a Green Party politician named Claude Turmes, a yoga teacher and environmental activist from Luxembourg. He recalled Hodson as "Enemy Number One," a shrewd operator with a flair for making his opponents sound radical. He reminded Turmes of a victory-obsessed rugby captain—all over the field, playing rough, making things happen.

"He knew how to move the show," Turmes said.

An engineer in the commission's environment ministry once challenged a series of errors in a Hodson presentation, including an equation that conflated mass and energy. He waved off her concerns as if she were a precocious girl who didn't understand grown-up politics. "I know you think the math is wrong," he said, "but the result is correct." He mocked another official's "exaggerated belief in the truth," gloating that perception mattered more than reality in government work. He kept warning legislators that attaching an ILUC test to the mandate would kill biofuels, an implicit admission they would flunk the test.

The food crisis did help Turmes push surprisingly tough conditions through Parliament that September—shrinking the mandate to 5 percent, limiting it to biofuels that dramatically reduced emissions, and attaching an ILUC provision that looked strong enough to kill crop-based fuels. "I don't think you could have come out any better," Searchinger exulted to his European friends. Still, the final deal would hinge on a closed-door multinational negotiation, so he also quoted his mother's advice: Celebrate today, for tomorrow things will fall apart.

"Your mother must have been an environmentalist as well," one of his fellow fighters replied. "That's definitely our credo."

• • •

Things did fall apart.

Lehman Brothers collapsed three days later, sparking a global financial panic that made climate issues seem less urgent. The biofuels mandate was part of a sweeping directive requiring 20 percent renewable energy throughout Europe by 2020, and some countries wanted to put it all on hold. For climate-hawk politicians like Turmes, seizing a generational opportunity to jump-start solar, wind, and other renewables was a higher priority than making sure the fine print of a biofuels mandate reflected controversial new science from an American lawyer.

French President Nicolas Sarkozy was serving as E.U. president, and he also wanted to do something big on climate. But when it came to biofuels, France and its muscular farm lobby wanted the original 10 percent mandate, with none of Parliament's green strings. Hodson threatened negotiators that imposing any conditions on biofuels could doom the entire renewable energy directive.

Sure enough, the final deal in December 2008 restored the 10 percent mandate and shrank the required emissions reductions. It also stripped out ILUC rules, although Turmes did secure an ILUC study as a consolation prize. Searchinger's European allies were so angry they began plotting a media hit on Turmes for selling them out. But as worried as Searchinger was about the coming renewable fuel boom, he thought it was silly to blame a sympathetic politician for caving to reality—especially when the deal ensured a renewable electricity boom that could actually reduce emissions.

"It's almost never a good idea to burn your bridges with your friends," he advised the group.

He had mellowed enough in middle age to see the problem wasn't Turmes or even Hodson. The problem was structural. Farm interests had too much power. ILUC science was too contentious. And renewable fuels sounded too appealing.

Unfortunately, the same structural problems that saved biofuels in Brussels persisted in Washington and Sacramento.

SYSTEMATIC OPTIMISM BIAS

Congress and California, unlike Europe, had required biofuels to pass ILUC analyses. But they hadn't required the analyses to be any good. Their early drafts reminded Searchinger of Army Corps studies of Army Corps projects: All the questionable assumptions made biofuels look better.

The EPA and California models both conveniently assumed many crops diverted to biofuels wouldn't need to be replaced because higher crop prices would reduce food demand, rewarding biofuels for making grain more expensive so that the poor would eat less food and literally exhale less carbon dioxide. The models then assumed other crops diverted to biofuels wouldn't need to be replaced because farmers would simply grow more food on their existing land, reviving the Miracle Yield theory he thought he had already debunked.

The models also minimized the carbon impact of any crops that did need to be replaced. Even as Indonesian peatlands were being drained for palm oil, EPA's assumed new farmland could never replace wetlands; even though ILUC was tearing up the Amazon, California's assumed almost no forest conversion. How could regulators evaluate the potential damage from land-use change when their models practically ruled out the most damaging changes?

Searchinger got a sense politics might drive California's results when he reached out to Dan Sperling, the gray-bearded policy maven who had designed the Low Carbon Fuel Standard with the science-obsessed Alex Farrell. Sperling was also a scientist, but when Searchinger urged him to rule out biofuels grown on good farmland, he wrote back that "the real question" was whether that would be "politically palatable."

Searchinger's suspicions deepened after an exasperating exchange about Miracle Yield theory with Thomas Hertel, the Purdue agricultural economist overseeing California's land-use model. The Purdue model assumed a huge "yield elasticity" of 0.4, which meant if corn prices doubled, corn yields would soar 40 percent, dramatically reducing the pressure to clear land. There's an old joke about an economist stranded on a desert island with a can of food who solves his problem by assuming a can opener, and yield elasticity sounded like a similarly fortuitous can opener.

When Searchinger dug a bit, he saw one of the studies Hertel had based his 0.4 value on had found no yield elasticity whatsoever. He wrote a sharp note calling 0.4 "basically made up," and Hertel agreed to reduce it to 0.27, which he said was the result in a more recent study.

That was wrong, too. That study didn't find any relationship between prices and yields. Hertel again conceded his error, yet defended his numbers. He believed prices drove yields, and insisted anyone who didn't had his head in the sand: "Just look at what has been happening to corn yields in the last two years!"

Searchinger looked. Nothing had been happening to corn yields in the last two years, even though prices had soared.

In short, more bullshit. Searchinger considered Hertel relatively thoughtful for a biofuels researcher, but he was still flailing around for evidence to confirm his priors. When Yale's Berry did a comprehensive literature review, he concluded yield elasticity was definitely below 0.1, probably closer to zero. Hertel still went with 0.25—not just for U.S. corn, for all crops everywhere. There was no reason to think farmers outside the U.S. with less access to fertilizer and easier access to additional land would make similar choices, but that one number derived from a misreading of one study of one crop in one nation drove California's entire analysis of global land-use change.

Hertel defended his work in an interview, accusing Searchinger of turning every interaction into a litigation, defending his team's performance in a "hyperpoliticized" situation.

"Searchinger said our numbers were too low, the industry said too high, nobody was happy," Hertel said. "It was a circus."

But the fact that the numbers had critics on both sides did not make the numbers reasonable, and the critics warning California regulators that the Purdue analysis drastically understated ILUC were repeatedly no-no-la-la'ed. "We'd wave our hands around at meetings and explain it didn't make sense," said Berkeley's Michael O'Hare. "The numbers didn't change." The charade reminded him of the classic economist lament about a beautiful theory tragically murdered by a vicious gang of facts—except the theory that biofuels were low-carbon survived.

Dan Sperling was trying to prevent the murder of the entire Low Carbon Fuel Standard. It was his baby, and he feared overly aggressive ILUC results would get it strangled in its cradle. He already faced intense political pressure, from farm interests pushing crop-based biofuels as well as advanced biofuels investors lobbying Governor Schwarzenegger to ditch ILUC entirely. Sperling told Searchinger he had a "compelling case" on the merits, but not the politics.

"We had to be realistic," Sperling told me. "Tim had astronomical numbers that would've ruled out crop-based biofuels. That would've been a political disaster."

Clearly, the California process was wired to approve farm-grown fuels. Remarkably, it wasn't yet wired enough. In April 2009, Hertel settled on an ILUC value that was only one-fourth of Searchinger's, but still large enough to make ethanol look worse for the climate than gasoline. Growing booze for fuel was too inefficient to withstand even blatantly lenient assumptions.

But biofuels interests erupted in protest, and Hertel says California officials began pressuring him to "do endless work to refine my results." It was no mystery how they wanted those results refined. Hertel had the typical farm-friendly biases of an agricultural economist at a Corn Belt school, but he was honest enough to quit rather than submit.

"It was time to pass the baton," he said.

Politics was circling the science in Washington, too.

New biofuels groups called Growth Energy, fronted by retired four-star general Wesley Clark, and the Alliance for Abundant Food and Energy, formed by agribusinesses like Archer Daniels Midland and Monsanto, furiously lobbied the Bush administration to ditch ILUC. The EPA emissions model was designed by another agricultural economist from another land-grant university, and it was full of farm-friendly assumptions, but *Inside EPA* reported amusingly frantic concerns from industry insiders that any ILUC analysis would find biofuels worse than gasoline, "making it difficult to promote them as clean-burning renewable fuels."

The empire was striking back. Republican Senator Charles Grassley of Iowa, an avuncular corn farmer, delivered a thunderous floor speech accusing moneyed interests of backstabbing biofuels. Not only did he keep a straight face as he portrayed his Big Ag donors as bullying victims, he seemed genuinely offended anyone would oppose farm-grown energy: "Never before have the virtuous benefits of ethanol and renewable fuels been so questioned and criticized!"

Obama's election signaled a sharp departure from the Bush era on foreign policy, economic policy, and climate policy, but not biofuels policy. He was a Corn Belt politician who had always backed ethanol. His agriculture secretary, former Iowa Governor Tom Vilsack, was an ethanol diehard, and his energy secretary, Nobel Prize–winning physicist Steven Chu, was just as enthusiastic about advanced biofuels. Obama quickly delivered for the industry, announcing billions of dollars for new advanced biofuel refineries and upgraded ethanol plants, along with a new cabinet-level task force to promote the domestic biofuels industry. However, Obama's EPA concluded that because of indirect land-use change, corn ethanol's emissions were 5 percent higher than gasoline's—another whitewash compared with Searchinger's numbers, but not enough of a whitewash for new plants to meet the required 20 percent reduction.

"And then," recalled Margo Oge, a veteran regulator who ran the EPA transportation and air quality office, "all hell broke loose."

Oge said ILUC became the most politically charged issue she faced in three decades at the agency. Growth Energy accused EPA of using "speculative models to blame American farmers for deforestation in Brazil." Monsanto lobbyists bombarded her staff with inventive rationales for massaging the numbers. Farm interests dubbed her Margo Ogre, and House Agriculture Chairman Collin Peterson, a gruff rural Minnesota Democrat, accused his own party's EPA of sabotaging ethanol.

"I don't trust anybody anymore!" he shouted.

At a Senate hearing, when Grassley grilled Oge about her agricultural experience, she had to admit she had never been to a corn farm, which became his go-to anecdote for mocking out-of-touch anti-ethanol bureaucrats.

When Oge later flew to Des Moines to kiss Grassley's ring, she told the man in the next seat she worked for EPA.

"Oh, you must be the woman who never visited a farm!" he said. "You're famous in Iowa!"

Farm groups had even more leverage than usual in 2009, because Obama was pushing a landmark cap-and-trade bill to start reining in emissions, and they had the power to kill it. The bill already exempted farms from its emissions cap, but Chairman Peterson still vowed that no Democrats on his committee would support it unless it banned EPA from even considering indirect land-use change. Since few Republicans would support any Obama bill, much less an Obama climate bill—politics had changed in the two years since Gingrich and Pelosi sat on that love seat—the defections would have doomed it. Its House sponsors, Henry Waxman and Ed Markey, had to insert language prohibiting EPA from even looking at ILUC for five years.

Ultimately, the maneuvers didn't matter. Cap-and-trade died in the Senate, after conservative media shredded it all summer as a communist cap-and-tax scheme. Rod Snyder, the corn lobby's policy director, recalls the futility of selling it to farmers while Rush Limbaugh, Fox News, and the Tea Party movement were slagging it as an ecoterrorist plot to destroy America.

"They just heard it was Obama's climate bill and shut off," said Snyder, who later became the top agriculture adviser at Biden's EPA.

The death of cap-and-trade not only delayed U.S. climate action, it doomed the Copenhagen summit. Democratic Senator Tom Harkin of Iowa threatened to resurrect the ILUC ban anyway, until EPA officials promised to subject ILUC science to a new "uncertainty analysis." There wasn't much uncertainty about what the new analysis would conclude.

In the fine print of a spreadsheet buried in an appendix to EPA's revised analysis, corn ethanol performed a miracle.

The agency's model found that increased ethanol production would predictably increase deforestation in 2012 and 2017. But in 2022, devoting more farmland to fuel would somehow *reduce* the need to clear forests for new farmland. The model then forecast increased deforestation again for 2027. For some reason, it only expected things to get funky in 2022.

The miracle then compounded: EPA decided to base its entire analysis on that 2022 anomaly. Lo and behold, limiting the analysis to 2022 nudged corn ethanol's emissions 21 percent below gasoline's, barely clearing the 20 percent threshold. The cherry-picking of 2022 was as pristine an example of Reverse Murphy's Law as the math error that salvaged the Army Corps boondoggle in Missouri, stark evidence that the beautiful theory of biofuels could defeat any vicious gang of land-use facts. And it ensured that 10 percent ethanol blends would become the standard at American gas stations.

Now that ethanol was a reality, Dan Sperling wanted California to nudge reality in greener directions—for example, by offering bonuses to ethanol plants that ditched coal. His problem was that the Purdue modeling, while extraordinarily favorable to crop-based biofuels, wasn't quite favorable enough to qualify them for the Low Carbon Fuel Standard.

"We never liked crop-based fuels, but we couldn't just say: 'Go away, we don't want you here,'" Sperling said.

So the revisions to the Purdue model conveniently pushed crop-based biofuels above the necessary threshold. A paper in the *Journal of Cleaner Production* later chronicled their "systematic optimism bias," an academic way of saying Hertel's less principled successors jammed their thumbs on the scale. They repeatedly tweaked assumptions "based on extremely questionable logic," manipulating their model "almost beyond recognition." A paper in an obscure technical journal couldn't change the science or the politics, but it demonstrated the fix was in. Even Hertel agrees his Purdue colleagues went easy on biofuels, finding ways to produce the less damaging ILUC estimates that California needed to get ethanol to yes.

"They got a lot of positive feedback whenever they came in with lower numbers," Hertel told me. "These things can gather momentum."

Searchinger got unwitting confirmation of that feedback loop at a conference at Purdue, when an assistant dean seated next to him at dinner, unaware of his stance on biofuels, confided that the school faced heavy pressure from farm interests after its model spit out bad results for ethanol.

"Fortunately, the results changed," the administrator said.

Searchinger drew two conclusions from these battles, and his first was that models like this were usually bullshit, assumptions masquerading as results. Their inherent uncertainty about the future made them inherently susceptible to bias and wish-casting.

It was impossible to know precisely what would happen when millions of acres of food crops became fuel crops, but common sense suggested millions of acres of natural land would become farmland to replace the food. It was also impossible to predict the precise carbon impact, but common sense suggested a decent opportunity-cost comparison would be the amount of carbon that land would store if it reverted to nature. The models seemed like elaborate exercises in obscuring common sense, with results depending on random variables dredged up from sketchy studies that ratified the inclinations of the modelers.

Searchinger's other takeaway was that he had been naive to think a few lines of congressional text requiring ILUC analysis might derail the biofuel train. Too much money was at stake for policymakers to say, Oh well, if some model says something called ILUC might cause problems in some other country, forget biofuels. It was quaint to think books wouldn't be cooked and bureaucrats wouldn't be pressured to get rid of that something called ILUC.

Still, he was upset—not only about the cover-ups for corn ethanol, but about the congealing consensus that advanced biofuels would fix everything. Many public servants who recognized the dangers of crop-based fuels—Sperling in California, Oge at EPA, Turmes in Europe—genuinely believed noncrop biofuels would avoid those dangers. And models that downplayed ILUC enough to make corn ethanol look plausible made advanced biofuels look wonderful. EPA claimed cellulosic ethanol from switchgrass would cut emissions 124 percent.

Ugh. Using good farmland to grow energy crops might not trigger agricultural expansion in systematically optimistic models, but in the real world, land was not free. Searchinger thought the Obama and Schwarzenegger administrations were sending a dangerous message that land didn't matter.

"The significance of California's effort is what it teaches the rest of the world," he emailed another state official. "It's got to have the courage to lead correctly."

The official replied that she doubted biofuels could be more damaging than fossil fuels. But the climate didn't care about her doubts.

The biofuels frenzy that Searchinger feared quickly became a reality.

U.S. ethanol production doubled in two years, approaching its 15-billion-gallon cap by 2009. More than one-third of the U.S. corn crop was shifted to fuel, diverting enough calories to feed every American away from food production. Biofuel production also doubled in Europe, while palm oil imports tripled, contributing to unprecedented deforestation in Southeast Asia. Oregon and British Columbia passed low-carbon fuel mandates like California's, while India, Indonesia, and other nations crafted mandates of their own. And investors began throwing cash at any vegetation that looked like potential fuel, snapping up millions of acres in Africa to grow a shrub called *Jatropha* for biodiesel.

The biofuel train had left the station, and it looked like a runaway train.

BEYOND BIOFUELS

It was frustrating to face so many attacks on his work. It inflamed the competitive spirit that made him so annoying in childhood games of Monopoly. It was also frustrating to spend so much time defending his work, chasing down wild assumptions and sketchy data to prove dead parrots were dead. The bullshit asymmetry principle was costing him a lot of family time, reading time, and sleep.

But one nice thing Searchinger learned from all the battles over his work was that it was the work he wanted to do.

He hadn't enlisted in the biofuels wars to publish in *Science* or land an academic job. He just wanted to avert a disaster. He had planned to do his research, prove biofuels were climate menaces, then decide the next phase of his career. But at some point, as he struggled to get the world to accept truths he considered almost self-evident, as he found himself doing more low-profile, high-stakes work nobody else would, he decided this would be the next phase of his career.

He'd practice science without a license at the dangerous intersection of science and politics, bringing lawyer logic and introductory-economics-student common sense to the blinkered world of emissions analysis. He'd stay in the land-use fight, applying his figuring-out skills to new almost-lost causes. He'd continue his crusade against biofuels, but the land-use issues he was crusading about applied to food as well as fuel—and he had seen how often land use was misunderstood. Biofuels were only a small part of the agriculture story, and he wanted to figure out how agriculture fit into the climate story.

He had passed for an agriculture expert among U.S. enviros, but he still had gaping blind spots about which crops grew in which soils, how different livestock were raised in different regions, which styles of manure management were the most sustainable. Once again, he needed to learn about shit. And academia felt like a candy store for a synthesizer, so he was delighted when Princeton extended his position after his paper got so much attention. He liked the attention, too, even when it came from a cottage industry of biofuels defenders who kept churning out press releases attacking his work. He even liked teaching Ivy League students why land is not free.

"In what other context would intelligent people sit and listen respectfully to me pontificate for 40 hours?" he wrote a friend.

He got to spend half his time on a picturesque campus, wearing tweed jackets and corduroy pants rather than suits. His family moved to a bigger house on the same Takoma Park street, so he could still take the Metro to Amtrak to Princeton, reading all the way. He also moved to a new office in a neo-Gothic building with an allosaurus skeleton in its atrium, which made him feel like a real scientist. In the *Curious George* books, scientists always seemed to have dinosaur bones around.

Mostly, though, Searchinger liked being useful, and the wrongness around biofuels had demonstrated the usefulness of his wrongness-identifying talents. There had never been a more vital time to figure out what was good for the climate and what only sounded good, because climate policy was finally happening. Obama's economic stimulus was blasting $90 billion into green energy—not only biofuels but solar, wind, electric cars, and more. Climate hawks were even more excited about the

E.U.'s renewable energy directive—not its renewable fuels mandate, which he had exposed as a terrible threat to forests, but its larger renewable energy mandate, which looked like an excellent threat to coal plants.

But Searchinger didn't think it looked so excellent. It soon lured him into another crusade that would distract him from his day job investigating food and agriculture.

FOUR
DESTROYING THE CLIMATE IN ORDER TO SAVE IT

THE BIOMASS LOOPHOLE

If a tree falls in a forest—and then it's driven to a mill, where it's chopped, chipped, and compressed into wood pellets, which are driven to a port and shipped across the Atlantic to be burned for electricity in a European power plant—does it emit any carbon?

Duuuhhyee, of course clear-cutting and combusting trees releases their carbon. Wood-burning power plants actually send more carbon up their smokestacks than coal plants, because wood burns less efficiently than coal. And that doesn't even include the carbon emitted by logging, debarking, drying, transporting, and pelletizing the trees before incinerating them. Since trees also absorb carbon from the atmosphere when they're not incinerated, it's bizarre to think incinerating them wouldn't hurt the climate.

Yet the European Union's renewable energy directive had decreed "biomass power" inherently carbon-neutral. It was legally a zero-emissions alternative to coal and gas, just like solar and wind. The climate wasn't supposed to notice it. And that was perfectly consistent with IPCC accounting rules, which did not require countries to report carbon emitted by biomass plants as carbon emissions.

The rationale was familiar: Burning trees emits carbon, but growing trees absorbs carbon, so as long as burned trees get regrown, the amount of carbon in the atmosphere shouldn't change. You could have your trees and burn them, too. The case for ignoring smokestack emissions from biomass resembled the case for ignoring tailpipe emissions from biofuels: Unlike fossil energy, which liberates long-buried carbon, bioenergy merely recycles aboveground carbon.

But land is not free!

In early 2009, when Searchinger read the fine print of the renewables directive, he realized it could spark a European biomass boom. Solar and wind farms were still risky new ventures, especially during a brutal recession—expensive to build, hard to connect to the grid, producing only intermittent power—but wood already generated half the continent's renewable electricity, and it produced 24-hour power. His inconvenient land-use question was: Where would all the wood come from? If the biofuels frenzy posed a bank-shot threat to forests through ILUC, a biomass frenzy threatened to obliterate forests directly.

He had planned to pivot from bioenergy to food and agriculture, but biomass felt like Biofuels 2.0—another renewable alternative that would hike deforestation emissions a lot in order to cut fossil fuel emissions a little, another climate disaster that only penciled out as a climate solution if you ignored the opportunity cost of land. Just as analyses of using farms to grow fuel ignored the road not taken of growing food, the assumption that using forests to grow electricity had no carbon impact ignored the road not taken of leaving forests alone. Yes, growing a tree might eventually recapture the carbon released by burning a tree. But it didn't keep as much carbon out of the atmosphere as growing a tree and *not* burning it.

You had to ask: What if the tree didn't fall in the forest?

Biomass also had a familiar carbon debt problem. Burning a tree releases carbon instantly, while regrowing the tree reabsorbs the carbon slowly, and it isn't fully reabsorbed until the tree is fully regrown. How could that be carbon-neutral? Buying a house isn't money-neutral, even though you might sell it someday for the purchase price. The up-front carbon debts from burning trees would take decades to repay, when the world needed to reduce emissions now.

The more Searchinger read about biomass, the more he feared it would inspire a global race to plunder forests, another sacrifice of the land sector to try to green the energy sector. Less than two centuries after fossil fuels displaced trees as the world's leading energy source, it seemed crazy to turn back to trees, especially now that we needed them to soak up the carbon our fossil fuels had spewed into the sky.

What really scared him, as banal as it sounds, was the accounting. He liked to say that if you want to screw the world up a little, do a bad thing, but if you want to screw the world up a lot, set up accounting rules that incentivize everyone to do the bad thing. He wonk-joked that nobody in history had wasted more tax dollars than whoever devised Medicare's reimbursement rates, which made fancy specialists more attractive than preventive care. In the same way, carbon rules that declared tree-burning carbon-neutral would make burning trees more attractive than not burning them.

Now that governments were finally contemplating climate action, it was vital to get the banal rules right. Otherwise, carbon regulations, carbon markets, and other climate policies could wind up encouraging the destruction of carbon sinks. And it looked like it could happen here.

Gene Searchinger, the question-everything thinker who shaped Tim's life, died that April. The next day, Tim compartmentalized his grief to send a hair-on-fire email to his Princeton colleague Michael Oppenheimer: "Please read this memo right away." The emergency, he explained, was that the latest version of Obama's cap-and-trade bill had declared biomass carbon-neutral, which would turbocharge deforestation and possibly doom humanity. He needed to log off because of his dad, but he didn't want America's first emissions limits to create an emissions disaster, so he pleaded with Oppenheimer to help mobilize scientific opposition.

His memo synthesized the literature about bioenergy's potential for explosive growth in dry prose that only sounded like a scream for help when read carefully. But the studies it cited all quantified how accounting rules that treated land as free could help bioenergy devour the earth. A recent paper in *Science* had estimated that if bioenergy was treated as carbon-neutral, it could expand enough to wipe out *nearly all forests and savannas* by 2065. A more industry-friendly study suggested bioenergy would only require a land mass the size of Brazil, which sounded less awful, except the world didn't have a spare Brazil.

Searchinger didn't believe in letting the perfect be the enemy of the good, but an emissions cap with a forest-shredding loophole didn't seem good. He feared it would just encourage utilities to burn wood in antiquated coal plants rather than retire them. An accounting mistake that could wipe out nearly all forests and savannas seemed like a pretty big mistake.

"If we don't solve this problem," he wrote, "we are going to destroy the world!"

Since cap-and-trade didn't ultimately become law, Congress didn't create incentives to game the law by burning wood, so the world wasn't destroyed in 2009. But the carbon-neutrality loophole that made biomass so attractive endured—in Europe, in two dozen U.S. states with renewable power targets, and in the global accounting rules, which created incentives for countries to exempt biomass from their own rules.

The crazy loophole had surprisingly sane origins.

When the IPCC was created in 1988, its scientists knew that logging and burning trees emitted carbon. But if they had required countries to report emissions from biomass when the carbon was removed from the forest and again when it was released up the smokestack, the emissions would've been double-counted. So they only required the carbon to be counted in the land-sector account covering forests, exempting biomass from the energy-sector account covering power plants. In other words, if a tree fell in a U.S. forest to be burned in a European biomass plant, only the U.S. would count the emissions, so the global accounting would balance.

The problems only began after the biomass exemption made it into the Kyoto Protocol, which required countries to cut emissions, not just count them. That created perverse incentives no one noticed at the time.

The U.S. didn't join Kyoto, and developing nations were exempt from its emissions rules. But countries that burned biomass didn't have to count smokestack emissions even if the wood was logged in countries that didn't report forest emissions, so a system designed to avoid double-counting often ended up zero-counting. And the loophole incentivized the burning of imported biomass no matter where it came from. Searchinger pointed out that a European country could level the Amazon, import the wood, burn it for electricity, and count the entire process as a national emissions reduction.

It was an honest mistake, and when he explained it to Sir Robert Watson, the British climate scientist and former IPCC head, Watson gasped: "We did that?"

Searchinger had a knack for catching Army Corps–style mistakes hidden in plain sight, and gaming out how a law's bad incentives or a model's bad assumptions could drive bad behavior. Lawyers were trained to think that way. In his experience, most scientists didn't understand law, while most lawyers didn't understand science, and he tried to bridge the gap.

But once the phrases "biomass power" and "carbon-neutral" began appearing in close proximity, wood-burning acquired a green halo that transcended science as well as law. One civil servant who helped craft the E.U. carbon trading system told me his team made biomass eligible for lucrative credits because everyone assumed it was carbon-neutral.

"Whoops," the official said. "We've learned a lot since then."

Searchinger was now a quasi-scientist with Ivy League business cards, but he thought of himself as the kind of academic who does, not just thinks; he joked that in a less frivolous world he'd inspire a tweed-jacketed bobble-head action figure. He wanted to close the loophole, not just study it, so he decided to write a paper explaining it. Maybe he could shake up the debate over biomass the way his ILUC paper had shaken up the debate over biofuels—and maybe this time, policymakers would side with forests.

Once again, *Science* accepted his contrarian take. But once again, the bioenergy boom he feared was already becoming a reality.

Europe's renewable energy directive created a rush to burn trees for heat and power as frantic as the rush to burn crops for fuel, with hundreds of new plants coming online across the continent. European coal plants also began shifting to biomass, aided by subsidies from their home countries. This wasn't rocket science or complex climate science. It was Economics 101: When governments made it profitable to burn wood, wood got burned.

The U.K. led the way, giving a power company with the movie-villain name Drax Group billions of pounds to retrofit its largest coal plant to burn wood. The Drax compound in Yorkshire—where steam billowed out of cooling towers taller than Big Ben and smoke billowed out a chimney twice as tall—was an unlikely vanguard for the green-energy transition. But just as Willie Sutton robbed banks because that's where the money was, governments looking to cut carbon looked where the carbon was, and Drax was Britain's worst emitter. Soon it was rebranding as a green renewable biomass company, "enabling a zero-carbon energy future."

A new clean-tech industry was springing to life, and whether or not the tech was truly clean, a U.S. government report predicted biomass would soon be the fastest-growing source of new electricity. The IPCC and other global institutions projected that by 2050, it would produce one-fifth to one-half of the world's energy.

But all that biomass did have to come from somewhere. Europe's increased wood demand was already improving the economics of logging, from the primeval mountain forests of Slovakia and Romania, home to the lynx and wolves of folktales, to the boreal spruce forests of Canada, breeding grounds for billions of migratory birds. Drax imported most of its wood from pine and hardwood forests in the American South, where a new pellet industry was racing to build mega-mills as fast as it could secure permits.

The forecasters predicting spectacular increases in bioenergy were energy experts, not land-use experts, and they rarely extrapolated how much land would be needed to grow it. But Searchinger did the math, and it inspired him to write a sky-is-falling email to an environmental reporter at *The Washington Post*, apologizing for "sounding like the usual calamitous environmentalist" but begging for coverage of the new threat to forests.

His math showed that meeting the various bioenergy forecasts would require doubling to tripling the global biomass harvest. For every ounce of vegetation used for food, feed, cardboard, telephone poles, or anything else, two or three additional ounces would have to be harvested for bioenergy. If he sounded like a calamitous environmentalist, it was because that would be an environmental calamity. We were already using three-quarters of the planet's vegetated land for agriculture and forestry, so meeting all those bioenergy forecasts would require another planet. He felt a frequent urge to yell: DO YOU REALIZE THE IMPLICATIONS OF WHAT YOU'RE SAYING?

Burning trees in power plants to meet renewable electricity mandates looked like an even bigger threat to nature and the climate than burning crops in engines to meet renewable fuel mandates—and politically, it felt like another runaway train.

Until one government decided to get off.

THE NEXT REVOLUTION

Massachusetts was a progressive state with an ambitious Democratic governor, Deval Patrick, and an aggressive energy and environment secretary, Ian Bowles. They were Harvard-educated go-getters who envisioned the cradle of the American Revolution as the cradle of a climate revolution. Massachusetts was already a hub for knowledge industries like higher education and biotech, and they saw clean energy as its next economic engine. A climate-smart economy could be a beacon for the world, like the colonial city on a hill.

"Our mindset was: We don't have to wait for someone else to be out front," Bowles said.

Bowles always had a green streak. He grew up in the Cape Cod hamlet of Woods Hole, where his father, an ecologist, and mother, a science writer, worked on early climate projects at the institute that later helped Searchinger model biofuels. His first job was as a congressional aide working on America's first climate legislation. He went on to Conservation International and the Clinton White House before becoming Patrick's point man on environmental issues. When the governor signed a Global Warming Solutions Act even tougher than California's, *The Boston Globe* noted a "lanky" and "excitable" cheerleader celebrating at the statehouse, screaming, "First in the nation! First in the nation!" That was Bowles.

His goal was to make Massachusetts a laboratory for fossil alternatives, and biomass seemed like an enticing alternative for a state that was 90 percent forest. He was thrilled when developers attracted by green subsidies proposed three biomass plants in western Massachusetts. He was annoyed when activists resisted, flocking to hearings dressed as trees and inhalers to rant about clear-cuts and air pollution; one protester in a policewoman costume confronted his deputy, David Cash, identifying herself as Officer Spew-Not and ticketing him for crimes against the planet.

"Ian's a do-big-things kind of guy," Cash said. "His reaction was: What the hell? We're on the side of progress!"

Bowles thought many enviros would say no to motherhood and apple pie if it displaced a few twigs. He was a yes-enviro who wanted to get green stuff built. He was already waging a losing war to approve America's first offshore wind farm, against Cape Cod elites who liked their views of Nantucket Sound, and he was ready to fight biomass opponents with similar not-in-my-backyard attitudes.

He was particularly irritated when a no-no-no-enviro named Meg Sheehan founded a group called Stop Spewing Carbon to push a ballot initiative banning biomass subsidies. Bowles saw her as a rich progressive opponent of progress who fought zero-emissions nuclear reactors, hydroelectric dams, and even solar farms, not just fossil plants. It was as if she thought electricity grew on trees—and now that it could, she was fighting that, too. Bowles also thought special interests used ballot initiatives to rile up uninformed voters and accelerate the politics of no.

Then he realized he was an uninformed voter.

Bowles didn't care if biomass plants intruded on mountain views or routed logging trucks through bucolic Berkshires towns. Those seemed like fair prices to pay for climate progress. He did care about climate progress, though, and he began to wonder if the Stop Spewing Carbon gripes might have merit.

A western Massachusetts ecologist named Mary Booth had raised early questions in a *Globe* op-ed, noting "it takes a minute to burn a tree and 70 years to grow it back." Bowles ignored Booth, a whip-smart but abrasive gadfly who was notorious for going out of her way to pick policy fights. At one state forestry meeting, she had confronted a climate scientist, William Moomaw—the genial Tufts professor whose IPCC chapter had implied the world's carbon sinks could be sacrificed for bioenergy—and yelled at him to do the fucking math.

But after Searchinger's October 2009 *Science* article about the biomass loophole suggested everyone needed to do the math, Booth's questions got harder to ignore. Searchinger also cowrote a *Globe* op-ed with Vinod Khosla, the tech billionaire who had attacked his biofuels paper, but now shared his concern that declaring wood-burning carbon-neutral would make forests worth more dead than alive. Their column began: "Although the very term 'accounting rules' may cause most people to turn the page..."

Bowles did not turn the page. He met with Booth and Searchinger, and found their critical arguments disturbingly plausible. He had assumed the carbon from burning and regrowing trees would even out in the long run, but Massachusetts had committed to cut emissions in the short run, and huge up-front carbon losses from burning trees could put its targets out of reach. So Bowles commissioned an independent study and promised to follow the science wherever it led.

Robert Perschel, a Yale-trained forester who worked on the study, had also assumed biomass power was carbon-neutral when he accepted the assignment. But as Searchinger explained in *Science*, that wasn't true when the biomass had other valuable uses. Perschel soon realized he had overlooked the opportunity cost of using whole trees for energy when they could otherwise go to paper mills, inducing the logging of more trees to feed the mills. Burning branches, treetops, and other logging residues for electricity really would reduce emissions, just like brewing crop residues and other waste products into fuel, but Perschel knew there weren't nearly enough residues in Massachusetts to feed large-scale biomass plants.

"The industry kept saying 'No, no, we won't cut trees,' and it was clearly BS," Perschel said.

In June 2010, the independent study concluded that burning wood was even worse for the climate than burning coal. True to his word, Bowles suspended biomass subsidies. Searchinger sent a thank-you note to David Cash, expressing delight that a government had finally recognized the biomass emperor had no clothes.

"This action will have great value not just for Massachusetts but as an example to the world," he wrote.

"Your intellectual fingerprints are on it!" Cash wrote back.

The no-enviro Meg Sheehan met with Bowles hours before the filing deadline for her ballot initiative, and for once she took yes for an answer, leaving her boxes of signatures in the car. Some of her fellow fighters felt betrayed, but as Searchinger learned during his high school foray into activism, some activists enjoy fighting The Man more than persuading The Man.

"It just felt like Ian had come around, why keep fighting?" Sheehan said.

Bowles showed that public servants can adopt new policies when confronted with new evidence, just as Alex Farrell and Nathanael Greene had showed for scientists and environmentalists. It hurt to admit the no-no-no whiners in dopey costumes were right, but Bowles wouldn't keep pushing bad climate policies just because they were his policies.

Scrapping a few plants that would have provided 0.003 percent of global power capacity wasn't earthshaking on its own, but Searchinger hoped Massachusetts would inspire other governments to reconsider biomass subsidies. That could transform the energy landscape and the actual landscape, because burning wood without subsidies made no economic sense.

"I'm trying to balance out my usual role as Chicken Little," he wrote in a cheery update to environmentalists.

Growing trees, like growing crops, was an inefficient way to make energy. He calculated that even if America diverted its entire wood harvest to power plants, it would only supply 3.4 percent of American energy. The algae that invented photosynthesis had been too dumb to recognize its inefficiencies, but he thought people, armed with facts, would be smarter.

In retrospect, he thinks his optimism that one liberal state's evidence-based decision could avert a global biomass boom was as naive as his optimism that a congressional ILUC provision could snuff out biofuels. The biomass battle wasn't even over in Massachusetts; it would keep flaring up for more than a decade, and while the industry never succeeded in getting its subsidies back, Searchinger clearly got too excited about one science-friendly state official embracing his cause. He should've listened to his inner Chicken Little.

"It was incredible how hard we had to fight to get Massachusetts to do the right thing," he says. "That should have been a lesson."

The lesson was: As goes Massachusetts, so goes Massachusetts.

"SUBOPTIMAL CARBON SCENARIOS"

Obama's EPA momentarily embraced the Massachusetts model in 2011, announcing a bold proposal to crack down on biomass emissions. But it withdrew the proposal a week later, after the forestry industry's friends in Congress threatened to gut its authority to regulate any carbon if it tried to regulate biomass carbon. Forestry interests had almost as much Washington influence as farm interests, and they envisioned biomass power plants propping up wood prices just as biofuels propped up grain prices. For Searchinger, the result was another extended game of Whac-A-Mole, of refuting flimsy arguments only to see new flimsy arguments pop up.

The draft plan EPA unveiled that year for biomass—for promoting it, not regulating it—acknowledged that burning trees for energy wasn't inherently carbon-neutral, but said it was carbon-neutral in regions where carbon stocks were increasing. While that sounded less extreme than blanket carbon-neutrality, the logic was strange. Even if a region's forests were expanding overall, why would incinerating some of its trees have no impact? It was like saying that if a region's companies were profitable overall, all its companies were profitable, even the ones losing money. Would burning Olympic National Park's old-growth forests for electricity create no emissions if carbon stocks happened to be increasing elsewhere in the Pacific Northwest?

Obviously not. There was also an obvious alternative to burning trees from a carbon-positive region: not burning them, which would make the region even more carbon-positive. EPA was making an anyway error, giving the biomass industry credit for the growth of trees it didn't harvest, even if the trees would've kept growing anyway. It was like giving Ted Bundy credit for women he didn't murder. "Even for someone with long experience with Corps of Engineers science, this document is extraordinary," Searchinger emailed Mary Booth.

Booth went to Washington to appeal to Jennifer Jenkins, the forest scientist who led EPA's biomass study. A blizzard hit the capital that day, so Booth trekked through the snow to EPA headquarters to make the case that the carbon emitted by burning trees didn't vanish just because trees were growing somewhere else. Jenkins, who later became an executive at Enviva, America's largest pellet manufacturer, responded with a glare as icy as the sidewalks outside.

"She gave me the stink-eye the whole time," Booth recalled.

Searchinger got even more worried about the EPA's direction when its bioenergy advisory panel recommended analyzing emissions over a 100-year horizon, as if cutting carbon over a century was as good for the climate as cutting carbon right away. That was an implicit argument for delaying all climate action—and, if you assumed the trees would grow back, for clear-cutting the Amazon for bioenergy. It was like suggesting that confiscating all your money would make you no poorer if the confiscator pinkie-promised to return the money a century later.

The panel was stacked with academics from agriculture and forestry departments, but the driving force behind the 100-year horizon turned out to be Harvard geologist Daniel Schrag, director of the university's Center for the Environment and winner of a MacArthur genius grant. Schrag was not a shill or an idiot. But when Searchinger corralled him at a 2012 meeting, he matter-of-factly explained: Look, my models say the world will need to clear a couple hundred million hectares in the tropics for biofuels. We might as well start now.

Yikes. A certified genius wanted to clear a Mexico-sized swath of forest for energy, because the up-front emissions might be offset by 2112? Schrag just didn't think emissions over the next few decades mattered much in the long run. Geologists tend to think in geologic time, and Schrag, who was later disciplined by Harvard for bullying students, was not overburdened with intellectual humility.

Searchinger didn't think the world could wait a century to start cutting emissions. Time, like land, was not free. Bioenergy's up-front carbon losses could trigger doom loops that could desiccate the Amazon and ravage the climate long before the hypothetical long-term reductions ever arrived. He suggested that cutting carbon now with innovations like solar and electric cars might help stave off some of the deforestation Schrag considered inevitable. Maybe those models that said the tropics would eventually need to be sacrificed would turn out to be wrong. Schrag's approach violated the planetary Hippocratic oath: First, do no harm.

Schrag shot back that even if all cars went electric, biofuels would still be needed for trucking and aviation.

Searchinger replied: Why not wait and see? There's a reason you wouldn't buy your next computer decades in advance. Humanity needed to buy time, and accelerating bioenergy was like selling time.

The chat didn't change anyone's mind. Schrag saw himself as a pragmatic grown-up, charting a sensible middle ground between ideological extremists who thought bioenergy would destroy the world and self-serving industry types who claimed all bioenergy was carbon-neutral.

"Both sides were ridiculous," Schrag said.

Searchinger didn't see what was ideological about his peer-reviewed articles in *Science*, or what was pragmatic about the middle ground between right and flat-out wrong. Hardly anyone was watching this bureaucratic dispute, but he refused to let the 100-year proposal stand. He eventually pestered EPA's science board into overruling its own bioenergy panel, an unheard-of reversal, and he continued to harangue EPA officials that endorsing biomass power would encourage the world to double tree harvests and pulverize rainforests. But he sensed a catch-22: They assumed that if the problem was as hideous as he kept screaming it was, he wouldn't be the only expert screaming about it. Biomass wasn't an obvious climate menace like coal; the industry portrayed it as a climate savior. Enviva's website opened to a forested vista with a green ribbon and the slogan: "Displace Coal. Grow More Trees. Fight Climate Change."

He knew he sounded like Cassandra, but hadn't Cassandra been right to warn about Greeks bearing gifts? He again begged his better-known and better-liked colleague Michael Oppenheimer to join his fight, in a plea that doubled as a professional manifesto:

> There is a natural and generally accurate tendency to dismiss warnings of this kind as at least partially Chicken Little. It is also very hard to believe people of good will could actually be responsible for the levels of harm I am warning about. We also know that anyone working on an issue, in my case bioenergy, has a natural tendency to exaggerate its importance (although in my case, I would give my right arm to be able to abandon all work on bioenergy). On top of all that, in a large world, it is hard to believe any one of us has a necessary role to play.
>
> But sometimes a fire does break out, or a dike begins to break, and only a few people are in position to alert others of the consequences before it becomes too late. And sometimes, even Chicken Little does not convey the true extent to which the sky will in fact fall.

He saw biomass as a comically simple issue: burning trees bad, growing trees good. But it was worrisome that a genius like Schrag disagreed. Unlike the broader global warming debate, bioenergy wasn't a few well-paid cranks disagreeing with 99 percent of scientists. There really were grown-ups on both sides. At a Princeton conference, Searchinger watched a plant scientist named Steve Long argue that biofuels wouldn't induce deforestation even if they increased demand for new farmland, an argument so illogical Searchinger assumed he must be a bit slow, before learning he was a brilliant expert on photosynthesis. Searchinger tried to remind Booth—and maybe himself—that not everyone who overlooked opportunity costs was stupid or corrupt.

"Even though the errors are obvious to you and me," he wrote in an email, "they are not obvious to others."

Searchinger spent almost as much time trying to correct errors in Europe. He helped an E.U. science board with a report warning that biomass power placed "enormous pressures on the Earth's land-based ecosystems." He sounded similar alarms in a paper in *Energy Policy*, recruiting scientists from a dozen E.U. nations as coauthors to maximize its impact. He also threw himself into the fight against the Drax biomass subsidies, sending the U.K. energy and climate ministry a long critique of its support for wood-burning plants, then contributing to a Greenpeace U.K. report titled *Dirtier Than Coal?* that used the ministry's own data to answer yes, about 50 percent dirtier.

The ministry's head of renewable energy, a former BP executive named Bernard Bulkin, released a wildly misleading response to the critics, gushing about the benefits of biomass while burying the pitfalls in word salad. The response floated multiple scenarios where burning wood reduced emissions, all involving waste wood, then only briefly conceded the benefits vanished with whole trees. But the critics hadn't objected to burning waste wood, only whole trees. The misdirection reminded Searchinger of the confession scene in *Moonstruck* where Cher's character tries to race through a list of minor sins: Twice she took the Lord's name in vain, once she slept with her fiancé's brother, once she accidentally bounced a check . . .

Searchinger felt like the priest who interjects: Wait, what did you say about your fiancé's brother?

The queen had named Bulkin an Officer of the Order of the British Empire, an unlikely honor for an American Jew who grew up in outerborough New York, got a doctorate in chemistry at Purdue, and became BP's chief scientist in London. Now in his early seventies, he had a wispy gray mustache and precise diction that evoked an American actor playing an English nobleman. He was a member of the Reform Club, the Victorian marvel that hosted the fencing scene in the Bond movie *Die Another Day*, and he once took Searchinger to lunch there along with the ministry's chief scientist, the celebrated physicist and author Sir David MacKay.

As they dined beneath Corinthian columns and a gilded ceiling, Bulkin insisted that while it might take decades to repay the carbon debt from burning one tree, there's no overall carbon debt when the overall forest is expanding. He basically accused Searchinger of missing the forest for the trees. "You're thinking about this all wrong," Bulkin said.

MacKay remained diplomatically silent, but Searchinger had already discussed the biomass loophole with him, and knew he considered it a mistake. If I'm thinking about this wrong, Searchinger said, then so is Dr. MacKay! Why don't you listen to your chief scientist?

Bulkin replied that he was a scientist, too.

"David MacKay was not my boss, and he never went to my boss to push back," Bulkin told me later. "He did think Searchinger's views had merit, but I had different views."

Bulkin's actual boss later became a biomass lobbyist—and honestly, Bulkin wasn't interested in Searchinger's views. Like Schrag, he thought he was charting a responsible path between boosters convinced biomass was perfect and haters convinced it was Armageddon, and Searchinger reminded him of the unreasonable eco-radicals who harassed him during his time at BP. He said MacKay, who died of cancer a few years later at age 48, never objected to his middle-ground strategy.

Searchinger thought Bulkin was the unreasonable one, hiding behind bioenergy targets that would require a second earth to achieve. He seemed like another wish-casting energy expert who knew nothing about land except that it could help solve energy problems. He was relying on biomass to meet his agency's renewable energy goals, so he wouldn't entertain the possibility that it could undermine global climate goals. He was in no-no-la-la mode.

It was hard to get people to grasp things they preferred not to grasp, and the E.U. was on track to use more wood for energy than all other purposes combined. The U.S. shipped six million tons of pellets to Europe in 2012, a 30-fold increase from just a decade earlier. And the official view of the scientific establishment, despite Searchinger's tireless Chicken Little-ing, was that even more would be even better.

The IPCC was the global arbiter of climate science, bringing together thousands of experts under the UN umbrella. It had voice-of-God influence even before its Nobel, but it didn't do research; it only summarized existing research. And most of its bioenergy experts had contributed to the existing research Searchinger had challenged.

William Moomaw, who did join the fight against biomass in Massachusetts after Mary Booth shamed him into doing the fucking math, got a sense of the IPCC's politics while coordinating its 2011 report on renewables. When he suggested that the authors of a bioenergy chapter should incorporate Searchinger's land-use insights, he was shot down by Dutch scientist André Faaij, a bioenergy adviser to the European Commission, International Energy Agency, and a dozen other institutions. "The guy was a bioenergy fanatic," Moomaw said. "He was so hostile: 'How dare you question this?'" Faaij refused my interview request, explaining in an email that journalists like me have hurt the climate by criticizing bioenergy—and also that I'd probably claim Searchinger was a better scientist just because he published in *Science*.

Moomaw's overview chapter did include the IPCC's first warning that using land to grow fuel and power "may even result in increased greenhouse gas emissions." But his brief words of caution were overshadowed by the extravagantly bullish bioenergy chapter written by Faaij and his allies. It suggested bioenergy could supply 300 annual exajoules, which exceeded the energy potential of every crop and tree harvested on Earth that year. Unless human beings stopped eating and using wood for other purposes, they'd have to double their use of land for agriculture and forestry to produce all that energy, which the chapter didn't even flag as a problem.

When the IPCC reviewed bioenergy again in 2014, Felix Creutzig, a young German physicist without preconceived notions, oversaw a fractious team that included a few skeptics like Richard Plevin and Mark Delucchi as well as evangelists like Faaij. Creutzig tried to steer the group toward a compromise that emphasized bioenergy's land-use downsides, but the boosters resisted any implication that their baby might be a little ugly.

"It's a huge bias in science," Creutzig told me. "People make a career investment in studying a technology, then they defend that technology."

In the end, IPCC officials sided with the evangelists, so much so that Creutzig joined the skeptics to publish an independent critique of rosy bioenergy analyses that he couldn't get into his own IPCC chapter. The chapter did briefly acknowledge that assuming carbon-neutrality was problematic, but still projected a future with 300 exajoules of bioenergy, relying on papers by Faaij and others that assumed carbon-neutrality. And a separate IPCC modeling chapter suggested that limiting warming to 2 degrees would require an all-out bioenergy push.

The modelers were particularly gung-ho about "bioenergy with carbon capture and storage," a strategy known as BECCS that did not yet exist in the real world but had free-lunch appeal in model world. With the world overshooting its emissions targets, modelers were looking for sources of "negative emissions"—and if you assumed burning wood was carbon-neutral, a biomass plant that captured carbon at its smokestack would be carbon-negative. The more wood you burned, the cleaner the atmosphere would get. But since burning wood is actually worse than burning coal or gas, BECCS is worse than capturing carbon from a coal or gas plant.

"The models loved bioenergy, and the modelers didn't know how to reach their goals without bioenergy," Creutzig said.

The IPCC used bioenergy as what Searchinger called a "magic asterisk," relying on it to supply as much clean energy as its models needed to meet global goals. Yet it subtly admitted in a *Moonstruck* way that there were "shortcomings" to its assumption that bioenergy was clean energy—and if bioenergy wasn't clean, the magic asterisk had no magic. It seemed absurd that the scientific community would bless energy sources that sacrificed trees now as indispensable for reducing emissions in the future. But as the *Times* reported: "The argument for caution has so far mostly fallen on deaf ears."

He seriously doubted he could convince the world bioenergy was a climate disaster if the IPCC kept calling it a climate necessity. Who would trust a lawyer's science over the scientific establishment's science? The establishment was usually right. The did-my-own-research types who denied global warming were liars and kooks. Still, it's an enduring theme in scientific history—from the idea that the earth revolves around the sun to the germ theory of disease to the long-ridiculed notion of plate tectonics—that when the establishment does get things wrong, it's slow to fess up. He figured he just had to keep whacking away at the truth.

But there were costs to picking fights with the in-crowd of bioenergy experts. One scientist with a prominent IPCC position told Searchinger he agreed with 90 percent of the criticisms, but even if he had agreed with 100 percent, he wouldn't say so publicly, because he wanted to remain "part of the conversation." In other words, he didn't want to develop a reputation for controversy like Searchinger, who was never invited to join IPCC committees.

As long as the science of biomass remained disputed, the politics of reform would be daunting. In the end, what Searchinger's incessant Chicken Little-ing about the biomass loophole helped persuade the Obama administration to do was . . . nothing. It ultimately declined to take a stand on the climate impact of burning wood, kicking the issue to the next administration.

Of course, that turned out to be the first Trump administration, which opposed any climate action. It was clear extractive interests would get their way under Trump, so it wasn't surprising when his cartoonishly anti-environmental EPA head, Scott Pruitt, announced the agency would provide "much-needed certainty" to forestry interests by declaring biomass power carbon-neutral. But since the Trump team had no interest in restricting carbon, its opinions about the carbon impact of biomass had no substantive effect, and no influence with climate-conscious officials abroad. As much as Searchinger despised Trump's reactionary climate policies, they didn't do much to advance fake climate solutions like bioenergy.

Congress also declared biomass inherently carbon-neutral after Republican Senator Susan Collins of Maine inserted some legislative language drafted by the forest industry into an energy bill; Collins cribbed her floor speech explaining her amendment from a forest industry website. But legally and scientifically, the Collins amendment evoked Abraham Lincoln's riddle about how many legs a dog has if you call a tail a leg: four, because a tail is still a tail. Biomass didn't become carbon-neutral just because Congress declared it carbon-neutral.

Outside the IPCC process, Searchinger was persuading more and more scientists that biomass was a climate disaster. He got 800 researchers to sign a letter urging the E.U. to exclude biomass from its renewable energy directive, impressive support for a Cassandra warning that "the fates of much of the world's forests and the climate are literally at stake." The E.U. did not heed that warning, but as biomass power plants became more common, their climate impact became an empirical question, not a political one. It depended on whether the pellets that got burned were made of "waste wood," forest residues with no other valuable uses, or "roundwood," whole trees that could otherwise be converted into paper or Amazon boxes or the fluff inside diapers. It was a question of opportunity cost, and it was an answerable question.

That's why Searchinger found himself standing outside Enviva's pellet mill in Northampton County, North Carolina, watching a steady stream of logging trucks come and go. He didn't have to be an ecologist or an arborist to identify the cargo the trucks were all carrying to the factory.

"Those," he helpfully pointed out, "are whole trees."

They were pines and hardwoods, stripped of their branches and cut into 20-foot-long logs. They weren't top-quality timber logs that could become houses or furniture; as one Enviva official later told me, they were Walmart wood, not Gucci wood. But they weren't waste wood, which meant they were driving deforestation, inducing the logging of more trees for pulpwood. We later spent an hour watching trucks carrying the same kind of logs to a nearby paper mill.

The biomass industry talks about using residues, but there's abundant evidence in America's wood basket that it mostly uses whole trees. Enormous log piles sit outside its mills, and satellite photos show forest cover vanishing around them. A North Carolina forest advocacy group, Dogwood Alliance, documented that the five largest U.S. pellet exporters to Europe all relied mostly on roundwood; four had photos on their websites of large logs at their mills. Even an industry analysis commissioned by the American Forest & Paper Association found 76 percent of the wood processed into pellets was roundwood, only 12 percent residues.

This was all damning evidence of European renewable energy rules shrinking American carbon sinks. Enviva inadvertently provided more evidence when it introduced me to several local civic leaders, who all praised Enviva's mill for reviving depressed wood markets so forest owners could harvest and sell their trees. But the climate prefers trees to remain standing. Enviva also took me to visit an 85-year-old North Carolina tree farmer, Gene Brown, who was growing half a million loblolly pines in orderly rows on his neatly maintained plantation. Brown made a similar case that the pellet industry was helping his bottom line, which, again, is not the same as helping the climate. Enviva's demand for low-grade wood made it profitable for Brown to cut his trees after 30 rather than 40 years, and some of his competitors were cutting after just 15 years.

"Enviva changed the incentives, in a good way," Brown said.

The problem for the climate is that the faster trees get harvested, the less carbon they store. I discussed this problem with Jennifer Jenkins, the forest scientist who left EPA to be Enviva's head of sustainability. She acknowledged that her company relied on whole trees, but claimed that by providing financial rewards for clear-cutters, it was actually encouraging landowners to grow more trees by increasing demand for pulpwood.

This industry argument that increasing demand for dead trees somehow increases the supply of live trees makes Searchinger want to scream. The whole point of recycling paper is to *decrease* demand for pulpwood; why would we want to burn the trees our conscientious sorting is saving? Jenkins was essentially arguing that there would be more trees standing if we stopped recycling and used more paper.

She also argued that since the carbon in Southeast forests was increasing, Enviva's pellets were by definition carbon-neutral. But one reason forest carbon was increasing regionally was that pine plantations were replacing pastures and farms, creating an ILUC problem where lost food production had to be replaced by clearing nature elsewhere. Another reason for the increase was that the atmosphere was full of so much carbon for trees to absorb, fertilizing their growth with fossil fuel pollution, a troubling phenomenon that did not seem to justify more clear-cutting. But the larger point was that the Southeast's forest carbon would have increased even more if less of it had been fed to Enviva's mills. Jenkins repeated her argument that biomass was carbon-neutral if the overall forest was expanding so often that I finally asked whether setting a forest fire would be carbon-neutral if the overall forest was expanding. I saw her press handler, who quit the next month, stifle a giggle.

"I'm not suggesting you should burn a forest," Jenkins said curtly.

As long as respected scientists like Jenkins and Schrag and Bulkin were willing to throw argumentative spaghetti against Searchinger's wall, there was never going to be a consensus that biomass power was a problem. Like biofuels or bio-anything-else, it sounded clean and green if you didn't think too hard about opportunity cost, and the politics of forests usually revolved around the money that could be made logging them. By 2019, not a single state or national government had followed the trail that Massachusetts blazed.

Nevertheless, Searchinger's fears that bioenergy would destroy the world had not come true.

Biomass power was still growing, but no longer at dizzying rates. It was constrained, ironically, by the world's failure to take action on climate; since most nations outside Europe hadn't imposed limits on carbon, the carbon-neutrality loophole hadn't created incentives for them to burn wood to evade those limits. And in countries that did take climate action, solar had become more attractive than biomass, so it had the dizzying growth rates; unlike photosynthesis-dependent biomass, it was highly efficient. Even Europe was generating less than 6 exajoules from biomass, a long way from the IPCC's global forecasts of 300 exajoules.

The biofuels train was also losing speed. In America, ethanol plateaued at 15 billion gallons, unable to compete beyond its mandate. In Europe, the ILUC study that got slipped into the renewable energy directive as a consolation prize for biofuels skeptics ended up confirming Searchinger's dire warnings and shaking up biofuels politics; E.U. leaders decided to scale back their 10 percent mandate, capping crop-based fuels at 7 percent. Electric vehicles now looked like the future of green transportation. Not only was solar 100 times as efficient as ethanol, electric cars were three times as efficient as gas models, so an acre of solar panels could move an electric car 300 times as far as an acre of corn could move a conventional car.

As for the advanced biofuels that were supposed to replace crop-based fuels someday, those miracle elixirs just weren't panning out. Vinod Khosla had predicted to Congress a decade earlier that cellulosic ethanol would fulfill the entire 35-billion-gallon renewable fuels mandate by now, but it was about 35 billion gallons short, and several Khosla-backed cellulosic startups had failed. It had become a stale joke that second-generation biofuels were five years away and always would be. *Jatropha* also fizzled, and the land-grabbers who cleared millions of acres in Africa to grow them quickly abandoned them. The only advanced biofuels that gained traction in the U.S. were diesel substitutes made from recycled cooking oil in California, where they earned extra subsidies because they didn't induce land-use change.

This flatlining of bioenergy helped save a lot of forests. Honestly, it was lucky that electric vehicles got attractive in a hurry, and that advanced biofuels turned out to be a much tougher economic proposition than many smart people expected. But Searchinger's science and rhetoric also helped problematize bioenergy, just as he problematized the Army Corps. His critiques pierced its aura of inevitability, buying time for better renewables. There's no way to know how many more forests would have been cleared if he had never stumbled across that Argonne ethanol study, because there's no Earth Two to test counterfactuals like that. But he helped drag most of the environmental community and some of the scientific community off the biofuels and biomass trains, stigmatizing solutions that once looked like win-win-wins.

"Who knows how bad it would've gotten if Tim hadn't exposed how bad it was?" asked his Princeton colleague David Wilcove.

But even though bioenergy looked less existential a threat in 2019, it was still a threat. Searchinger estimated that meeting the world's biofuels goals would consume another Bolivia's worth of land—way better than another Brazil's worth of land, but the equivalent of adding another Japan's worth of emissions. There were now much greener ways to make energy with much less land, and the main thing he had learned at his day job studying agriculture was that the world's land needed to grow a lot more food and store a lot more carbon.

This was the tragedy of bioenergy: It made the global land squeeze worse, eating up scarce acreage when humanity desperately needed to free up more acreage to feed the world without frying it.

The scariest thing Searchinger had learned in his research at Princeton was how incidental bioenergy was to that land squeeze. While he had shown in excruciating detail how expanding bioenergy could destroy the world, he warned in a report that July that getting rid of bioenergy would achieve just 3 percent of the land-sector emission reductions necessary to meet the Paris 2050 goals. A bioenergy boom could make the eating-the-earth problem unsolvable, but a bioenergy bust wouldn't come close to solving it.

In any case, bioenergy was now a relatively straightforward problem, an ineffectual approach to the fossil fuel problem that made the land problem worse. Searchinger had done what he could to keep it under control, even though he hadn't persuaded the world to solve it. But he had said he'd give his right arm to stop working on it—because he already understood it, and he wanted to understand the harder problem of how humanity could avoid eating the earth.

That was the focus of that July report, his 564-page nerd manifesto called *Creating a Sustainable Food Future*. Only a small portion was about bioenergy, because bioenergy was only a small portion of the eating-the-earth problem. But someone had to devise a plan to fix it—not just 3 percent of it, all of it—and as the patron saint of almost-lost causes, he figured he was that guy.

FIVE
THE MENU

FEEDING A HOT AND HUNGRY PLANET

Bioenergy was a gigantic problem with an obvious solution: Stop growing bioenergy. The politics would be ugly, but if we decided as a species that it was suicidal to use precious land and resources to grow relatively trivial amounts of fuel and electricity, we could stop.

We couldn't stop growing food, though. Back in 2007, when Searchinger began plowing through the agriculture and climate literature, he began to see that food was a far more gigantic problem, with no obvious solution. Really, it was three gigantic problems.

First, farmers needed to grow more food to fill more stomachs. Nearly 1 billion of the 8 billion people on Earth were already hungry, and the global population was expected to approach 10 billion by 2050.

At the same time, agriculture needed to generate much fewer greenhouse gases to avoid climate chaos. Emissions from farms, pastures, and the deforestation that made room for them were already three times higher than the Paris targets for 2050, so they'd have to come way down even as production ramped way up.

Finally, agriculture's footprint needed to stop expanding, so that nature's footprint could stop shrinking. Farmland already covered a land mass six times the size of the continental U.S., while the part of it growing biofuels was only the size of Texas. Somehow, farmers had to produce far more calories and protein without clearing more forests.

This was like saying a basketball team had to grab far more offensive rebounds without giving up more transition baskets. Normally, focusing on one of those goals detracts from the other. Normally, "do more with less" is a strategy signifying a lack of strategy.

But the world did need to do more with less, or the world would burn. So even as he continued his fight against bioenergy, Searchinger decided the broader eating-the-earth problem would be the challenge of his career.

It felt like the challenge he was born for. Food needed an intellectual grinder who could dig into agronomy, biology, chemistry, data science, economics, and the rest of the alphabet; a synthesizer who could see connections across all those disciplines; and an independent thinker who would follow the facts rather than the herd. In other words, it needed the kind of obsessive who would become an amateur scientist to stop a White House assault on wetlands, an amateur economist to expose shenanigans by the Army Corps, and an amateur climate analyst to stop biofuels. Now that he was a professional climate analyst, his special sauce was still his willingness to learn everything there was to learn about topics he knew nothing about. He liked coming in cold and coming out with evidence-based hot takes.

Searchinger's initial take on food and climate was that researchers were missing how gigantic a problem it was, because most of them assumed land was free. Land was not free, and using more land for agriculture was not sustainable. Somehow, the world needed to expand its supply of food while reducing its demand for land.

His take on the supply side of the problem was that it was mostly a yield problem. If farmers could make more food per acre, they wouldn't need as many acres to make food. More intensive agriculture could be less extensive.

This was the beauty of the high-tech seeds, fertilizers, and other Green

Revolution innovations that had helped triple global harvests since the 1960s. Jacked-up yields produced 90 percent of the additional food, vindicating the pro-growth techno-optimists (dubbed "wizards" by the author Charles Mann) who had scoffed at gloomy forecasts of perpetual famine. Farmers still cleared a land mass larger than all U.S. cropland to produce the other 10 percent, a sobering reminder of the limits of sustainable intensification, and a little-noticed point in favor of the anti-industrial eco-pessimists Mann calls "prophets." But it was even more sobering to think how few forests would still be standing without a half century of heroic yield growth.

Now the world needed even higher yields, because it needed to produce even more calories over the next 30 years than it had produced over the previous 12,000. But as Searchinger began devouring studies, he found many of them simply assumed the Green Revolution's heroic yield growth would continue indefinitely, conveniently supplying all the extra food the world would need without requiring farmers to clear extra land. He kept seeing more Miracle Yield forecasts, as if sustainable intensification would automatically ward off hunger and deforestation. Many studies, including the IPCC's, treated heroic yield growth as a given, an antigravitational force that always pushed upward no matter what humans did.

One problem with those upbeat forecasts was that much of the Green Revolution had already spread across most of the planet, so the lowest-hanging yield fruit had already been picked. For example, irrigated farmland had doubled worldwide since 1962, but now that the most accessible rivers and aquifers were already tapped, it was only expected to expand another 7 percent by 2050. Amping up yields would be much harder in the future than it had been when most farmland was unirrigated, unfertilized, and relatively unproductive.

Another problem was climate change. More extreme droughts, heat waves, and other semi-natural disasters were expected to stress and kill more crops and livestock, which would make it even harder to replicate the Green Revolution's productivity gains. Climate scientists were warning that by 2050, maize yields could fall by a third, so it was strange that so many experts expected the future to reprise the past.

Searchinger did have faith in humanity's ability to feed itself. In the two centuries since the Debbie Downer economist Thomas Malthus first warned that population growth would inevitably overwhelm food production, Malthusian predictions had been reliably wrong. He suspected the techno-wizards were correct that if we again faced food shortages, we'd again find a way to avert perennial famines.

But the easiest way would be to convert more native habitat into farmland, and the eco-prophets were correct that releasing all that carbon would shred the climate. The hard part of feeding the world would be doing it without frying the world—and he again found experts hand-waving the dilemma away with implausibly sunny forecasts. Most IPCC scenarios assumed emissions from deforestation would vanish by 2050, with one scenario hitting zero by 2020. The cockeyed optimism reminded him of President Bush's speech in front of a Mission Accomplished banner early in the Iraq war, except Bush at least had progress to report when he prematurely declared victory.

The IPCC scenarios assumed not only that miracle yields would make most land-clearing unnecessary, but that farmers who did clear land would conveniently avoid forests and wetlands, only clearing lands designated "other" with supposedly minimal carbon costs. But he already knew those "other" lands included tropical pastures and savannas with significant carbon costs. In model world, there always seemed to be plenty of surplus land, and little cost to using it, but in the real world, land wasn't free even when it wasn't cropped or forested. Land use always involved trade-offs. He fantasized about shooting a video from space that would zoom in closer and closer on Earth's allegedly marginal land, until it became clear almost none of it was truly marginal.

It could feel strange to think of land as a scarce resource, especially in America, which first got its reputation as a land of opportunity from the New World's bountiful and seemingly boundless land itself. The developed areas where most humans spent most of their time were still a tiny fraction of the earth. But we had entered an age of limits. We were using more than half the world's ice-free land for farming and forestry, while the land we weren't using was soaking up carbon, sheltering wildlife, and performing other ecological services. It all had work to do, and we needed to use it as efficiently as possible.

Searchinger described his new credo with a twist on Bush's "No Child Left Behind" education slogan: no acre left behind.

His take on the demand side of the problem was that it was mostly a meat and dairy problem. Nearly 80 percent of the world's agricultural land supported livestock. If we could eat more plants, we'd eat less of the earth.

Diet for a Small Planet, a 1971 bestseller by a 26-year-old community organizer named Frances Moore Lappé, first popularized the idea that animal agriculture was unsustainable, and that eating lower on the food chain could ease pressure on nature. Lappé grew up eating meat in the cow town of Fort Worth, and she had no nutritional or environmental background. But she got curious about why people were going hungry in a world with abundant farmland, and after a few months in a library with a slide rule, she concluded we were producing more than enough grain to feed the world. We were just feeding too much to animals instead of people.

Lappé's tract was often dismissed as a hippie cookbook—her publisher made her include vegetarian recipes—but it was stuffed with visionary insights about the opportunity cost of meat. "Imagine sitting down to an eight-ounce steak, then imagine the room filled with 45 to 50 people with empty bowls," she wrote. "For the feed cost of your steak, each of their bowls could be filled with a cup of cooked grains." This was why livestock drove deforestation: They were an inefficient mechanism for converting plants into human nutrition. Lappé didn't link this inefficiency to the climate, but hardly anyone was thinking about the climate in 1971.

"It's the same concept, though: Our diets wipe out nature," said Lappé, who is still a vegan activist.

Searchinger hadn't read Lappé, and the studies he did read about meat's climate impacts routinely short-changed land use, only assessing direct emissions from meat production. They often parroted industry arguments that grazing added value to grasslands, because cattle used otherwise worthless land to convert grass we couldn't digest into meat we could. But almost three billion acres of grazing lands were former woodlands, a deforested area larger than the continental U.S. The other five billion acres of pastures also had carbon and biodiversity value. Using land to graze cattle had an opportunity cost, just like using land to grow fuel. Some grazing lands were more sustainably managed than others, but none of those lands were free.

His wake-up call came from *Livestock's Long Shadow*, a 2006 FAO report that had attributed more than half of all emissions from agriculture to animal agriculture. Its reputation had suffered after the UN retracted its claim that livestock created more emissions than transportation, prompting snarky headlines mocking its general theme that meat was warming the planet. But the methodological error that led to the retraction involved transportation, not livestock, so it didn't refute the report's findings that pigs, chickens, and especially cud-chewing ruminants like cattle and sheep used far more land and produced far more emissions than beans or lentils. As Lappé had noticed, eating plants was a much more efficient use of the sun's photosynthetic energy than eating animals that ate plants. It cut out the middlemen.

One *Livestock's Long Shadow* coauthor actually critiqued Searchinger's biofuels paper in *Science*, arguing that livestock were a much worse threat to the climate. Searchinger hadn't claimed biofuels were the worst climate threat, only a serious threat that shouldn't be portrayed as a climate fix, but it was true that animal agriculture was worse, occupying 30 times more land than energy crops. He had identified the danger that bioenergy mandates could convert the earth into an energy farm, but much of the earth was already an animal farm. And meat consumption was expected to keep rising as more poor people joined the global middle class, which would induce more destruction of nature to grow grass and grain for livestock.

The question was what to do about it. Nobody seemed to have a plausible strategy for how to shorten livestock's long shadow or stop eating the earth. Farms and ranches needed to produce quadrillions of extra calories while shrinking their carbon footprint by billions of tons and their physical footprint by hundreds of millions of acres—and the world didn't have a plan.

That's what inspired *Creating a Sustainable Food Future*, his doorstop report, the first comprehensive strategy for feeding the world without frying it. Searchinger hoped to do for agriculture and climate what he once did for wetlands, assembling the best science into a definitive guide to a complex issue. He got a Packard Foundation grant in 2008, hosted a Princeton conference in 2009, and basically grasped the problem by the spring of 2010, when he delivered a keynote titled "Feeding a Hot and Hungry Planet" at a conference in New Zealand. At a workshop that year in San Diego, he met Craig Hanson of the World Resources Institute, an action-oriented research organization with the motto "Making Big Ideas Happen." Hanson already wanted WRI to do a major food and climate report, so he hired Searchinger as a senior fellow to write his opus.

It took nine more years to finish. Contrary to Searchinger's Third Law, its completion would give him satisfaction as well as relief. But it would also inspire his WRI colleague and coauthor, Richard Waite, to suggest a fourth law: For some projects, by the time you get them done, you question all your life choices.

INTO THE WEEDS

You don't want to hear every detail of why it took so long. There's a reason thrillers don't get filmed about the writing of reports. What mattered about this one was its substance, not the laborious process that brought it into the world.

Still, it's worth noting that Searchinger's lawyerly compulsion to bulletproof every fact and anticipate every objection, as if he might be cross-examined about all 564 pages, did not speed up the process. He was perpetually dissatisfied with agricultural data, and spent weeks ground-truthing minutiae it was hard to imagine anyone noticing. He refuted dozens of studies it was hard to imagine anyone reading, cataloguing their flawed assumptions in sidebars and footnotes. He became so cynical about the guesswork in global agriculture models that he enlisted a land-use modeler from France's leading agricultural research institute to help him build a new one for WRI, which added several years to the work.

In fairness, he was laying out a blueprint for a habitable earth. He didn't want to give imprecise advice. And since he was new to the field, he had to do inordinate amounts of work, at times creating inordinate amounts of pain in the asses of others, to make sure his advice was rigorous.

Some of the work was just pulling together existing research. For example, since feeding humanity would be easier with fewer mouths to feed, he inhaled the scholarship on population growth. This was a touchy topic; WRI's leaders didn't want to come off as rich Americans lecturing developing countries about their excessive baby-making. But he was determined to learn what drove birth rates, and whether there were morally acceptable ways to reduce them.

What he learned was that most of the world was already on its way to achieving replacement-level fertility rates around two babies per woman, with only sub-Saharan Africa nowhere close. Poverty wasn't destiny, though. There was strong evidence that three strategies reduced rates everywhere. First, reduce child mortality, so parents don't have to conceive lots of kids to ensure a few survive. Second, educate girls; the longer girls stay in school, the later they start having kids and fewer kids they have. Third, expand access to contraception and other reproductive health services. That was the formula. Birth rates plummeted in every country that checked those boxes, including Botswana and Rwanda in sub-Saharan Africa. Typically, those countries then enjoyed economic growth spurts.

All this was common knowledge in population circles; it just hadn't infiltrated climate circles. And as nervous as Searchinger's bosses had been about a brash white guy with no development background lecturing Africans about overpopulation, they were happy for him to recommend keeping babies alive and girls in school. Since sub-Saharan Africa also had the world's highest hunger rates, lowest farm yields, and deepest dependence on firewood for cooking and heating, the formula could help avoid a lot of suffering, deforestation, and emissions.

Population growth was a relatively simple issue, since all he had to do once he finished his reading was graft an existing development agenda onto his climate agenda. On more obscure issues, like methane emissions from rice farms, he had to create a brand-new agenda.

He again came in ignorant, so he was surprised to learn that methane-producing microbes on flooded rice fields generated 10 percent of agricultural emissions. He was then pleased to find a rich vein of studies evaluating techniques to limit methane by limiting flooding, like "midseason drawdown" and "alternate wetting and drying." But he found almost no data or planning about where or how the techniques should be deployed. When he attended a rice science workshop in the Philippines with two dozen leading researchers, he was struck by the lack of focus on translating their diligent field experiments into action.

Searchinger doubted farmers or bureaucrats would slog through the science to analyze the best practices or policies, so he did it for them. He spent months getting deeply and sometimes literally into the weeds, learning how rice growers used pumps, mulch, and reservoirs in Punjab, Sichuan, and Arkansas. He repeated his learn-everything *How Wet Is a Wetland?* approach, relentlessly grilling the best scientists he found—one American, one Indian, one Chinese, one German based in the Philippines—until he had a plan to promote lower-methane rice.

One key to his plan was a strategy the scientists had overlooked: increasing yields. Emissions from a paddy mostly depended on its size, not its output, so producing more rice on the same paddy would mean less methane per bushel. Again, *duh*. Land mattered, so obviously yields mattered. But like the blind men who only focused on parts of the elephant, the rice experts had focused on their specific methane mitigation techniques. Searchinger again saw the elephant.

The dispiriting news was that even if all his rice recommendations were adopted globally and worked perfectly, they'd generate only 7 percent of the emissions reductions needed to meet the Paris targets for agriculture. And reducing birth rates to replacement levels worldwide would only reduce the 2050 population by 4 percent, so it would barely dent food demand. The developing world already had so many women of childbearing age that much of the population growth projected for the next few decades was almost inevitable.

This was the consistent theme throughout his research: It would be incredibly hard to stop eating the earth. It would require all kinds of daunting changes to food consumption and production. Energy analysts often said there was no silver-bullet solution to the fossil fuel problem, just silver buckshot, but he thought the food and land problem would require copper bullets; they'd have to be more powerful than buckshot to succeed, and cheaper than silver to scale. And it would take tremendous amounts of work just to figure out what they were.

Creating a Sustainable Food Future was the absurdly detailed result of his decision to start doing that work. The bibliography alone was 51 pages. It was a team effort, with three coauthors and dozens of contributors, but he was the lead author and animating force, stuffing it with esoterica about harvest losses from primitive threshing tools in Senegal, reforestation opportunities on sloped Brazilian pastures, and more than any normal person would want to know about shit.

To summarize the eating-the-earth problem, the report quantified the food, emissions, and land challenges as "gaps"—the chasms between what they were on track to look like in 2050 and what they'd need to look like to end hunger, climate change, and deforestation.

The food gap was clear. According to FAO, farmers produced 13.1 quadrillion calories in 2010, and they would need to produce 20.5 quadrillion calories to feed the world in 2050—a gap of 7,400,000,000,000,000 calories. If you visualize pizza better than strings of zeroes, that would require increasing production by a pepperoni pie's worth of calories every day for everyone on Earth.

The emissions gap was even clearer. Agricultural emissions were on track to hit 15 gigatons by 2050, and they needed to shrink to 4 gigatons to meet the Paris goals. That was a gap of 11 gigatons, equivalent to the carbon footprint of every factory on Earth.

The land gap, the extra farm acreage needed to close the food gap, was more complex, because it depended on how much extra food could be produced per acre. WRI's model found that even if yields kept growing through 2050 at the same heroic rate they had grown since 1960, agriculture was on track to expand 1.5 billion acres, a gap nearly twice the size of India. It would be like giving every adult on Earth a forested quarter-acre lot, then chopping down the trees.

As alarming as that was, Searchinger thought it wasn't alarming enough, because of its wishful-thinking assumption that the Green Revolution's yield growth would continue. It had already slowed a bit in recent decades, and merely tweaking the model to reflect that expanded the land gap to three Indias.

He worried that baking heroic yield growth into the business-as-usual baseline could create an impression that the hard work of achieving it had already been done. A friendly WRI research assistant named Austin suggested showing what would happen with zero yield growth, to make clear how much work actually needed to be done. Searchinger and his colleagues called this hypothetical the Austin Scenario—the joke was naming doomsday after such a sweet guy—and it produced a land gap of *seven* Indias.

Zero yield growth was an overly pessimistic assumption, but the Austin Scenario was a reminder that assuming an eternal Green Revolution was wildly optimistic. Even that wildly optimistic baseline left canyon-sized food, emissions, and land gaps—and forests would keep falling until the gaps were closed.

The rest of the report explored ways to hold back the chain saws. Searchinger hoped to show the world not only that the eating-the-earth problem was a spectacularly consequential problem, but that potential solutions were just waiting to be studied, funded, and scaled. WRI pitched it as a "five-course meal" with 22 specific "menu items," from reducing cow-burp emissions to improving land-based fish farms to reforesting abandoned farmland.

Searchinger loved learning new things about soils and microbes and grasses, and what he learned often filled him with hope. Outside Cali, Colombia, he had an almost religious experience seeing how "silvopasture" systems on grazing lands could ratchet up beef yields; that religion's god would have been a super-shrub called *Leucaena* that provided protein-rich leaves and shade for cattle while sucking nitrogen out of the air that fertilized nearby grasses. He had another intellectual thrill when, after spending half his career fighting for wetlands without even knowing they were valuable carbon sinks, he realized that restoring peatlands, the wettest wetlands, could be one of the land sector's most cost-effective climate solutions.

What made him hopeful was the potential for change, because not much change was happening yet. The silvopasture program he saw in Cali had only spread across 15 square miles, or 0.0001 percent of the world's grazing lands, and far more peatlands were still being drained than restored. But maybe his research into agroforestry, agrochemicals, and agro-everything-else could shake up debates over food, just as he had done with bioenergy. To continue the culinary metaphors, he was at least putting a soup-to-nuts plan on the table.

WANTING LESS

Just as reducing demand for oil reduces drilling, reducing demand for farmland reduces deforestation. That's why WRI's demand-side "course" included menu items on growing less bioenergy and reducing fertility rates, ideas that could ease the global land squeeze.

It also explored reducing food waste, because one-fourth of our food didn't make it to our mouths, which meant one-fourth of all farmland and farm inputs were wasted, too. It seemed pointlessly cruel to butcher billions of animals we didn't eat. It seemed perversely destructive to use a land mass larger than China to feed our garbage bins. The UN goal of cutting food waste in half by 2030 seemed more sensible, and even the Trump administration had endorsed it.

Still, Searchinger eye-rolled when advocates claimed the world could erase its climate and hunger problems by erasing food waste, as if the world could snap its worldly fingers and optimize the actions of billions of farmers, retailers, and eaters. He hated to sound like the hyperrational economist who wouldn't pick up cash lying on the sidewalk, but if eliminating a billion annual tons of food waste was such a no-brainer, why hadn't it happened already? Behavioral change was hard. He suspected the smallest measurable unit of time might be the nanoinstant between his return home from the market with a slightly soft avocado and Brigitte throwing it out.

Anyway, profligate behavior by well-off consumers who let groceries spoil in their fridges or failed to clean their plates was only part of the problem. Rich countries mostly wasted food near the fork, in homes, restaurants, and stores, but most food losses in poor countries were near the farm, during harvest, storage, and transport. And structural change was often even harder than behavioral change.

He illustrated the problem with gari, a Nigerian flour milled from cassava. Nearly half of gari was lost before it got anywhere near a fork, but not at one chokepoint. Modest losses were spread all across the supply chain, from antiquated harvesting equipment that left cassava to rot in fields; to shoddy processing; decrepit storage facilities; and spoilage while moving product to distant markets on crumbling roads. Even though the cascading inefficiencies added up to an irrational mess, no one had a strong enough incentive to invest in fixing any one of them.

Still, the fact that rich nations had minimal food losses near the farm while poor nations had little food waste near the fork indicated room for improvement. And he got less skeptical after a British scientist showed him evidence that a "Love Food Hate Waste" campaign helped the U.K. eliminate one-fifth of its household waste. One study found that every pound spent raising awareness in London generated 90 pounds in household savings. Cash really was lying on sidewalks. A food waste team at WRI came up with dozens of other promising strategies, from low-cost "evaporative coolers" for regions without cold storage to tray-free or pay-by-the-ounce cafeterias where diners are less likely to heap too much on their plate.

Searchinger also met Apeel Sciences founder James Rogers, a material scientist who opened his eyes to technology's potential to curb food waste. Rogers got the idea for his Santa Barbara startup while driving past some lush farmland, when he heard a radio story about food shortages. Just like Frances Moore Lappé, he got obsessed with the idea of deprivation amid abundance, and soon began developing invisible, tasteless, odorless biotech peels that prevent fruits and vegetables from going bad. Rogers was a compelling salesman—he sent one potential investor two pallets of avocados, one protected by Apeel, along with a note that said "Watch Me"—and he helped persuade Searchinger that innovation could prevent a lot of spoilage.

Searchinger just worried that all the rhetoric about waste could encourage silver-bullet thinking, as if we wouldn't need to change how we produce and consume food if we could just pledge not to throw so much away. He worried about anything that made any of this seem easy.

He was also skeptical that people would eat less animal protein, because people, including him, loved animal protein.

Americans clearly ate way more Bacon, Egg & Cheese McGriddles than was good for our bodies or the planet. But he had little patience for vegan utopians who argued that we should all just stop being such heartless gluttons and eat lower on the food chain. How likely was that? It was no coincidence that Republicans were trashing the Green New Deal promoted by left-wing Democrats as a plot to ban meat and cows, even though the legislation didn't mention meat or cows. Framing it as anti-meat framed it as anti-American, because meat was as deeply ingrained in American culture as cars or guns.

Anyway, most future increases in demand were expected in fast-growing countries like China and India, where billions of poor people ate almost no meat. The first lifestyle change most people make when they stop being poor is to start eating meat, and he didn't think it was realistic or fair to tell them not to.

But he learned two things that tempered his skepticism about dietary change as well. The first was that even though researchers often minimized disparities among different kinds of meats, ruminants generated far more emissions than other livestock. While it wasn't exactly good news that cattle and sheep were climate outliers, it provided a narrower target for action.

WRI's model found that chicken and pork used three times more land and produced three times more emissions per calorie or gram of protein than beans. Dairy was even worse. But beef and lamb were at least six times worse than dairy. Searchinger calculated that eating five pounds of beef had the climate impact of driving an SUV cross-country. Ruminant burps and farts were a problem, but the main problem was that beef cattle used nearly half the world's agricultural land to produce just 3 percent of its calories. Compared with pigs or poultry, they needed five to ten times as many calories of feed to produce a calorie of meat.

Searchinger couldn't even imagine the world trying to cut out meat. But when he crunched the numbers, almost all the climate benefits of vegetarian diets came from cutting out ruminant meat. Cutting out chicken and pork barely affected emissions, because vegetarians tended to replace the protein with dairy. Getting people to eat less beef wouldn't be easy, either—it was, as the slogan said, what's for dinner—but he could at least imagine it.

In fact, it was already happening. That was the other thing he learned: America's per capita beef consumption had declined by a third since the 1970s. We didn't shift to plants. We shifted to chicken, which was cheaper and perceived as healthier. But that shift inadvertently spared millions of acres of land, probably reducing more U.S. emissions than any other shift before the rise of solar. The progress was real, if you weren't a chicken, and since U.S. beef consumption was still four times the global average, there was room for more progress.

The world was still on track to eat at least 50 percent more meat by 2050, and even that forecast assumed six billion poor people would still eat almost no meat. But the well-off regions that ate the most beef and lamb could help immensely by continuing to shift toward chicken and pork. Reducing the rich world's ruminant consumption 50 percent could close much of the land gap.

Still, getting people to eat less meat they loved to eat would be even harder than getting them to waste less food they preferred not to waste. The report included a suite of marketing strategies proven to steer dietary choices, like giving vegetarian dishes appealing descriptions like "twisted citrus-glazed carrots," refraining from describing them as "vegetarian dishes," and making them the default option at cafeterias. But even successful efforts to change behaviors are more like turning ocean liners than flipping switches. The fossil fuel experience suggested consumers would keep eating meat until they had attractive substitutes.

The only available substitutes when Searchinger started his research were hockey-puck veggie burgers, the oddity that was Tofurky, and other unconvincing knockoffs aimed at vegans who didn't know what they were missing. But by 2019, a slew of buzzy biotech startups were trying to engineer plant-based meat for the carnivores whose diets were straining the planet. He visited Silicon Valley to check out the buzziest, Impossible Foods, which was providing beefless beef for plant-based Whoppers at Burger King. Its founder and CEO, the pioneering biochemist Patrick Brown, exuded mad-scientist confidence about its inevitable triumph over factory farms; when Searchinger mentioned that reducing beef consumption and rewilding pastures would be the most direct way to close the land gap, Brown smiled and said: "We'll take care of getting rid of beef for you." When he heard about the 22 changes in WRI's report, he scoffed: "One change! If we can just get everyone to eat plants, you don't have to disrupt everything else."

Searchinger doubted anyone could get everyone to eat plants. Meatless Monday was barely a thing; how would a universal Meatless Every Day become a thing? Still, the average American ate three burgers a week, and replacing one of them with an Impossible Burger could save a Massachusetts-sized land mass every year. Many environmentalists distrusted technology, especially when it came to what we put in our mouths, but as he waded deeper into the food and farming culture wars, he found himself closer to the futuristic wizards than the old-school prophets. Humanity had a better record of solving problems with innovation than self-sacrifice.

In any case, it was clear our big-brained species was at least capable of making demand-side changes. We knew how to eat less beef, waste less food, educate more girls, and repeal bioenergy mandates. The question was whether we'd do it. The challenge of reducing demand evoked the joke about how many psychiatrists it takes to change a light bulb:

Only one, but the bulb really has to want to change.

MAKING MORE

Even if the world learned to consume food more efficiently, it would still need to produce food more efficiently. Searchinger repeatedly reminded his team that higher yields were the indispensable supply-side solution: *We need incredible yield growth, or we're incredibly screwed.* Incredible improvements would be necessary just to maintain the unacceptable business-as-usual baseline.

The first menu item in his supply-side course was increasing livestock productivity, because the yield literature focused far more on the crops that covered one-third of agricultural land than the pastures that covered two-thirds, and he wanted to send a message that animal agriculture mattered. He called pasture the Rodney Dangerfield of land use: It got no respect. It seemed like everyone who wanted to claim there was abundant land available to grow more fuel or plant more trees assumed grazing lands could be converted to croplands or woodlands without cost. Since vegan activists wanted to abolish animal agriculture, while the meat industry liked the status quo, it was hard to drum up interest in improving grazing productivity. But he believed humanity's future depended on how much more meat and milk per acre and per animal its pastoralists and agribusinesses could produce.

Big changes were clearly needed. The livestock sector already used a third of the habitable land on Earth to feed 1 billion hogs, 4 billion cows, sheep, and other ruminants, and 60 billion chickens. A no-yield-growth Austin Scenario for livestock would deforest two more South Americas. Even the wishful-thinking baseline would require another Russia's worth of pastures.

But Big Ag had proved big changes were possible. The U.S. cattle industry had doubled beef and milk production per head since 1970, creating a high-efficiency protein delivery system that would have baffled the cowboys of the open range. It featured advanced breeding of faster-growing beef cattle and higher-production dairy cows; scientifically optimized nutrition and veterinary care; and automated feedlots, milkers, and slaughterhouses that achieved industrial economies of scale. It was a brave moo world, and while it wasn't romantic—even bovine liaisons were replaced by artificial insemination—it was extremely productive.

America's pork and poultry industries achieved similar yield gains with even harder-core efficiency. Pigs were transferred into climate-controlled barns where they didn't have to expend energy braving the elements, so they could instead focus on getting absurdly fat absurdly fast. U.S. chickens—called "broilers" to emphasize their utility for people rather than their unenviable lives as birds—were also freakishly oversized marvels of modern genetics, bred to grow so breast-heavy so quickly they could barely stand. Searchinger remembered the first chicken fillet he tasted as a teenager in the 1970s was considered fancy cuisine; now it was fast food, and Americans cared more about cheap chicken sandwiches than humane treatment of broilers.

It gradually dawned on him that Big Ag's industrial efficiency, while motivated by profits, reduced emissions by reducing demand for farmland, an unsettling challenge to his environmental worldview. At EDF, he had associated factory farms with river-fouling manure pits, not forest-sparing productivity. It never occurred to him while he was litigating to keep pesticides and fertilizers out of the air and the Gulf of Mexico that they also protected nature by boosting yields. But even though elements of Big Ag still repelled him, he was willing to change his mind to accommodate new facts—and he had learned that without Big Ag's productivity, few forests would remain standing. Now that the world needed way more food with less deforestation, he was developing strange new respect for its ability to manufacture so much food.

It wouldn't be easy to manufacture even more. Industrial pork and poultry was already so hyperefficient that it no longer seemed biologically or humanely possible to extract much more meat from hogs or broilers in the U.S. or Europe. They were approaching their physical limits. There didn't seem to be any way to cram them much tighter, fatten them much faster, or breed them much top-heavier.

But there was still room to improve cattle efficiency. Cattle were the Draxes of agriculture, the biggest sources of emissions and biggest opportunities for emissions reductions. They grazed so much of the earth that even minor efficiency improvements could transform landscapes and climate math. The most advanced beef systems were 20 times more efficient than systems in the Global South, so there was still enormous potential to spare forests through better breeding, better feeding, and other livestock intensification strategies.

Climate activists tend to focus on shutting down dirty coal plants, not making them less dirty. They often take the same no-no-no approach to beef, which helps explain pasture's Rodney Dangerfield problem. But Searchinger believed better beef, or at least less disastrous beef, would be as important as less beef.

• • •

Crop yields were important, too. Searchinger was especially eager to see the Green Revolution spread throughout Africa and other left-behind regions. But boosting global yields would also require a new Green Revolution, a truly green one, and the modern revolutions in information technology and synthetic biology could help.

Farmers were already investing in "precision agriculture," using drones, weed-whacking robots, and smart tractors equipped with GPS and modern data analytics to improve their efficiency. It also seemed possible that as 21st-century biological innovations arrived in farm country, they could repeat what 20th-century chemical advances had done for productivity.

Searchinger thought the biggest yield opportunity, ironically, was improving crop breeding. Borlaug's cross-bred wheat launched the Green Revolution, but his genetic innovations weren't that different from Gregor Mendel's mating of pea plants during the Austro-Hungarian Empire. Now higher-tech advances were reshaping plant biology, and newer innovations that would have blown Borlaug's mind were inspiring dreams of new supercrops with super-yields.

The most controversial advance was "genetically modified organisms," crops reengineered with DNA from other species. Introduced in the 1990s, GMOs already dominated America's breadbasket, and their wizard-defenders hailed them as yield miracles. But GMOs were illegal in much of the world, and prophet-critics still attacked "Frankenfood" as a Monsanto plot to hook humanity on carcinogens. Despite his allergy to the common cop-out that the truth must be in the middle, Searchinger actually did think the truth about GMOs was in the middle.

Just because they were unnatural did not mean they were unsafe, and he saw no credible evidence they were harmful to people or the planet. In fact, GMO crops infused with the natural insecticide Bt reduced impacts to health and the environment by displacing chemical insecticides. "Roundup-ready" crops engineered to be sprayed with the herbicide glyphosate were more controversial, but even though Monsanto lost several big Roundup lawsuits, most research suggested glyphosate was also less damaging than the chemicals it replaced.

Then again, he didn't see much evidence linking GMOs to big yield gains, either. Most studies suggesting otherwise highlighted data showing yield growth in GMO crops without proving the genetic modifications had much to do with it. The most credible U.S. studies found no yield gains for Roundup-ready crops and only modest gains for Bt crops. Pests and weeds were also starting to develop immunity to Roundup and Bt, a potential drag on future yields.

But even if wizards overstated the yield impact of existing GMO traits, the impact wasn't zero. Bt cotton provided a lift in India, where pesticides were less common. Hawaii's papaya trees were saved from extinction by gene transplants from the virus that was killing them. And Searchinger saw enormous potential for new breakthroughs. Early GMOs were created by haphazardly injecting hundreds of copies of genes into plants in the hope that one might happen to express itself in helpful ways. Now scientists were unveiling faster, cheaper, more precise ways to identify and deliver useful genes.

The most exciting was the gene editing tool CRISPR, which could adjust, insert, or delete specific genes in a cell, or program them to turn on and off. It had the potential to unleash mega-yielding designer crops—from drought-tolerant maize to potatoes that poisoned aphids but not us—and even designer livestock. Even if it raised thorny sci-fi questions about parents ordering super-babies and Frankenbeasts creating *Jurassic Park* situations, it could help make more food with less land. "These techniques offer too much opportunity to ignore," he wrote.

He couldn't say which offered the most opportunity. He thought anything with the potential to produce more food with less land was worth investigating: "agroforestry" projects that planted soil-enhancing trees and shrubs alongside crops, "double-cropping" strategies to get two harvests out of the same field in the same year, even better wild fisheries management and more-efficient fish farms. No strategy was too obscure; he tracked down studies that found small yield gains in Mali and Burkina Faso from "rainwater harvesting," a tactic smallholder farmers used to fight erosion by capturing runoff in pits ("zai") and half-moon-shaped dikes ("demi-lunes").

His top recommendation for almost every strategy was: more research. He tried to figure out the right questions, but there wasn't enough information available to figure out the right answers.

He was also appalled by the quality of the information that was available. Researchers couldn't even agree whether pastures were expanding globally. After reading one report that claimed they were shrinking, he discovered in the fine print that Australia has caused much of the decline by simply redefining "pasture." Brazil reported 12 million fewer acres of pasture since 1985, but satellite images later showed its pasture expanded by nearly 70 million acres. In an era of unlimited data, it was frustrating how much of agriculture remained a statistical black box.

This information gap made the eating-the-earth mess even messier. As hard as it would be to electrify the world and run it on clean electricity, the energy sector at least knew what it needed to do. The land sector didn't even know what it needed to know. "More data needed," "more research needed," and "more money needed" were classic academic platitudes, but for Searchinger they were emergency responses, the equivalent of "more water needed" during a fire.

Privately, he put it in less academic terms: The world needed to *figure shit out*, and also *hurry the fuck up*.

BEYOND SUPPLY AND DEMAND

The food and land problem was mostly a demand and supply problem. If consumers could consume less land-intensive food, and producers could produce more food with less land, we'd eat less of the earth. More efficient consumption and production could close the food and land gaps.

But it was a little more complicated than that.

For one thing, even if the world's agricultural footprint stopped expanding, farmers and ranchers would still need to downsize their carbon, methane, and nitrous oxide footprints to close the emissions gap. Direct farming emissions were a bit tangential to the land-use changes that lured Searchinger into climate issues, but he still found it perplexing that IPCC reports, which often included lengthy discussions of green energy technologies, barely mentioned green agriculture technologies, as if farming emissions were an immutable fact of life.

As he learned when he studied how to reduce methane from rice fields, there were opportunities for improvement if you looked. For example, he found several promising approaches that dairies and feedlots could use to reduce methane from manure, some as simple as using gates or chemicals to separate solids from liquids. The problem would be getting them adopted, since governments weren't requiring dairies or feedlots to reduce their methane emissions. In North Carolina, state researchers developed a low-cost wastewater system for hog farms that dramatically reduced odors and pollution as well as emissions, but state regulators didn't mandate it, because it would have increased hog-farming costs 2 percent.

Theoretically, farmers had stronger incentives to embrace strategies to clean up nitrogen from their fertilizer—specifically, to make sure more of their nitrogen helped crops grow, and less ended up as nitrous oxide emissions that created global warming or nitrate pollution that fouled water bodies. Half their nitrogen was escaping to the environment, hurting their profits as well as the planet. But they lived in perpetual fear that using less fertilizer would endanger their yields; it was like the joke about how half of all advertising is wasted, but nobody knows which half. So they kept applying "insurance N," because underapplying and losing yield was much worse for their bottom lines than overapplying and overspending on fertilizer.

In the U.S., some farmers did use smart tractors that only applied fertilizer where it was needed, or "controlled-release" fertilizers that reduced nitrogen losses to the air and water. Searchinger hoped to see those technologies spread, and he was hopeful about new technologies. Still, it would be tough to get farmers to reduce their nitrogen until governments cracked down on nitrogen pollution, and the politics of farm crackdowns were always dismal.

Ultimately, he thought limiting emissions from fertilizer—and from rice, manure, burps, and farming in general—would depend on a combination of carrots and sticks to spur innovation and action. In his darker moments, when he tried to imagine politicians resisting pressure from farm interests and doing the right thing, it felt like another cartwheeling-across-the-country fantasy. But he was at least showing them some of the right things to do—even if, as Winston Churchill supposedly said of Americans, they tended to try every other option first.

Some critics of chemical fertilizers, pesticides, and industrial agriculture in general wanted humanity to embrace a totally different approach to the land, a natural approach that would revive a sick planet by breathing life into its degraded dirt. They envisioned a global transition from "extractive agriculture" that bludgeoned and poisoned soils to "regenerative agriculture" that nurtured and rejuvenated soils. Not only would farming in harmony with nature prevent Dust Bowls by saving our soils, it would sequester enough carbon in those soils to transform agriculture from a climate problem to a climate solution.

These revolutionary calls for a kinder and gentler new paradigm for our relationship to the land, unlike Searchinger's technocratic recommendations for incremental farming improvements, had powerful political and financial support—from the UN and other global institutions, much of the environmental and philanthropic communities, and even many agribusinesses and food conglomerates. Crops do need healthy soil, and Searchinger had seen agrochemicals ravage the Everglades and Chesapeake Bay, so he took regenerative agriculture seriously. But once he did his homework, he concluded that regenerative practices tended to inspire agricultural and environmental silver-bullet fantasies.

His first concern, as usual, was yields. Regenerative advocates liked to say there were more organisms in a teaspoon of healthy soil than humans on Earth, and natural farms did have the potential to support more biodiversity than industrial farms. But if they made less food per acre, they would induce the clearing of natural landscapes that supported far more biodiversity than any kind of farms. And while some advocates insisted less intensive farming that rebuilt soil would eventually match conventional yields, he didn't see much evidence. In Zambia, after the government subsidized regenerative practices, FAO found no yield gains, and 95 percent of the farmers eventually abandoned the practices. U.S. studies found that organic farming consistently reduced yields, which made sense; for all their faults, fertilizers and pesticides do boost crop growth and kill pests that stunt crop growth.

Other advocates argued that even if regenerative practices did depress yields, they could still save the climate by repatriating billions of tons of carbon from sky to soil. The European Union's science council called them "the planet's greatest untapped treasure for dealing with the climate crisis." Again, though, Searchinger didn't see much credible evidence to justify the global enthusiasm for "carbon farming."

The most common carbon farming practice was "no-till," where seeds are drilled directly into the ground without plowing. Early studies suggested it could sequester a little extra carbon in topsoils, so he hadn't objected when his old nemesis Sara Hessenflow Harper and the bosses at EDF wanted to offer farmers no-till carbon credits to bribe them to back cap-and-trade. (He had only opposed adding biofuel mandates to the bribe.) But more recent studies found no-till did not actually increase overall carbon sequestration when the soil was measured a meter deep. And even if a farmer did manage to sequester a little extra soil carbon by not tilling, most of it would escape back to the atmosphere when he resumed tilling, which most no-till farmers eventually did.

Searchinger didn't find much evidence that diverse crop rotations, winter cover crops, or other regenerative practices sequestered much soil carbon, either. Most of the literature merely emphasized the technical capacity of soils to store vast amounts of carbon, glossing over the practical challenges of getting it there and keeping it there. It was like emphasizing the capacity of a bank's vault to store vast amounts of money as evidence of its wise financial strategy. "Carbon farming" sounded climate-friendly, and "regenerative" sounded wholesome, but he was leery of betting the planet on a faith-based transformation of food production.

The cool kids in Searchinger's new world liked to pontificate about transforming the broken food system—with regenerative ag, or vegan diets, or various silver-bullet technologies. The latest hotness was the MIT OpenAg Initiative's "personal food computers," mini-greenhouses featured on *60 Minutes* as the vanguard of an at-home agricultural revolution, destined to replace the chaotic uncertainties of the outdoors with hydroponics and robotics.

But the OpenAg Initiative turned out to be a scam. There weren't many miracles in the agronomic literature. And while the food system was undeniably flawed, it was really good at producing the calories and protein the world needed even more of. He didn't think blowing it up made sense, even if that limited his appeal at conferences on reimagining food. What made sense was reducing impacts, by breeding lower-methane rice and higher-yield cattle, by developing tastier meat substitutes and less damaging fertilizers, by limiting agricultural expansion to natural land that didn't store much carbon and rewilding farmland that didn't produce much food.

His basic message was "Produce, Reduce, Protect, and Restore": We needed to produce more food and reduce our demand for land so we could protect and restore natural ecosystems. As for the specifics, well, we needed to figure shit out, because the math was unforgiving. Even if we hit the report's 2050 targets of eating one-third less ruminant meat, cutting food waste in half, and increasing global farm yields by two-thirds, we still wouldn't close the three gaps. We'd have to do even more—not All of the Above, because not every menu item would pan out, but probably Most of the Above. And it was time to hurry the fuck up and figure out which of the above.

"I'll admit up front that we won't cover everything," Searchinger's boss, Craig Hanson, announced at the Washington launch of *Creating a Sustainable Food Future*. "We just don't have enough time." It was a three-hour event, but as WRI president Andrew Steer pointed out, it was a 564-page report.

"This is one of the great challenges facing civilization, ever," Steer said. "We don't apologize for the report being that long."

It felt like a turning point in the food and climate fight, not only because WRI was releasing the first battle plan, but because the UN Development Programme, UN Environment Programme, and World Bank were all endorsing it as an urgent and hopeful blueprint for action. When it was Searchinger's turn, he talked about how inspiring it had been to meet so many unsung researchers figuring out ways to make meat without killing, prevent food from spoiling, and breed wheat that reduced fertilizer pollution.

"There were times doing this report when some of us wanted to throw ourselves off the roof in despair," he said. "One thing that made me not do that was for every challenge, we discovered small groups of scientists with limited budgets doing extraordinary things."

The report got virtually no press, other than a *Times* story on page A8. It was hard to express three gaps, 22 menu items, and a slow-moving crisis in a news story. Searchinger feared that in the absence of coverage, insiders who didn't read the report would just assume it confirmed their priors. The head of the UN Development Programme praised it for highlighting how "we must shift from land degradation to regenerative agriculture," when if he had skimmed chapter 30, he would have noticed its skepticism of regenerative agriculture. A major bank's head of sustainability told Searchinger she loved the report, then announced a new regenerative ag initiative. He felt like the Warren Beatty character in *Reds* who gives a speech in Baku calling for a class war, then is mystified when the Azerbaijani crowd goes wild with excitement—until he learns his translator had said "holy war."

Searchinger knew people were unlikely to go wild with excitement about the war he was proposing, a nerd war of nitrogen-use efficiency research, silvopasture subsidies, and peatland restoration projects. It wouldn't be as inspiring as a holy war against industrial agriculture, carnivores, or capitalism. And his own holy war against bioenergy had not exactly proved the power of science and logic. There was no reason to think the food world would be less prone to wishful-thinking bias, conflicts of interest, or an inability to understand that land is not free.

Still, if humanity wanted to solve the eating-the-earth problem, he had laid out some action items. Now it would be up to others to act. He had shown the world a multitude of copper bullets. The rest of this book is about the early efforts to figure out which ones can hit their targets.

SIX
IT'S THE FOOD THAT NEEDS TO CHANGE

A RADICAL'S STORY

One morning early in 2023, Searchinger hosted another sustainable food expert from Takoma Park for breakfast. His guest, Bruce Friedrich, biked over with vegan donuts. Searchinger served cappuccino with oat milk. They sat on his back porch, in the shadow of the towering maple and oak trees he loves so much, and discussed whether fake meat could save the world.

After publishing the only guide to decarbonizing the land sector and a half dozen influential articles in *Science* and *Nature*, while producing the intellectual ammunition for the global battles against biofuels and biomass power, Searchinger no longer felt like he was playing a scientist on TV. He was in his 15th year at Princeton, where his daughter Chloe was a senior. WRI had promoted him to technical director for agriculture, forestry, and ecosystems. He was even advising the deep-pocketed do-gooders at the Gates Foundation and the Bezos Earth Fund.

He was now a man to see in land-sector circles, and Friedrich came to see him to make the case that alternative proteins were the indispensable solution to the eating-the-earth problem, not just one item on a long menu.

Like Pat Brown of Impossible Foods, Friedrich burned with conviction that humanity's future depended on meatless meat for the masses. If the world didn't have to use so much agricultural land to feed livestock, the eating-the-earth problem wouldn't be such a problem.

Searchinger had written the long menu, and Friedrich had been warned about his reputation for disagreeing disagreeably. But he seemed genial, even jovial, and the two wonks clicked. They shared a mix of intensely logical intellectualism and intensely passionate idealism—and as they eased into a jaunty Socratic banter, they found they also shared some common ground.

"Do you agree meat consumption needs to go down, or else meeting the Paris goals will be impossible?" Friedrich asked.

Absolutely, Searchinger replied.

Humans now ate 350 million annual tons of meat, nearly a thousand Empire State Buildings in carcass weight. Animal agriculture used a land mass the size of Africa. If we kept eating more meat, especially beef and other ruminant meat, we'd have no realistic way to close the land or emissions gaps.

"Do you also agree that nothing other than alternative proteins can make meat consumption go down?" Friedrich continued.

Maybe, Searchinger said.

He didn't expect carnivores to switch to lentils to protect the planet, or politicians to risk the wrath of voters and cattlemen by making meat or dairy less ubiquitous or more expensive. He didn't think humanity was any more likely to stop eating animals than it was to stop driving or flying. But Tesla was proving that if you built a better alternative mousetrap, at least some people would buy it. Alternative energy was starting to grab market share from fossil fuels, and he hoped alternative proteins could displace some animal proteins, too.

"So we agree alternative proteins are our only hope!" Friedrich said.

Searchinger laughed. Meat and dairy substitutes did have the potential to liberate some farmland from the livestock-industrial complex. But our only hope?

Friedrich had no doubt. Sure, more efficient feedlots and pastures could reduce emissions from cattle, but only beef and dairy substitutes could reduce the need for feedlots and pastures—and for cattle, too. He believed that trying to make animal agriculture less damaging, instead of trying to make a world with less of it, would be fiddling while the planet burned.

"It seems to me that if we can't make alternative proteins work," Friedrich said, "we're colossally fucked."

It was a radical conclusion. But Friedrich had traveled a spectacularly circuitous path to reach it. And he had always been a radical guy.

Feed the hungry, Jesus said.

It wasn't a suggestion. He made that clear in Matthew 25, the Works of Mercy chapter: Give food and drink to the needy, clothe the naked, and welcome the stranger, or else you're damned to eternal fire. You must help the vulnerable to join me in eternal life, because whatever you do for the least of my brothers and sisters, you do for me.

Friedrich learned the Works of Mercy in 1982, in Christian confirmation class in Norman, Oklahoma, and it inspired him to dedicate his life to the less fortunate. He was only 13, but the idea that what matters most to Jesus is what you do for those in need spoke to his analytical brain as well as his tender soul. How could he sleepwalk through a middle-class existence while millions of his brothers and sisters were starving in Ethiopia? Intellectually and spiritually, fighting poverty and hunger felt like the right way to live.

His journey continued at Grinnell College in Iowa, where he was organizing fasts for Oxfam and volunteering at a soup kitchen when he read *Diet for a Small Planet*, that original exposé of how animal agriculture was overrunning the planet and hurting the hungry. He was inspired again, this time to go vegan for the poor. He had been raised on Oklahoma's holy trinity of Big Macs, Dairy Queen, and bacon, but once he grasped how eating animals that ate plants instead of simply eating plants wasted ungodly amounts of land, water, and other resources, he couldn't justify eating above the bottom of the food chain. Plant-based diets just seemed like common sense and common decency. If we had to feed 50 calories' worth of plants to a cow to make one calorie of meat for us, why involve the animal?

As a young man, he didn't sacrifice for his fellow humans in the humble spirit of Christ. He was more a prophet Jeremiah type, haranguing his fellow humans to stop sinning. He even browbeat his mother into going vegan. And after leading his county's Young Democrats in high school, he was losing faith that politics could produce moral outcomes. Inspired by the rabble-rousing homeless advocate Mitch Snyder, another Jeremiah type who visited Grinnell to scream at the students to fight the system, he dropped out and moved to Washington, D.C., to work at a Catholic Worker homeless shelter. It felt like another correct and Christian thing to do.

Friedrich lived the Works of Mercy for the next six years, feeding the hungry and welcoming the stranger in a poor African American neighborhood. He slept in a closet in the shelter's basement, taking a salary of five dollars a week, wearing the same donated clothes as his guests. He converted to Catholicism, swayed by the Catholic Worker vision of Jesus as a nonviolent revolutionary for social justice. He even looked like Jesus, with a bushy brown beard and long hippie hair.

He also got arrested enough to qualify as a nonviolent revolutionary in his own right. On Columbus Day 1991, he crashed a "500 Years of Faith" ceremony to spray-paint "500 Years of Genocide" on a Columbus statue, just as then–Deputy Attorney General William Barr (the son of Searchinger's old-school headmaster) was laying a wreath. On Pearl Harbor Day 1993, Friedrich and three comrades in the Christian pacifist Plowshares movement sneaked into an air force base and vandalized a fighter jet with hammers, a stunt that landed him in prison for a year.

He then read yet another life-altering book: *Christianity and the Rights of Animals*, which persuaded him the Works of Mercy directive to care for the vulnerable shouldn't only apply to vulnerable humans. The more he learned about factory farms—chickens stuffed into filthy cages, pigs crammed into crates so tiny they couldn't turn around, the routinized cruelty of slaughterhouses—the more he wanted to be a voice for voiceless livestock. He decided they were truly the least of his brothers and sisters, and he would devote his career to their defense.

• • •

This time, Friedrich didn't take a low-profile job at a shelter. He became head of campaigns for the in-your-face vegan group People for the Ethical Treatment of Animals, masterminding some of PETA's most provocative protests.

He threw fake blood at a fur-wearing model at a New York fashion show. He streaked outside Buckingham Palace before George W. Bush ate lunch with the queen. He orchestrated campaigns against Murder King and Wicked Wendy's; handed out anti-meat leaflets at a Southern Baptist convention alongside an intern dressed like Jesus; and called on Oklahoma City bomber Timothy McVeigh to go vegan in prison. ("Good job getting attention to your cause," McVeigh wrote back.) He and his wife, Alka Chandna, who quit a tenured job as a math professor to work for PETA, spent one Christmas in Kentucky dressed as Santa Claus and an elf, delivering coal to naughty KFC executives until one called the cops.

He fit the stereotype of a vegan extremist, working 80-hour weeks "informing everyone in the world," as the Pulitzer Prize–winning *Post* writer Gene Weingarten put it in one of his periodic humor columns about Friedrich, "that they are bloodthirsty monsters." Weingarten came to consider him a friend, but still described him in print as PETA's "obnoxious, argumentative, sanctimonious national spokesperson." Weingarten was an animal lover who loved eating animals, and Friedrich made him feel lousy about the torture behind his meals.

Still, Friedrich tried to remember he was in the business of changing hearts and minds. He calculated that if every vegan converted one carnivore every five years, the world would be vegan in a couple of generations, a concept he called "the vegan multiplier." He tried to enhance his appeal by cutting his hair, shaving his beard, and bathing regularly, transforming from a Jesus lookalike to, amusingly, a President Bush lookalike. In an essay titled "Effective Advocacy: Stealing from the Corporate Playbook," he argued that a hen trapped in a cage wouldn't want her defenders to be smelly and unkempt. *Details* magazine wouldn't have ranked him fifth on a list of influential Generation Xers, ahead of Leonardo DiCaprio, if he had stuck with his scraggly look.

"I guarantee that since I began sporting a conservative appearance, I've persuaded many more people to become vegans," he wrote.

He was losing faith in extremism's power to persuade. Even the radical filmmaker Michael Moore had groused that PETA's radicalism made him want to kick his dog. Friedrich's friend Paul Shapiro, who founded the group Compassion Over Killing as a teenager, liked to say that two half-time vegetarians saved as many animals as one full-time vegetarian. Friedrich began to temper his radicalism with pragmatism, negotiating with McDonald's executives to improve animal welfare in their supply chains even as he protested outside McDonald's franchises wielding bloody plastic chickens. He joined Compassion Over Killing's practical young activists as they displayed gruesome images of factory-farm cruelty on a big-screen "FaunaVision," then politely urged passersby to eat more plants instead of screaming that meat was murder. He still saw himself as a fool for Jesus, but he tried to maximize the marginal utility of his efforts on His behalf.

"Bruce never forgot that the goal was to reduce suffering," said Josh Balk, a Compassion Over Killing staffer who later became a top executive at the Humane Society of the United States.

Friedrich started advising colleagues to read *How to Win Friends and Influence People*, to behave like salespeople peddling a kinder world. Before handing out leaflets or appearing on television, he'd smile in the mirror to remind himself to project positivity. He went about his work with such a sunny Mary Poppins demeanor that fellow activists called him Bruce Poppins.

"There's a tendency in animal protection to demonize people we don't agree with," says Milo Runkle, the former PETA intern who dressed like Jesus at that Southern Baptist convention and later founded the advocacy group Mercy for Animals. "Bruce had this rare and radical belief that most people did the best they could with what they knew."

Paul McCartney once said that if slaughterhouses had glass walls, everyone would be vegetarian. Friedrich felt the same way: If only carnivores knew that broilers were bred to grow so fast their legs sometimes broke, so fast a human baby growing at the same rate would gain 600 pounds in two months! He tried to fill these knowledge gaps by making films like *Meet Your Meat*, featuring Alec Baldwin narrating footage of piglets castrated without painkillers and clumps of hens squeezed into cages the size of desk drawers. He produced a similar film with McCartney called *Glass Walls*. The idea that it's best not to see how laws or sausages are made is usually trotted out as commentary on lawmaking, but to Friedrich it highlighted the urgency of exposing sausage-making. Almost all Americans told pollsters they wanted animals protected from abuse, so he assumed that once they knew the truth, they'd change their diets to align with their values.

But they didn't change.

In Friedrich's 13 years at PETA, the vegan multiplier never multiplied. Vegetarianism held steady around 3 percent of the U.S. population, veganism around 1 percent. Friedrich wasn't focused on the environment, so even the one big shift in meat-eating, from beef toward industrial chicken, seemed like a big step backward. One cow can produce as much meat as 100 chickens, so far more animals were suffering and dying on their way to the center of our plates.

Friedrich started telling a joke about an Israeli woman who visited the Wailing Wall every day for decades to pray for Middle East peace, before a reporter asked her how it felt to keep at it for so long.

"Like I'm talking to a fucking wall," she replied.

Friedrich got so weary of talking to his wall that after turning 40 in 2009, he left PETA to teach in an inner-city Baltimore high school. He found the education system even more dysfunctional than the food system, so he returned to animal protection in 2011. Again, though, little changed. U.S. meat consumption, which finally dipped during his sabbatical from activism, resumed its steady rise as the pain of the Great Recession subsided, confirming the rule that when people have enough money to eat meat, they eat meat. It didn't matter that a new wave of "flexitarians" said they wanted to eat less meat, or that nearly half of Americans said slaughterhouses should be banned. Slaughterhouses kept slaughtering.

Friedrich gradually concluded persuasion would never reform the system. People loved meat too much.

"Our movement had been educating people and shaming people and yelling at people to go vegan for 50 years," Friedrich said. "It wasn't working!"

He had failed to change people's hearts and minds. But what if he could help change their food?

BEYOND BEAN CURD

One running gag in Weingarten's columns about Friedrich was the abysmal quality of the vegan restaurants where they met. The menus overused quotation marks, as in "meat"-ball subs and grilled "cheese." The dishes challenged his descriptive talents, as in "indefinable vegetation in a soup of what appears to be regurgitated spinach" or "brown spores in slime." Friedrich enjoyed Gardenburgers, Quorn's fungi-based faux chicken, and other fake meats of the day, so he was always disappointed when his friend drew conclusions like: "These foodstuffs all seemed to be made of the same substance, possibly turnip."

But Friedrich wasn't delusional. Meat-eaters clearly didn't think the imitations tasted like the real thing, or, for that matter, good.

The first commercial meat substitute, a peanut-based mutton knockoff, was invented in 1896 by Dr. John Kellogg, a eugenicist and anti-masturbation activist who also invented cornflakes, the flagship product for the company that still bears his family name. Kellogg was a character—Anthony Hopkins played him in *The Road to Wellville*—but what's relevant here is that he was a vegetarian who made meat substitutes for his fellow vegetarians. Ever since some Chinese Buddhist nuns were reportedly served "vegetarian food made to look like meat" at a Ming dynasty banquet, fake meat had always been crafted for people who didn't eat real meat. Friedrich got excited when Bill Clinton started eating Boca Burgers in the White House, but even Boca's founder admitted they were "not like your backyard burger, with juices flowing."

That's why veggie burgers had never broken out of the vegan aisle. Not all of them looked like hockey pucks or tasted like cardboard, but none of them looked or tasted much like meat. Vegan cheeses and ice creams were niche products, too—fine if you had sworn off the real thing, but lacking dairy's stretchy, melty, creamy functionalities. They inspired in-jokes like: *Hear about the fire in the vegan cheese factory? Still didn't melt.*

Margarine did outsell butter, because it was cheaper, perceived as healthier, and tasted similar. New soy milks that no longer evoked liquefied paste were also making inroads with non-vegans. The celebrity chef Tal

Ronnen, a PETA alum, often told Friedrich the path to a better food system was feeding people better meat and dairy substitutes, not guilting people for hating lousy ones.

But of course a vegan chef would say that. Friedrich never seriously considered animal-free alternatives for animal-eaters before the Humane Society's Josh Balk helped him move back to D.C. after his teaching stint in Baltimore. As Balk drove Friedrich back to the animal rights fray in a U-Haul, he argued that innovation could succeed where education and demonization were failing. We'll never get everyone to share our ethics, he said. But if we can make ethical eating cheap and appealing, we can save a lot of innocent animals.

"This is how we *win*," Balk said.

Balk had worked undercover in factory farms to film hideous abuses of chickens, but he had an even sunnier disposition than Friedrich, with an engaging smile he didn't need to practice in the mirror. He now led campaigns to persuade corporations to use cage-free eggs, and he believed most food executives were decent people who would gladly sell less cruel products if they were more economical. He also believed most carnivores were decent people who would gladly eat those products if they were delicious and affordable.

Balk figured consumers loved cheeseburgers, not the grotesque system that produced cheeseburgers. Even his father, a kindhearted teacher who had marched for civil rights, ate animal products at almost every meal. But systemic change would never happen in the vegan ghettos of natural food stores. Bean curd wasn't the answer to any question that mattered. The solution was animal-free meat, dairy, and eggs for mainstream consumers like his dad.

Balk was trying to help make that happen.

One frigid afternoon in Minneapolis, after a fruitless meeting with a General Mills executive who couldn't make cage-free eggs pencil out, he thought of a twist on the which-came-first cliché: What if we could make eggs without chickens? He proposed the idea—first gently, then relentlessly—to his best friend, Josh Tetrick, a rudderless searcher who spoke with great fluency and frequency about using capitalism to help the world, but had never done it.

Now the two Joshes were starting a company called Hampton Creek—named for Balk's cuddle-monster of a Saint Bernard—to engineer egg substitutes from plants. (They first called it Beyond Eggs, until Savage River, the only other biotech startup exploring alternative proteins in 2011, rebranded as Beyond Meat.) They planned to begin by replacing the eggs used as ingredients in mayonnaise and cookie dough, because mimicking the unseen glop that emulsifies mayo and adds structure to cookies seemed easier than mimicking entire eggs. But the Joshes had just turned 30, and neither had any background in science, engineering, or food. Balk was staying at the Humane Society, so Tetrick, a former college linebacker with no job or tangible record of achievement, would be in charge of building a brand-new kind of business.

Tetrick did fit a certain startup-founder profile in the age of Uber, a cocky, bro-ish, ultra-competitive pitchman who could sell water to the ocean. He had a compelling story about being raised in Alabama by a single mom who used food stamps. He had gone on to Cornell and Michigan Law, helped African street kids as a Fulbright Scholar, and built an impressive-sounding résumé that included stints as an adviser to Liberia's president, associate at Citigroup, and attorney at a big law firm. He already gave motivational speeches in his honeyed drawl about mission-driven capitalism, about finding purpose in life by making money and a difference.

But Tetrick had not found much purpose or profit in his own life. He had a speaking agency and a schtick, but no real career. He had never held down any of those prestigious jobs for an entire year. "When I gave those talks, I was really talking to myself," he said. He hated his corporate law job, and got fired after denouncing "cruel and inhumane" factory farms in a newspaper column about social entrepreneurship, a clear act of self-sabotage at a firm that represented the pork giant Smithfield.

Now he was trying to build a company in a nonexistent industry while crashing on an ex-girlfriend's couch, using the kitchenette in her Los Angeles studio as a lab, asking volunteers on the street to taste-test cookies. He knew nothing about foaming, binding, gelation, or any of the other functionalities that make eggs so good at what they do. Still, Balk believed in him, and believed Hampton Creek would help repair the world. And Balk's confidence got Friedrich thinking about animal-free animal foods.

"I was like, 'Huh, he isn't crazy, and it isn't a crazy idea,'" Friedrich said.

Friedrich began to think more seriously about this still-theoretical industry after meeting Ethan Brown, the vegan entrepreneur behind Beyond Meat. As a boy, Brown (no relation to Pat Brown of Impossible Foods) was fascinated by the Holsteins on his family's dairy farm in Maryland, along the Savage River that inspired Beyond's original name. He always wrestled with the morality of exploiting animals, which endeared him to Friedrich.

But Brown's mission was more about saving the climate, a mission more aligned with the Obama era. He also seemed more like a grownup, a six-foot, five-inch family man with a neatly trimmed beard, a Columbia MBA, years of experience as a fuel-cell executive, and a calmly analytical way of describing his wildly speculative venture. He'd explain in his soothing baritone that a food-tech version of the clean-tech boom was inevitable, and plant-based meats, unlike the fuel cells he had failed to make happen, could get cheap fast. It enhanced his credibility that he had burned his boats behind him, quitting his clean-tech job and emptying his 401(k) to license a University of Missouri patent for processing imitation chicken. In the non-vegan business proverb about the ham-and-egg sandwich—the chicken is involved, the pig is committed—he was the pig, fully committed to reinventing meat.

His spiel began with the idea that most humans would never quit meat, because meat helped make us human: When our ancestors began eating it two million years ago, they began developing the smaller stomachs and bigger brains that set them apart from apes. Humanity could always be counted on to procreate, fight wars, and eat meat, so instead of trying to undo evolutionary biology, why not give carnivores the meat they craved without the deforestation, pollution, and other downsides of animal agriculture? Thanks to meat, we had big brains that could recognize those downsides—and learn to make meat without animals.

Friedrich's favorite part of Brown's rap was his insight that meat was amino acids, lipids, and minerals, while plants were . . . amino acids, lipids, and minerals. So why couldn't big-brained scientists reconfigure plants into the architecture of meat? And why couldn't an extrusion process that converted plants directly into meat cost less than laundering plants through livestock that had to be cared for and cleaned up after? Animals were the bottlenecks of the food system, expensive to maintain and inefficient at converting their feed into our food. Why not bypass them, like Tesla was bypassing petroleum?

"What Ethan was saying made tremendous intuitive sense," Friedrich said.

Brown wasn't interested in making better knockoffs for his fellow vegans. They weren't causing the problems he wanted to solve. He hoped to shake up the trillion-dollar meat market by reverse engineering substitutes that tasted, smelled, seared, and chewed like the real thing, without the damage to the climate, forests, human arteries, and sentient animals. It was a classic tech-disruption pitch, and the Silicon Valley venture fund Kleiner Perkins, which had backed Amazon and Google, led Beyond's first funding round.

Vegan meat for non-vegans was a novel idea, and when Beyond debuted its first plant-based chicken strips in 2012, a *Slate* columnist called them "Fake Meat So Good It Will Freak You Out." Food writer Mark Bittman failed to distinguish them from chicken in a taste test. Bill Gates described them as "a taste of the future of food," and invested a bit of his own fortune in Beyond.

The venture firm founded by Vinod Khosla, whose bets on cellulosic biofuels were going south, cast another early vote of confidence in alternative proteins, giving Hampton Creek enough seed money for Tetrick to leave his ex's studio, set up shop in a San Francisco garage that previously housed a biker gang, and hire a few scientists to try to make plants emulsify and scramble. By 2015, Tetrick had raised $120 million from a Who's Who of billionaires, including PayPal's Peter Thiel, Salesforce's Marc Benioff, and the Hong Kong tycoon Li Ka-shing.

Tetrick had a talent for separating rich men from their money. But he also had a powerful pitch about revolutionizing a food system that overnourished the kids he grew up with in Alabama, undernourished the kids he worked with in Africa, tortured animals, and bulldozed forests. He looked investors in the eye and said: "Food sucks." He described Hampton Creek as a "leapfrog technology" that could bypass chicken farms just as mobile phones bypassed landlines. His pitch deck had a slide comparing the way Amazon ("a technology company pioneering e-commerce") took on Barnes & Noble, Toys "R" Us, then all retail, to the way Hampton Creek ("a technology company pioneering food") would take on mayo, cookie dough, then Every Conventional Egg.

Tetrick's deck didn't mention, as he told me later, that "objectively, I didn't know what the hell I was doing." It didn't explain how a food company, which would have to make stuff, could grow like Amazon, which sold stuff.

But why complicate the vision? The food system did suck. And his team did manage to make a creamy, shelf-stable mayo from yellow peas. Maybe the journey to a more sustainable world could start with condiments.

Pat Brown's pitch was even more audacious: Impossible would eradicate animal agriculture in two decades. And he was a renowned Stanford biochemistry professor, not an unemployed, couch-surfing dreamer.

Brown was best known for inventing the "DNA microarray" that helped launch the 21st-century revolutions in genetics and synthetic biology, work that revamped cancer detection and unlocked mysteries of the human gut. He also had disruption experience as cofounder of the nonprofit Public Library of Science, which upended academic publishing by making research free online. He lived in Silicon Valley—he biked to meetings with venture capitalists—and shared the local ethos of moving fast and breaking things. He had the unwavering intensity of a marathon runner, which he was, and the messianic fervor of an evangelist, which he basically also was. When he preached about the planet's catastrophic trajectory in his high-pitched nasal voice, it felt a sermon from an unhinged Kermit the Frog, until it began to feel like time to check Zillow for housing options on other planets.

Like Beyond's Ethan Brown, Pat Brown was a vegan who had concluded that climate change was humanity's worst problem and animal agriculture the worst part of it. He had tried to jump-start solutions by hosting a scientific workshop in Washington early in the Obama era, but after it failed to galvanize hope or change, he decided facts were too weak a tool to fix policies or diets. "I thought if we got together some objective people, the answers would be obvious and we'd convince everyone to change their ways," he told me, chuckling at his own naivete. Meat was just too central to civilization. Even environmental conferences served steak. Just as the Stone Age didn't end because we ran out of stones, the Meat Age wouldn't end until we found something better.

Brown was perfectly satisfied with the 1,500-square-foot condo his family had lived in since the '80s. But he made the boat-burning decision to leave academia and start a business, because he considered the free market the only mechanism subversive enough to disrupt livestock. His plan was to make meatless meat so irresistible it would sabotage the economics of animal agriculture, which would vanish and allow the rewilding of pastures, which would draw down carbon from the atmosphere and stabilize the climate.

If that sounded absurdly optimistic, he thought pathbreaking scientists had to be. And after all the work he had done to unlock the secrets of genetics, rearranging plant molecules into meat formations seemed quite doable. Meat was already made of plants, even if it was assembled inside animals that ate plants. He knew it would take time to bio-mimic whole cuts like chops and fillets, the complex slabs of marbled fat and muscle that fetch the highest prices, but he thought Impossible could reproduce ground meats like burgers and nuggets fairly soon.

The key challenge would be replicating the molecular magic that makes meat so awesome—the explosion of aromas on the grill, the chewiness, juiciness, and sheer meatiness that triggers something primal in our brains. He already suspected the key ingredient was a protein called heme, which gave blood its red color and was one of the most potent catalysts in nature. Once he was sure, it would just be a matter of getting the recipe and processing right.

By 2015, before selling his first burger, Brown had raised $180 million from investors like Gates, Khosla, Thiel, and Li Ka-shing—and rejected a

$300 million offer from Google to buy the company. He wasn't looking for a quick payday, and he doubted a search-engine giant would be a loyal steward of a mission-driven meat subsidiary.

Still, Friedrich saw Google's mega-offer for a food company that hadn't even started selling food as proof of concept for the alt-protein opportunity. The three California startups had raised more money in three years than farm animal charities had raised in their entire history. He had always raged against the capitalist machine, but he recognized its power.

Friedrich was getting interested in "effective altruism," a new utilitarian movement that aimed to bring data and rigor to the work of doing good, and he was losing confidence that activism was the most effective way to do good for livestock. America's first animal activists had mobilized to protest the abuse of overworked carriage horses, creating the American Society for the Prevention of Cruelty to Animals to demand mandatory rest and watering stations for literal workhorses. But it was Henry Ford, not the ASPCA, who had ended their exploitation, by creating a horse-free form of transportation that consumers liked better.

Friedrich wondered if modern Fords could create animal-free forms of consumption that consumers liked better than animal foods. Imagine how much good a protein equivalent of the Model T could do, not only for livestock but for people and the earth. If modern technologists could wipe out tollbooths, video stores, and compact discs, why not meat, dairy, and eggs?

FILLING THE WHITE SPACE

Alternative energy already supported eight million jobs in 2015, while alternative proteins just supported a few thousand. And most of those jobs were with old-school brands like MorningStar and Boca (owned by Kellogg's and Kraft) as well as Gardein, LightLife, and Sweet Earth (soon to be owned by Conagra, Maple Leaf, and Nestlé) that sold vegan meat substitutes to vegans. They were side hustles for Big Food, serving a niche market.

Plant-based milks had seized 10 percent of the U.S. milk market, and dairy milk sales were actually declining, an encouraging example for plant-based meat. Still, Americans spent less in 2015 on almond milk, which had replaced soy atop the plant-based charts, than tickets to the *Star Wars* sequel *The Force Awakens*. It was pretty niche, too.

Hampton Creek, Impossible, and Beyond had much grander ambitions. Tetrick talked about building the world's largest food company. Pat Brown vowed that Impossible would be the most important company of any kind, ever. Ethan Brown named Beyond's science team "the Manhattan Beach Project," to evoke its World War II–style urgency as well as its Los Angeles–area geography.

So far, though, that Big Three didn't have much company. Two celebrity chefs in California, Miyoko Schinner and Friedrich's PETA pal Tal Ronnen, started vegan cheese companies, Miyoko's Creamery and Kite Hill, but thanks to Big Dairy's clout, they couldn't even use the word "cheese" on their labels. ("Cultured nut product" didn't have the same ring.) A few Silicon Valley software nerds founded Soylent, a quirky effort (named after the human-based wafers in the film *Soylent Green*) to simplify mealtime by cramming all the nutrients needed for human survival into plant-based powders and shakes. But they tasted about as appealing as one might expect of functional food created by engineers who didn't enjoy eating.

Then there was Muufri, as in moo-free, a Bay Area "precision fermentation" play that hoped to genetically engineer microbes to produce dairy proteins in steel vats. (It was renamed Perfect Day in honor of an odd study that claimed cows made more milk while listening to that Lou Reed song.) Precision fermentation already had a history of saving animals; cheese makers used it to make an enzyme called rennet that was previously extracted from calf stomachs, while drugmakers used it instead of pig pancreases to make insulin. Still, Perfect Day's founders were 22-year-old vegans fresh out of college, unlikely candidates to disrupt the trillion-dollar dairy industry.

And that was about it. Alt-proteins were still an out-there California fad. Steve Molino, a young finance guy who stopped eating meat and got obsessed with finding substitutes after watching the documentary *Cowspiracy*, says the sector was so nascent, he even got excited about Soylent's joy-free meal replacements.

"I'd try to tell people about the space, and they'd say: 'What space? There is no space!'" said Molino, now an alt-protein venture investor.

Arturo Elizondo, a Mexican American Harvard student who grew up eating eggs every morning and meat every day in a Texas border town, also went vegan and got obsessed with substitutes after interning in USDA's slaughterhouse division. After graduation, he impulsively backed out of a plum Obama administration appointment and booked a one-way ticket to San Francisco to try to break into the alt-protein business. But after leaving for his new adventure with no job and no place to stay, he discovered there was practically no business.

"The ecosystem just didn't exist," he says.

Blessed with TED Talk eloquence, type A determination, and a telenovela jawline, Elizondo got lucky at an early food-tech conference when Isha Datar, the director of a "cellular agriculture" research institute called New Harvest and the matchmaker who had connected Perfect Day's founders, introduced him to a biologist who wanted to make egg protein without chickens. Two weeks later, they launched a precision-fermentation startup called Clara Foods, which, because rebranding seems almost mandatory in the alternative protein sector, later became the EVERY Company. Suddenly, the Bay Area had another 22-year-old biotech CEO.

It was one thing for newly minted college grads to develop apps, quite another to rewire microbes to make proteins that livestock already made cheap. Still, Datar liked to quote Carl Sagan's line that to make an apple pie from scratch, you must first invent the universe. She knew there was no recipe or infrastructure for what EVERY and Perfect Day wanted to do. But at the relatively seasoned age of 26, she thought mission-driven youngsters were more likely than food-industry veterans to invent the universe necessary to end animal agriculture.

"They don't look for reasons not to do things," Datar said. "They're like, 'Why not?'"

The Big Three at least had executives with labor-market experience. But they weren't big yet, and they all had growing pains.

Beyond Meat's 2015 revenues were $8.8 million, three hours' worth of McDonald's revenues. And Ethan Brown and his star scientist, Harvard-trained biochemist Tim Geistlinger, were losing faith in the source of those revenues, the imitation chicken strips Bill Gates had dubbed the future of food. After the initial shock of verisimilitude wore off, meat-eaters weren't loving them.

"They were never going to be game changers," Geistlinger said.

Instead, Beyond was betting its future on its burger project, which wasn't going smoothly, either. Replicating beef with plants was tough. Scientists and engineers had to isolate the plant proteins in crops from the starch and fiber; use heat and pressure to extrude the protein "isolates" into meat-like molecules; and blend it all together with binders, coconut oil for fat-like qualities, and beet juice for color. Beyond's executive chef, Dave Anderson, recalls months of tinkering with various ingredients—and when something improved the texture, it often threw off the taste, requiring additional months of tinkering to recalibrate.

Beyond's marketing team also complicated the task by insisting the burger couldn't contain soy, because consumers had negative health perceptions of soy, especially GMO soy. Those were unfair perceptions, especially the myth that soy hormones create "man boobs." But Brown, a soy milk drinker, decided the customer was always right, even when the customer was wrong. He didn't want to bet Beyond's future on man-boob-related rationality. So his team took on the even tougher challenge of fabricating a burger from pea protein, which was harder to texturize and notorious for bitter off-tastes. "I pushed to keep working the soy angle, too, in case peas were too big a hill to climb, but the decision was peas," Anderson said. Geistlinger is still proud of their work, but he left Beyond to join the kids at Perfect Day after it was done.

Impossible's 2015 revenues were zero—partly because once its scientists concluded heme was indeed the mystery ingredient that made beef beefy, Pat Brown had them waste a year trying to extract it from soybean roots. That turned out to be impractical, the molecular equivalent of harvesting rice with tweezers.

"A cockamamie idea," he said.

Impossible eventually used precision fermentation to recode yeast microbes to make heme, just as Perfect Day and EVERY were trying to do for dairy and egg protein. But it still hadn't gotten a product to market, and

Brown often behaved more like a religious revivalist than a CEO. His aides spent a lot of time begging him to stop attacking the slaughter cartel and blocking his other cockamamie ideas—like refusing to let employees expense meals with animal products, refusing to let restaurants serve Impossible Burgers with non-vegan cheese, and refusing to submit products for testing in rats. All he cared about was the science and the mission, which he summarized as getting rid of the fucking cows.

"We'd have all-hands meetings where nothing would get done, he'd just preach about how meat was the devil," one former executive recalled.

Hampton Creek was also drowning in drama. Beyond chef Dave Anderson, initially a partner there, describes Tetrick as a mercurial and manipulative cult leader who forced him out during a family crisis. "Josh just liked chaos," Anderson said. Thomas Bowman, another chef who left Chicago's molecular gastronomy scene to be head of product development, had worried he'd miss the adrenaline rush of the kitchen, only to find even more pandemonium in the garage. Tetrick once gave Bowman three months to create a dozen new dressings he had rashly promised to a national food-service operation.

"We spent a lot of time in panic mode," Bowman said.

Hampton Creek was conceived as an ingredients company, but when Whole Foods offered to carry its branded consumer products, Tetrick abruptly pivoted, triggering a frantic scramble to identify co-manufacturers and build a distribution network on the fly. Then its first branded product, Just Mayo—as in, it's just mayo, and also mayo produced in a just way—began disintegrating in the Northeast, prompting another frantic scramble to find the flaw. It turned out that a winter freeze had thrown off the emulsion and separated the oil from the water.

"It was like, 'Oh, thank God, the product is fine, we just need insulated trucks,'" Bowman recalled. "On to the next mess!"

The next mess was a lawsuit by Unilever, maker of Hellmann's Mayonnaise, challenging Hampton Creek's right to call its vegan spread "mayo." Facing a probably unwinnable and possibly unsurvivable legal case—mayo was legally defined as an egg product—Tetrick fought back with a David-and-Goliath media strategy, portraying his startup as a victim of corporate bullying, generating a torrent of publicity that shamed Unilever into dropping its suit. It later came out that one American Egg Board executive sent a presumably facetious email asking if producers could "pool our money and put a hit on" Tetrick, prompting another to offer to "contact some of my old buddies in Brooklyn to pay Mr. Tetrick a visit."

Tetrick was making enemies, inspiring gossip, and spending investor money like he was reprising *Brewster's Millions*. One insider told me he spent six figures on hoodies and other Hampton Creek merch. Still, he was getting his mayo and cookies into thousands of outlets, and the media lionized him as a floppy-haired rebel in a T-shirt, waging guerrilla war against Big Food. "Mark my words: Josh Tetrick will win a Nobel Prize someday," one food writer predicted.

But the drama only deepened. Disgruntled Creekers began leaking to reporters about shenanigans in the garage, accusing Tetrick of mislabeling ingredients, tampering with contracts, promoting an aide he was sleeping with, and other shadiness. They described him as a grifter and gaslighter reminiscent of Trump, the ultimate Bay Area insult. They even complained he let his omnipresent golden retriever, Jake, wander around the lab. One executive quit after nine days, then warned investors to ignore Tetrick's soaring projections of future sales. It soon came out that Tetrick was also inflating current sales by having employees buy back jars of Just Mayo from stores, prompting a federal investigation that led to no charges but lots of embarrassment.

The press, after pumping Tetrick up as a swashbuckling giant slayer, began tearing him down as a flimflam artist who had conned Silicon Valley into funding edible vaporware. *Bloomberg Businessweek* ran a photo illustration of Tetrick with mayo smeared across his roguish face, while *Business Insider* headlined its hit piece "Sex, Lies and Eggless Mayonnaise: Something Is Rotten at Food Startup Hampton Creek." Even as he denounced the pile-on as exaggerated clickbait, he had to admit that "in working to change the food system, we screw up sometimes."

Tetrick's most pressing concern didn't even make news. His scientists had concocted a liquid egg from mung beans that scrambled beautifully in the lab, but his engineers couldn't make it work at scale, so Hampton Creek couldn't release its signature product. He had always rejected suggestions to narrow his focus, because they felt like insinuations he wasn't up to his job. He wanted to save all the birds, as soon as possible. But he finally decided to shut down Hampton Creek's work on pancake mixes, brownies, and other nice-to-have products. An egg company needed to get its egg right.

"I had to admit: We can't do this, unless we do *only* this," Tetrick says.

In the global context, none of the Big Three did much in 2015. The world ate 1.5 trillion eggs; Hampton Creek displaced a few million. Carnivores ate 500 billion pounds of meat; Beyond and Impossible weren't even a rounding error. Food's wannabe Henry Fords were hemorrhaging cash, while its buggy-whip manufacturers kept raking in profits.

The Good Food Institute was created to change that.

Milo Runkle, the former PETA intern who led Mercy for Animals, wanted to create an alt-protein division inside his organization. But after brainstorming with his deputy Nick Cooney and fellow vegan pragmatists Balk and Friedrich, they all agreed the world needed a new organization, with a mission transcending mercy for animals. Just as the Model T cleaned up manure-choked streets, alternative proteins could clean up the carbon-choked atmosphere—and also reduce habitat loss, water pollution, and other animal agriculture problems unrelated to animal abuse. A nonprofit unaffiliated with animal protection could have more mainstream credibility, and could also work on meatless products with conventional meat companies, which might be wary of a group that considered their businesses inherently barbaric.

"Everyone knew we were all about saving animals," said Cooney, who wrote an effective altruism book called *How to Be Great at Doing Good*. "This new thing couldn't have our baggage."

Before Mercy for Animals launched the institute with a $600,000 donation, the other brainstormers did insist on one piece of animal-rights baggage: Friedrich had to run it. The former college dropout, statue defacer, and jet vandalizer was now a savvy strategist with a Georgetown law degree. When Cooney told him activism was no longer the best and highest use of his talents, an effective altruism catchphrase, he had to agree.

He was still the same Works of Mercy disciple who streaked Buckingham

Palace, with the same Bruce Poppins hope for change. But pushing to make factory farms unnecessary now felt like a better and higher use of his talents than pushing to make factory farms marginally less grisly. He no longer wanted to fight for the food equivalent of water stations for carriage horses along their daily exploitation routes. He suspected the animal welfare movement's most important legacy would be inspiring entrepreneurs like the Browns and the Joshes.

His new mission was to build the ecosystem that hadn't been there for Arturo Elizondo after Harvard graduation. GFI would be part think tank and part business incubator, a one-stop marketing department, research hub, technical adviser, and social network for alt-protein entrepreneurs. It would help startups get started and help tell their stories. It was technically a charity, not a lobbying group—a tough-talking vegan lawyer named Michele Simon was starting a Plant Based Food Association with a lobbying focus—but it aimed to be the sector's voice.

Friedrich pitched GFI as an effective altruism play that was less about saving animals than preventing climate change, famines, antibiotic-resistant superbugs, and livestock-borne pandemics. Facebook cofounder Dustin Moskovitz's effective altruism foundation became his top donor. Friedrich also raised $250,000 from a then-obscure effective altruist named Sam Bankman-Fried, whose check cleared well before his cryptocurrency exchange imploded.

Friedrich's main argument, updating *Diet for a Small Planet* for the climate era, was the inefficiency argument that first attracted him to veganism. In a TED Talk, he gave an audience member one bowl of pasta and tossed eight bowls in the trash, to illustrate the folly of feeding chickens nine calories of grain to make one calorie of meat. Why waste feed, land, water, and fertilizer to grow beaks, hooves, feathers, and lungs with no gastronomic or economic purpose? Unlike fossil fuels, which were destructive but super-efficient, farm animals were destructive and inherently inefficient—which made animal farms inherently vulnerable to competitors who didn't have to pay for feed, farmland, or veterinary care.

Still, taking on animal agriculture would require a slew of new Beyonds, Impossibles, and Hampton Creeks. That's why the former Jesus lookalike who had worked in a shelter for five bucks a week decided to launch a venture capital fund.

One Wall Street vegan had secretly financed Friedrich's advocacy, Balk's advocacy, and much of the animal rights movement. The anonymous benefactor had been thinking about alternative proteins even before Friedrich called to ask him to invest in a $25 million vegan venture fund, a serendipity he ascribed to "the grace of the animals." He believed livestock were calling the cosmic shots, guiding him to help build a more humane world.

"The chicken in the factory farm was always my boss," he told me.

The donor told Friedrich to stop fundraising: He'd put up the entire $25 million for GFI's New Crop Capital fund. It was a lot to give an activist who knew nothing about money, but the funder told me he didn't sweat the uncertain returns; he's sunk $300 million into animal sanctuaries, anti-meat billboards, and other vegan causes with no returns. And while Friedrich now looked like a venture capitalist, wearing sports jackets without ties and yuppie-chic glasses, his first hire at GFI was an actual investor to run New Crop: Chris Kerr, a serial entrepreneur who had sold his companies to focus on animal welfare and his house full of stray cats.

"Chris was literally the only person I knew who understood capitalism," Friedrich said.

Kerr had been trying to help the Humane Society invest in animal-friendly companies like Beyond Meat. But it wasn't the Investment Society, and its financial overseers were never comfortable letting him risk donor dollars on biotech burgers. New Crop would be all about risk, and in the fall of 2015, Kerr and Friedrich began identifying "white spaces" where the anonymous funder's karmic cash could fill voids in the marketplace.

One glaring void was Asia, and Friedrich talked would-be entrepreneurs into founding a plant-based meat company in India and an alt-protein incubator in China. New Crop put the first money into both firms, planting the sector's flag where it was needed most. Another obvious white space was plant-based seafood, but Kerr was skeptical of the startups pitching themselves as Impossibles of the sea. "So I said: 'Fuck it,

I'll start my own,'" he says. He helped some friends in the business create a fish-free fish startup called Good Catch—and New Crop was its first investor.

Really, the sector was almost entirely white space, and Kerr's goal was to help a thousand flowers bloom to fill it. His strategy was "spray and pray"— spread cash around a variety of startups, then hope a few of them hit.

He did have a few easy calls, like Beyond, Miyoko's, and Kite Hill, whose founders included not only the vegan chef Tal Ronnen but Impossible's Pat Brown, who had invented a way to convert nut milk into cheese curds when he started tinkering with food at Stanford. Kerr also bet on longshots like Fora Foods, which made butter out of "aquafaba," the starchy liquid in chickpea cans; New Wave Foods, a plant-based shrimp startup led by two young women; Nobell Foods, another woman-led effort to reengineer crops to grow dairy proteins; and Alpha Foods, a plant-based frozen food company that made a vegan "Alpha Dog" he considered utterly mediocre. He usually considered taste a dealbreaker, but Alpha's leaders were seasoned executives, and he was afraid he was sending too many mission-driven newbies into the food jungle. Anyway, he figured consumers ate plenty of mediocre-tasting frozen food.

At the same time, GFI was building a virtual alt-protein community with monthly video calls and a lively Slack channel. It created a startup manual so every alt-protein entrepreneur wouldn't have to reinvent every wheel. Friedrich also persuaded a bored investment banker named Curt Albright, a recovering addict who had realized in therapy that there was more to life than toys, that the best and highest use of his talents would be to start another alt-protein fund.

"Bruce showed me the best way I could be of service," Albright said. "If I could help investors make gobs of money, people would notice this asset class."

Friedrich even encouraged his own employees to join the new industry. The first scientist he hired left to launch a plant-based chicken startup. His corporate engagement director quit to build a plant protein team for a food conglomerate. The innovation manager who wrote GFI's startup manual jumped to a vegan cheese venture. Another innovation specialist teamed up with Hampton Creek's culinary guru Thomas Bowman to make plant-based ice cream.

Kerr felt like every time he invested in a startup, Friedrich sent three new ones his way. "It was a pretty frigging chaotic time," he said.

After seeding four startups in seven years at the Humane Society, Kerr seeded 15 in his first year at New Crop. And some of them were gambles on a purely speculative industry without products to sell or the legal authority to sell them.

THE SECOND DOMESTICATION

In 2013, a Dutch scientist named Mark Post unveiled a burger grown in his lab from bovine stem cells, generating a flurry of publicity about "in vitro meat." It was also described as "cell-based meat," "lab-grown meat," and even "schmeat," but it arguably wasn't fake meat. It was real meat grown outside an animal, fibers of cow muscle smushed into a beef patty. The only hitch was that Post's five-ounce burger cost $330,000 to create.

In vitro meat's champions always noted that in 1931, a part-time futurist named Winston Churchill predicted that "fifty years hence . . . we shall escape the absurdity of growing a whole chicken in order to eat the breast or wing by growing these parts separately under a suitable medium." Instead of feeding birds, which expend energy clucking, pecking, and growing feathers, why not feed cells that direct all their energy into meat making? But 80 years hence, the meat industry was not alarmed by a $1.2 million-a-pound competitor. It still had huge hurdles to relevance, especially the exorbitant cost of the "growth media" the cells consumed, the difficulty of getting cells to multiply quickly, and uncertainties surrounding the bioreactors where the cells would grow. Friedrich hoped those problems would be solved someday, but he planned to focus on companies that could displace animal products sooner. He saw cell-based meat as a long-term science project.

He changed his mind after meeting Uma Valeti, a star University of Minnesota cardiologist who had also burned his boats, quitting medicine and moving to the Bay Area to grow slaughter-free chicken and beef. Valeti wasn't as bombastic as Tetrick or Pat Brown, but he shared their dream of dismantling animal agriculture, with a quieter confidence that inspired confidence in others.

Valeti's story began near his childhood home in India at a birthday party he attended when he was 12, a joyous occasion with guests celebrating in the front yard while cooks slaughtered chickens and goats in the back. He never forgot that creepy juxtaposition of birthday and death day. But he didn't go vegan for another decade, so he also understood the allure of animal flesh. He'd often tell patients after stenting their arteries that they needed to change their diets to save their lives—and often, they didn't. Humans were lousy at breaking habits like meat.

Valeti began to imagine a better way at the Mayo Clinic, where he used stem cells to grow muscle for damaged hearts. He wondered: Why not use animal stem cells to grow muscle for food? He knew it could be done, maybe even with tweaks to reduce red meat's contribution to the heart diseases that plagued his patients, and remarkable new advances in synthetic biology and supercomputing could make the trial-and-error process much faster and cheaper. In 2015, Valeti cofounded Memphis Meats, which later became Upside Foods, with a heart logo in honor of its compassionate mission as well as its cardiology roots.

Valeti was an enthusiastic early investor in Beyond, Impossible, and Hampton Creek, but when he met Friedrich for dinner that fall, he asked: What if plant-based meat can't make it across the chasm? He doubted carnivores would ever quit whole muscle cuts like rib eye steak, or settle for not-quite replacements made of soy or peas.

"We need to grow real meat, too," he told Friedrich. "Asking people to give up what they love doesn't work."

Valeti envisioned a "second domestication," when humanity would end its dominion over animals by taking dominion over cells. As a doctor, he could save a few thousand lives, but as a cellular agriculture pioneer, he could help save billions of human and animal lives, and the rainforests, too. He wanted to maximize his impact, an effective altruism concept Friedrich found relatable.

Valeti knew it would take at least a decade to get pharmaceutical-grade cell-culturing costs down to food-grade levels. The economics of growing life-saving organs was much easier than growing lunch. But he also knew cultured meat wasn't ten-years-away-and-always-will-be. And his calm bedside manner made his sci-fi fantasy sound like quadruple-bypass surgery—challenging, sure, but he's got this. Soon Friedrich was declaring publicly that cell-based meat was just a decade away—"and at that point, you'll see animal-based meat go the way of the horse and buggy."

"Uma convinced me this was not just an opportunity down the road, this was investable right now," Friedrich said.

Michele Simon of the Plant Based Food Association, who thought Friedrich even overhyped the plant-based products they were both paid to promote, found his cell-based propaganda laughable: "He was selling a research project disguised as an industry." Impossible's Pat Brown also considered cultured meat "scientifically ludicrous," like trying to replace horses with horse cells hooked up to gears and pulleys instead of automobiles. Even Isha Datar, whose nonprofit promoted cellular agriculture, thought Friedrich was undercutting its credibility with unrealistic promises, while blurring the lines between advanced synthetic biology and glorified veggie burgers.

But by the end of 2015, Valeti was growing ground beef in a small lab above a Walmart for $18,000 per pound—still $17,996 more than the meat counter downstairs, but 98 percent less than Post's proof-of-concept burger. It was sliding down the kind of cost curve associated with transistors and computers. After New Crop seeded Upside with $500,000, the company began attracting investors like Gates, General Electric CEO Jack Welch, Richard Branson, Elon Musk's foodie-entrepreneur brother Kimbal, and Khaled bin Alwaleed, a Saudi royal known as "the vegan prince" who had invested in Beyond and Hampton Creek as well.

Friedrich was also pitching meat companies on alt-proteins, urging them to be Canon, which embraced the transition to digital photography, not Kodak, which invented the digital camera but went bust after underestimating its impact. Cargill and Tyson agreed to invest in Upside, and Tyson's CEO publicly explained he didn't want to be Kodak. (Kerr sold Tyson half of New Crop's stake for $500,000.16, banking the first 16-cent profit on cell-based meat even if the other half went to zero.) Tyson also bought 5 percent of Beyond and started calling itself "the Protein Company," while Cargill dropped the word "animal" from its Animal Protein Division. Meat executives obviously didn't plan to quit meat; they saw substitutes as an "and" play, not an "or" play, providing options in a flexitarian moment.

"We don't want to be disrupted," Tyson's chief sustainability officer said at a GFI event. "We want to do the disrupting."

Kerr also began spraying and praying in the cell-based space. New Crop led seed rounds for Aleph Farms, an Israeli startup culturing steaks; Mosa Meat, a Dutch firm commercializing Mark Post's burger science; and BlueNalu, a San Diego cultured seafood company that Kerr cofounded to fill more oceanic white space. But they all faced a tough road to market. They'd have to culture cells in tanks tens of thousands of times larger than the flasks they were using in their labs—and they'd have to cut costs another 99.9 percent to approach parity. There were no suppliers to sell them cell lines, growth media, or other necessities, so they'd have to develop their own. It was as if Old MacDonald had to do his own genetics research to breed his livestock, figure out what they liked to eat and prepare it himself, then design and build his own tractors and slaughterhouses. And cell-based meat wasn't approved for sale anywhere on Earth.

Still, Kerr believed in taking shots on goal. He felt like he was investing in solar in the '60s, making small bets that might pay off big for the planet someday.

"We weren't chasing returns," Kerr says. "We were trying to build an industry that could change the world. We were trying to start a conversation."

Friedrich wanted to get that conversation right, because consumers would never eat meat grown in vats if they saw it as a freaky science experiment. In the 19th century, the industry that sold ice from lakes tried to smear new imitations made in freezers as "artificial ice," but the public quickly accepted them as ice, just as cellular phones are now known as phones. Ideally, the public wouldn't think of cellular meat as fake meat or lab meat—just meat.

He was haunted by the pariah status of GMOs, which were vilified as deadly Frankenfood even before they hit the market—and the consumers most likely to embrace meat alternatives because they cared about animals or the earth were often the same consumers who rejected food that wasn't "natural." Friedrich didn't see what was so natural about livestock bred for insanely fast growth, then pumped full of drugs and GMO grain in overcrowded barns. But the public saw animal-corpse meat as normal, and there was no point arguing with the public.

He wanted lab-grown meat to seem normal, too, so for starters it needed a more appetizing name than "lab-grown meat," which sounded like something a Bond villain would concoct to destroy civilization, and wouldn't even be accurate once the technology scaled out of the lab. "In vitro meat" was even worse, evoking test-tube babies on plates. Even "cell-based meat" had overtones of mush on a petri dish—and anyway, almost all food was cell-based.

GFI settled on "clean meat." It was a reminder that meat grown in breweries rather than farms would have no hormones or antibiotics, less risk of pathogens, and much less impact on forests and the climate. Friedrich loved the analogy to clean energy, since he believed ditching animal meat was as urgent as ditching fossil fuels. When the animal pragmatist Paul Shapiro left activism to write a book showcasing the industry, he titled it *Clean Meat*.

But the nomenclature question was not yet closed. The meat industry was enraged by the implication that its products were dirty, and the clean meat industry didn't want war before it even had products. When Valeti visited the National Cattlemen's Beef Association's offices near the White House to discuss thorny regulatory issues with the major livestock groups—known in D.C. as "the Barnyard"—they were angrier about "clean" than anything else on his agenda.

"They said they didn't realize 'clean meat' was offensive," recalled Danielle Beck, a lobbyist for the cattlemen. "Uh, what did that make us?"

Valeti promised a less disparaging term, and GFI eventually forged a new consensus around "cultivated meat," which was accurate without sounding icky or hostile. The Barnyard didn't care, as long as its theoretical competitors didn't make its actual products sound filthy. Cultivated meat was still so astronomically expensive in the lab that nobody expected it on the market anytime soon.

Nobody, that is, except the impatient leader of a turbulent startup that, by his own admission, had no sensible rationale for cultivating meat.

Josh Tetrick thought a lot about death, which explained his frenetic approach to life. In law school, after he was diagnosed with a heart disease that often killed young athletes, he got focused on making an impact before time expired. He programmed his phone to send him a macabre daily reminder: "Prepare to die today." His mortality obsession intensified after he lost his first loved one: his golden retriever Jake, his constant companion as he matured from aimless wanderer into purpose-driven businessman. The pain of Jake's passing left Tetrick feeling aimless again.

He had built a dynamic mayo business. His next funding round would value Hampton Creek as the first billion-dollar alt-protein unicorn. But so what? Even if it did become the Amazon of eggs, the food system would still suck.

"I need to do more," he whined to Balk. "What else should I do?"

How about meat from cells? Balk replied. Meat created far more slaughter, deforestation, and emissions than eggs. Why not tackle the biggest problem?

"You don't know how much time you have left to make an impact," said Balk, who knew how to touch his pal's emotional buttons.

As a business strategy, cultivating meat seemed deranged, like Uber pivoting to time travel. Tetrick had just decided to prioritize the liquid egg—"we can't do this, unless we do *only* this"—and it was still two years away from market. How could a cash-bleeding startup that hadn't released its core product afford a side project that would require multiple breakthroughs just to get out of the lab? It was the kind of hubris that got entrepreneurs booted off *Shark Tank*. And if he refused to stick to his knitting, wouldn't the sensible next act for a plant-based egg company with a high-powered team of plant scientists be plant-based meat?

"We hadn't even nailed the egg yet. Any normal strategic person would've said this was a bizarre idea," Tetrick said. "I was just taken by the bigness of the bet."

The timing was bizarre, too, since plant-based alternatives were finally escaping what Ethan Brown called the vegan penalty box. Supermarkets were selling Beyond Meat in refrigerated meat aisles, while the Impossible Burger was debuting at Momofuku Nishi, a Manhattan hotspot whose chef, David Chang, was such an aggressive carnivore (and found vegans so annoying) that he once pulled his vegetarian dishes off his menu. Just Mayo was achieving such crossover success that Unilever, in a classic can't-beat-'em-join-'em move, rolled out an eggless Hellmann's.

But when Tetrick thought about his childhood—the barbecues and tailgates, the wings his mom fed him after football practice—he doubted his Alabama buddies would ever give up real meat. It was baked into their identity and culture. He suspected that growing meat from cells was the only way to stop them from eating animals.

"If Option A is a 100 percent chance of building a nice profitable company with the egg, and Option B is really fucking going for it and having a 20 percent chance of helping to end this cruel system of animal agriculture, I'm never gonna feel motivated by Option A," he said. "It's not enough for me."

That was the origin of Project Jake, the cultivated meat lab Tetrick set up in a converted storage closet. Hardly anyone at Hampton Creek knew about his skunkworks. But in June 2017, after his team decided success was at least possible, Tetrick publicly vowed to sell the first cultivated meat in 2018, long before anyone thought it could be cost competitive. And then all hell—well, more hell—broke loose.

Tetrick had always careened from crisis to crisis, and as his focus whipsawed from ingredients to branded products to his never-quite-ready egg, the criticism of his cowboy style had only gotten louder. One hit piece asked if Hampton Creek was "the Theranos of Mayonnaise," as if his eggs were as bogus as the blood tests that sank that other Bay Area unicorn. But his lurch into cultivated meat seemed like next-level megalomania, and a security firm Tetrick hired soon discovered that three of his top lieutenants were plotting to oust him.

Tetrick fired them, purged his entire board of directors, and changed Hampton Creek's name to Eat Just, signaling a fresh, mission-driven start. He'd keep trying to make eggs from plants, but he'd also try to save animals and the climate by making meat from cells. It would be an "and" play, not an "or" play, and while it wouldn't make money for years, if ever, someday it might help revamp a food system that sucked for Alabamans, Africans, animals, and the earth.

"I get it, I was increasing the probability of the company blowing up," Tetrick said. "I was also increasing the probability of the company doing something really meaningful in the long run. Isn't that the fucking point?"

In November 2017, Friedrich invited the *Post*'s Gene Weingarten to a farm-to-table restaurant with real meat and dairy options, not grilled "cheese" or spores in slime. The occasion was the Washington debut of the Impossible Burger, which was being hyped as the first indistinguishable substitute, with headlines like "Silicon Valley's Bloody Plant Burger Smells, Tastes and Sizzles Like Meat." His own newspaper asked: "Is This the Beginning of the End of Meat?"

Weingarten hoped so. If the Impossible Burger could match the magnificence of beef, he could avoid a lot of guilt. He ordered one of each, blindfolded himself with napkins, and bit into the burger to his right.

"This," he said, "is not meat."

He chomped to his left.

"This," he said, "is meat."

He removed his blindfold, saw Friedrich was despondent, and knew he had aced the test. But his friend Rachel, who only ordered an Impossible Burger, had a different reaction: "Whoa! This satisfies my meat thing totally."

Friedrich was delighted, until Weingarten tried hers as well.

"This," he said, "is meat."

Sure enough, the waiter had accidentally brought beef.

Impossible's early formulation clearly wasn't indistinguishable to everyone. Still, as Pat Brown liked to point out, the first locomotive lost a race to a horse. The rationales for alternative proteins—cutting emissions and pollution, sparing forests and wetlands, feeding the hungry in accordance with the Works of Mercy—were only getting stronger. For Friedrich, the only questions were whether the products would keep improving like locomotives did, and whether people would be willing to eat them. And Weingarten realized he had been so focused on identifying the imposter, he had overlooked something:

"The Impossible Burger, whatever the hell it was, tasted pretty darn good."

SEVEN
THE FAKE MEAT HYPE CYCLE

THE PEAK

The Good Food Institute's conference in September 2019 drew 900 people to a San Francisco hotel. They were some fired-up people. I kept hearing phrases like "phenomenal trajectory," "perfect storm," and "trending up and to the right." In his welcoming remarks, Friedrich celebrated the sector's "colossal watersheds," "awesome news," and "stunning progress." His vision of alternative proteins that were no longer alternative no longer seemed implausible. The movement to swap out meat and dairy for less destructive substitutes was approaching critical mass.

"Next year, we'll have to move to a stadium!" he crowed.

Beyond Meat had just pulled off the 21st century's biggest-popping initial public offering, and plant-based meat was enjoying, in Ethan Brown's exuberant phrasing, "a time of unlimited growth." The media had crowned it "The Hottest Thing in Food," declaring its "Long-Awaited Tipping Point Has Finally Arrived." Plant-based diets were trending, as *The Game Changers,* featuring macho men like Schwarzenegger promoting them as athletic and even sexual performance enhancers, became the bestselling documentary ever on iTunes. The EAT-Lancet report, a landmark scientific investigation of healthy and sustainable food, also recommended sharp reductions in meat-eating.

The idea that "Fake Meat Can Save Us," as a *Times* column proclaimed, was especially hot in sustainability circles, as Beyond and Impossible shared the UN's prestigious Champions of the Earth award. Life-cycle analyses found both of their burgers used about 95 percent less land and generated 90 percent fewer emissions than beef. As historic wildfires ravaged California—the sky above the conference was hazy with ash—the earth seemed to be crying out for climate-friendlier forms of protein, and non-vegans were finally trying them.

Beyond was trending up and to the right, with products in 50,000 outlets, revenues up 30-fold in four years, and new partnerships with chains like Subway, Hardee's, KFC, and Dunkin' almost every month. It was suddenly valued at $10 billion, almost a third as much as Tyson—which unwisely unloaded its stake in Beyond before the IPO, missing a sevenfold spike in the stock price.

Impossible was growing just as explosively and raising even more cash from private investors, including Katy Perry and Serena Williams as well as Vinod Khosla and Li Ka-shing. It had released its "2.0" beef, like a software firm releasing an update, and Pat Brown was vowing to shove the entire meat industry into a death spiral within five years. As unhinged as that still sounded, Burger King had just provided the best validation yet of fake meat's mainstream potential by rolling out the Impossible Whopper nationwide—along with ads featuring the bleeped-out double-takes of real customers who couldn't believe it wasn't beef.

"This is a [bleep]ing cow!" one insisted.

"No [bleep]ing cow," the narrator deadpanned.

The Whopper was an American icon, and the Impossible Whopper signaled such instant cultural relevance that Arby's aired a parody ad for a meat-based fake vegetable. More signs of the times: A corporate front group called the Center for Consumer Freedom launched a "Fake Meat, Real Chemicals" campaign smearing plant-based burgers as processed poison, and several states with powerful livestock lobbies obediently criminalized descriptions of "meat" that didn't come from carcasses—even though nobody had ever criminalized hamburgers that contained no ham, hot dogs that contained no dogs, or Girl Scout cookies that contained no Girl Scouts. Fake trends rarely inspire real backlashes, and fake meat was a real trend. Its sales grew 18 percent in 2019, six times the growth of animal meat.

Hundreds of plant-based startups were vying to be the next Beyonds and Impossibles, just as Friedrich had envisioned four years earlier when there were only a few in California. Cash was also pouring into three dozen cultivated meat startups, up from just Upside in 2015. Another three dozen fermentation startups had joined the youngsters at Perfect Day and EVERY, which moved to the grown-up table with $60 million Series B rounds. GFI itself had 80 employees in six countries and a $20 million budget. Its conference was swarming with so many eager investors, I thought I might accidentally raise a Series A while waiting on line for the bar.

Alt-Protein-Land's capital was the Bay Area, where Silicon Valley tech met Central Valley ag, but the gold rush was global. The city-state of Singapore, on a mission to produce more of its own food on its limited land, sent a dozen bureaucrats to GFI. Israel, whose desert location surrounded by hostile neighbors had fueled a perennial focus on food tech, was also a fake meat hub, with a half dozen startups in the city of Rehovot alone. Redefine Meat CEO Eschar Ben-Shitrit, a former Hewlett-Packard executive whose Rehovot startup was 3D-printing plant-based beef, told me that after Beyond's IPO, an investor who had blown off his earlier request for $100,000 resurfaced to offer $3.5 million.

So much white space was being filled by so many startups that Friedrich's top priority was no longer promoting new ones. He had separated GFI from New Crop Capital, then folded his innovation division into corporate engagement, which was his new top priority. Sure enough, the four largest meat corporations—JBS, Cargill, Tyson, and Smithfield—were all launching plant-based lines. After spending years denouncing them as murder syndicates, Friedrich had forged close relationships with all of them. He was probably the only PETA alum ever introduced at a corporate gathering as a "Friend of Tyson."

"These guys aren't chasing a billion-dollar pie," Friedrich told the crowd. "They're making the pie bigger."

Actually, plant-based meat hadn't quite hit the billion-dollar mark in the U.S., lagging slightly behind esports and car washes. It was only a $3 billion pie worldwide, commanding just 0.2 percent of the global meat market. And cultivated meat hadn't even made it to market, except for a couple of pounds of lab-grown duck Josh Tetrick sold in Amsterdam in the waning hours of 2018 to fulfill his Project Jake pledge, a symbolic gesture that Dutch authorities subsequently ruled illegal.

But even though plant meat was still a mini-category and cultivated meat wasn't yet a category, analysts were giddy about their future. Barclays forecast $140 billion worth of vegan meat sales within a decade. One consultancy predicted fake meat would outsell conventional meat within two decades. "Technology may make animals obsolete," an AgFunder report proclaimed.

It was no accident that big law firms like Holland & Knight, accounting firms like Ernst & Young, and engineering firms like Black & Veatch were hawking their services at GFI. Alternative proteins were the new new thing. That first day, Perdue revealed it was devoting half its marketing budget to its half-plant, half-meat Chicken Plus blended nuggets, and Tyson announced it was buying a stake in the woman-led faux-shrimp venture New Wave Foods.

"It seems like every week something happens that would've been the biggest news of 2018," Friedrich said. "It just goes on and on and on."

I toured the original California startups that month, and they all had the same vibe: Gen Z geeks with tattoos under their lab coats. Shiny new extruders and fermenters doing the jobs of cows and chickens. Sterile glove boxes like the one Tom Cruise used to handle the killer virus in *Mission: Impossible II*. And mission-driven leaders whose talking points about reinventing food no longer sounded so dreamy.

My first stop was the Silicon Valley headquarters of Perfect Day, which had inspired a lot of eye-rolling when its kiddie-founders first proposed to reengineer yeast into mini-factories manufacturing whey protein. Now they were doing it. Their dairy-free ice cream tasted like ice cream, because the whey was molecularly identical to dairy whey. Compared with the dairy industry, the startup still looked as microscopic as its yeast, but CEO Ryan Pandya believed it had a long-term advantage: It didn't have to milk 2,000-pound animals that used too much land and burped too much methane.

"Our microbes are way more efficient, and they do what we tell them to do," Pandya said.

In its eighth-floor South San Francisco offices, across DNA Way from the Genentech labs that first used precision fermentation to make insulin, the EVERY Company was using it to make food. Arturo Elizondo gave me a few drops of a clear liquid from a tiny pipette, then told me I was the first reporter to try an animal-free egg-white protein. It didn't taste like much, but neither do egg whites. And while Elizondo still looked more likely to appear on the cover of *Tiger Beat* than *Fortune*, he had an ambitious plan to grow EVERY into an edible "Intel Inside," by optimizing its yeast cells until the ingredients they produced in fermenters were tastier and cheaper than the ingredients chickens produced in cages.

"Humanity won't be guilt-tripped into eating sustainable products," he said. "We can use technology to overcome the trade-offs."

Just had left its garage for a former chocolate-factory-turned-Pixar-studio in San Francisco's gentrifying Mission District, fitting symbolism for a food-tech firm in a sector entering adolescence. Tetrick had matured enough to update his phone reminder to the slightly less morbid "What if today is the day." His liquid egg was finally in stores, and he was cutting deals with his former egg-industry enemies to distribute it, which he had sworn he would never do, until it became the sensible thing to do. It was easy to forget how new and novel this industry still was. After I sampled a Just omelet, I told a Tetrick aide I had noticed an earthy off-taste. He reminded me that off-taste was cholesterol-free, cage-free, and antibiotic-free.

"Chefs say eggs are God's perfect creation, and we're not there yet," the aide said. "But you just ate a good omelet made from beans. That's something, right?"

My next dish at Just was really something, a chicken nugget cultivated from cells from a chicken that was still alive and clucking. Yes, it tasted like chicken. It did cost an unappetizing $1,000 a pound, when a 20-pack of McNuggets cost just five bucks, but that was an impressive 99.9 percent cheaper than Mark Post's experimental burger six years earlier.

Tetrick's new short-term goal was to make the first legal sale of cultivated meat in 2020 in Singapore, to show the world it wasn't just science fiction. His longer-term vision was reflected in a sketch on Just's wall: a meat brewery with glass walls. Paul McCartney had explained why slaughterhouses could never look like that, but the meat packers of the future would have nothing to hide.

I visited Beyond Meat the day the delivery service Blue Apron agreed to carry its products, which didn't seem like big news. Delivery services carry lots of products. While I was interviewing Beyond's chief growth officer, Chuck Muth, he peeked at his phone to see if the markets had noticed.

"Whoa," he gasped. "This is *crazy*."

Muth showed me a stock-price chart with a vertical spike: Blue Apron was up 50 percent in morning trading, simply because Beyond's name had appeared in its press release.

"Validating, I guess," Muth said with an uneasy laugh.

I then sat down with Ethan Brown, who exuded chill amid the fake-meat-mania in dad jeans, a sweatshirt, and a ball cap. A former Connecticut College basketball player, he adhered to the athletic philosophy of not getting too high about highs or low about lows. He had just wrapped a meeting with NBA star Chris Paul, a Beyond investor, and when I asked how the mega-IPO had changed his life, he sheepishly confessed he had partied the night before with Snoop Dogg, another investor. After a decade of unglamorous work, he could only laugh at the breathless coverage of Beyond's supposedly overnight success. His office had posters of early quotes mocking his product ("slightly better Tofurky") and predicting its failure, but rather than dunk on the doubters, he said Beyond was still only a blip in the meat market.

"Honestly, we're peeing in the wind," he said.

Brown's model was Tesla, another Wall Street darling taking on another high-emissions industry deeply rooted in the American psyche. He admired how Elon Musk, who had not yet become consumed with anti-woke politics, was helping the earth by giving customers great products, rather than nagging consumers to buy its products to help the earth. Brown had hired the architect of the "Got Milk?" campaign to broaden Beyond's appeal, and recruited a stable of athletes to talk up its benefits for their bodies. Research suggested that consumers subconsciously assumed detergents marketed as eco-friendly and jeans touted as child-labor-free must be inferior products, so Beyond avoided virtue-signaling words like "sustainable" or "vegan." Its slogan was "Eat What You Love," a message to carnivores that it wasn't about deprivation. And they were responding. One supermarket survey found 93 percent of Beyond buyers also had animal meat in their carts.

I thought Beyond Burgers had an off-putting grassy aftertaste—and on the grill, they smelled like someone literally peeing in the wind. When I was rude enough to tell him so, Brown readily agreed his multi-billion-dollar burger was "nowhere near where it needs to be." That's why his Manhattan Beach Project scientists were studying beef's nanostructure and biochemistry, running samples through an "e-tongue" to quantify bite resistance and an "e-nose" to isolate aromas, auditioning new plants that might replace pea proteins someday.

His case for optimism was that the cow was a pretty mature technology, while bioengineered meat wasn't.

"This is like the early days at Wang, when a computer was the size of an entire room," Brown told me.

I thought the Impossible Whopper basically tasted like a Whopper—greasy, overshadowed by sauce, yet satisfying in a guilty-pleasure way. Impossible's 2.0 beef, now mostly GMO soy, was so much meatier than the version Gene Weingarten confidently identified as plants that in an on-air taste test, right-wing shock jock Glenn Beck confidently misidentified it as beef. "That is insane!" Beck shouted. "I could go vegan!" Even a Missouri Farm Bureau official admitted after trying a plant-based Whopper that if he hadn't known, he wouldn't have guessed, and he warned farmers and ranchers not to dismiss it as a passing fad.

"This is not just another disgusting tofu burger that only a dedicated hippie could convince himself to eat," he wrote on the bureau's blog. "It's 95 percent of the way there, and the recipe is likely to only get better."

Impossible's scientists were already working on Beef 3.0—along with 1.0 chicken, pork, and fish—in its fancy Silicon Valley lab. Chief science officer David Lipman, a butcher's son who had run a National Institutes of Health biotechnology center, said his team was testing hundreds of prototypes every week. He showed me the notes on that day's compounds: "fresh melon," "less stinky," "tires!" Mix the right bit of this with the right bit of that, he said, and eventually you'll get meat, without the butchery or greenhouse gases.

"We'll continue to get better, because we don't have to deal with 20,000 genes interacting inside an animal," he said. "You want to cook beef well done but keep it juicy? You want a slightly porky flavor? We can make it happen."

Impossible's 2019 impact report noted that its production would have to double every year until 2035 to meet Pat Brown's goal of ending animal agriculture. That would be a 6 million percent increase, which seemed a tad ambitious. But the report also quoted one Timothy Searchinger declaring this "might be the year when true substitutes first became viable." Phenomenal trajectory seemed more important than minuscule volumes. Plant-based meat was getting better, while cattle would never escape their ruminant inefficiencies.

"They breathe, they fart, they do all kinds of things that don't produce food, and it's hard to improve them," Lipman said. "We're creating a new system that's simpler than animals."

No wonder the mood at the GFI conference was so euphoric.

The Saudi vegan prince Khaled bin Alwaleed, looking like a Brooklyn hipster in a hoodie and ripped jeans, gloated about the investors who had rejected his pitches for Beyond: "Are they kicking themselves in the ass right now? I'm sure they are!" A fast-food executive recounted her C-suite's first meeting with Ethan Brown, an "instant love affair" that quickly produced a deal to put Beyond in every Hardee's: "The energy was so great, we just *knew!*" The head of Archer Daniels Midland's venture arm delighted the crowd by predicting that once plant-based meat hit price parity, it would swiftly replace animal meat. Archer Daniels Midland! The head of innovation for JBS said it was about to cannonball into the plant-based pool. JBS! One panel of investors agreed it was becoming *too* easy for startups to get funded.

"There's a lot of dumb money right now," one warned.

I did endure one contrarian rant about overhyped plant-based crud from Alan Hahn, the nebbishy CEO of the fungi-based ingredient firm MycoTechnology—until he walked away mid-rant to trash-talk a competitor a foot taller and 25 years younger. That was Tyler Lorenzen, a chiseled former NFL tight end whose pea protein venture, Puris, had just secured $75 million from Cargill.

"Cargill owns you now!" Hahn taunted him.

Lorenzen, unaccustomed to playing defense, stammered that Puris remained independent. He looked confused. Three hundred pea protein products were hitting the market that year, and he had far more demand than he could supply. Why wouldn't he take on a deep-pocketed partner who could help him scale?

"You'll see," Hahn said. "I'll raise more money than you, anyway!"

Otherwise, jubilation reigned. George Peppou, the 26-year-old CEO of Vow, told me there was no reason to assume chicken, pigs, and cows were the best-tasting animals, which was why his Australian startup hoped to cultivate cells from exotic species like kangaroos, yaks, and endangered Galápagos tortoises—and maybe even extinct woolly mammoths. His main theme, the conference's main theme, was that the future didn't have to look like the past.

"Let's create brand-new experiences!" Peppou gushed.

I sneaked into a back room for a new experience with Thomas Jonas, whose Chicago startup Nature's Fynd made protein out of an "extremophile" microbe harvested from the volcanic springs beneath Yellowstone National Park. It was among the most efficient organisms on Earth, because the acidic hot springs in which it had evolved were among the harshest environments on Earth; his cofounder discovered it while working on a NASA project to explore life in settings as inhospitable as space. Now they were fermenting their badass fungus into "mycoprotein" that could form anything from burgers and sausages to yogurts and cream cheeses—and they had backing from ADM and Danone as well as Bill Gates's climate fund.

With a furtive smile, Jonas reached into his backpack and showed me a plastic-wrapped substance that looked like chicken breast.

"These are harvested microbes, fresh from the farm," he said.

The "farm" was stacks of trays where the fungi were fed cheap sugars and starches. Conveniently, they were some of the least picky eaters on Earth, thanks to their origins in a resource-deprived acid bath. Also conveniently, the farm did not need to be restocked from the Yellowstone springs, where a tourist had recently fallen in and dissolved. The fungi worked like reusable sourdough starter, so Nature's Fynd already had all the samples it would ever need in one freezer.

Just about everything about the microbes was more efficient than livestock. They doubled their weight every few hours, so one tray could grow as much meat as 30 chickens. All they did was consume feed and become food, all day every day. They didn't grow beaks or brains or waste energy breathing or pooping. And they didn't express proteins, like the yeast engineered by Perfect Day and EVERY; they *were* proteins, with all nine essential amino acids. That made Nature's Fynd's "whole biomass fermentation," a process first applied to vegan meat in the 1980s by the old-school fungi brand Quorn, much simpler than precision fermentation.

"We're growing our protein like you might grow a tomato," Jonas said.

He didn't let me try the chicken, and his chocolate mousse prototype was bland and insufficiently creamy. Still, it was mousse grown from a microbe that could survive in boiling acid, and that was something, right? Nature's Fynd had the potential to grow protein with negligible land use, the ultimate Anthropocene food hack. Jonas was building a pilot plant on an acre of Chicago's meatpacking district, the former epicenter of the livestock-industrial complex, that could produce enough mycoprotein to replace 7,000 acres of cattle pastures.

"We can return that land to nature and start rejuvenating the planet," he said.

Making slaughter-free meat in the former enclave of slaughter where Upton Sinclair set *The Jungle* was a perfect illustration of the power of innovation. It was as if Sinclair had concluded *The Jungle* with the triumph of mission-driven capitalism rather than utopian socialism.

"This kind of technology isn't nice-to-have, it's must-have," Jonas said. "If we're going to feed 10 billion people, this efficiency is absolutely necessary."

I heard that a lot at GFI, the idea that factory farms were so awful, and climate change so devastating, that the alt-protein revolution *had* to happen. Boosters kept riffing about grim realities—the fires burning California's forests into the sky, a swine fever wiping out half of China's pigs—as if they made meat substitutes that could feed the world without destroying it inevitable.

But if severe problems made solutions inevitable, they wouldn't be so problematic. Plant-based meat was nowhere close to feeding the world. Nature's Fynd's fungi weren't feeding anyone yet. And neither was cultivated meat. Aleph Farms showed a video of the world's first lab-grown steak, but the big reveal was a sad-looking clump of Steak-umm-like substance resembling the tongue of a beat-up work boot. I wrote in my notes: WHY SHOW THIS???

I had another hmm moment when *Clean Meat* author Paul Shapiro told me he had cofounded a startup, the Better Meat Company, and it wasn't a clean meat company. It made the plant-based filler for Perdue's blended Chicken Plus nuggets, the food version of a hybrid Prius rather than an all-electric Tesla. He admired the pioneers in his book, but he couldn't wait for $1,000-a-pound cultivated nuggets to slide down the cost curve. As a teenage activist, he had seen that two half-time vegetarians did as much good as one full-time vegetarian, and he figured Perdue could get far more than twice as many consumers to eat half-chicken nuggets as a cell-based startup could get to eat no-chicken nuggets. His goal was better meat, not perfect meat. "I want to make a difference *today*," he said. It just seemed concerning that after writing how clean meat would "Revolutionize Dinner and the World," he had partnered with corporate animal killers to make products he wouldn't even put in his own mouth rather than pursue clean meat.

It also seemed concerning how many plant-based products I put in my own mouth were gross. I spit out some scrambled eggs made from pumpkin seeds. I nearly broke a molar on some plant-based jerky. Ocean Hugger Foods, another New Crop play, served sushi made of tomatoes that looked like tuna nigiri but tasted like tomatoes. And comparing the faux bacon from Hooray Foods to cardboard would be unfair to cardboard. It tasted like dumb money.

I kept thinking about a cautionary note Friedrich dropped into his triumphal remarks: "This is all exciting, but not self-executing."

Friedrich was now the sector's unofficial dad,* and I kept encountering startups whose cofounders met through GFI, like Better Meat, or were inspired by a GFI pitch, like the alleged-bacon venture Hooray Foods, or previously worked for GFI. But when I asked Friedrich if he was worried that some startups his team had nurtured might fail, he flashed a guilty smile: "Oh, there will be blood."

Privately, he had even deeper doubts about the revolution he helped spark.

Before the conference, Friedrich shared his unvarnished take in an internal "Bruce's Musings" memo, labeled "Sensitive: Please do not share." He was sure meat substitutes could save the world, once they tasted as good and cost as little as animal meat. But his message was that they weren't there yet, it would be hard to get them there, and the industry needed to focus on getting them there instead of pretending they were already there.

"We are in a bizarre time where everywhere you go, there is an assumption that these technologies have succeeded," he wrote. "That is quite simply and unequivocally not true."

He was especially worried about Beyond, which was the plant-based poster child even though it lacked "a product that tastes like meat to a meat-eater or costs even close to animal-based meat." Ethan Brown had quietly pulled its unsatisfying chicken strips off the market, a mature but not exactly reassuring decision. And its wackadoodle rise in valuation—to

---

* He even started his speech with a dad joke: "Did you hear Bill Gates bought the *Seattle Times* this morning? Yeah, he buys it every morning."

40 times revenue, when less than five is standard for a startup—felt like a precursor to a Humpty-Dumpty fall.

"If Beyond crashes, that's going to be painful, and it very well could crash," Friedrich wrote.

Impossible performed much better in taste tests. And Friedrich was glad it was using GMO soy to improve its product, even if natural food purists squealed; taste was king, and soy wasn't harmful. But even though nutrition was an early selling point for Impossible and Beyond Burgers, which had no cholesterol and less saturated fat than beef, they weren't whole grains or spinach.

"Neither product is especially healthy," Friedrich warned.

That was a general vulnerability for plant-based alternatives, and the "Fake Meat, Real Chemicals" campaign was exploiting it. Conceived by the Center for Consumer Freedom's gleefully sinister director, Richard Berman, a Washington lobbyist whose industry-funded crusades against smoking bans and drunk driving laws inspired the satire *Thank You for Smoking*, it compared fake meat's long lists of nasty-sounding ingredients to dog food labels. Who would want to eat "niacinamide"? It sounded like a felony.

Of course, regulators had approved plant-based meats as safe, and neither their long multisyllabic ingredient lists nor the stigma around "processed foods" made them unsafe. As Pat Brown often said, a poisonous mushroom was one natural ingredient. Beef wasn't whole grains or spinach, either, and while it was marketed as one natural ingredient, it also contained chemicals like 4-hydroxy-5-methyl-3(2H)-furanone and, yes, niacinamide, a harmless form of vitamin B. What happened inside cows was arguably as icky a form of processing as what happened inside twin-screw extruders. Butchering animals and plastic-wrapping their meat onto polystyrene trays were processes, too.

Still, the CEO of Chipotle, home of the 1,200-calorie steak burrito, declared fake meat didn't fit his "food with integrity" principles. And even if Beyond and Impossible were being held to snobby standards that weren't applied to animal-corpse meat, it was true they didn't deserve health halos. Their corporate rhetoric was writing nutritional checks their ingredient lists couldn't cash.

"That may not matter long term," Friedrich wrote, "but it might while the products are more expensive."

Friedrich's concern was how little was being done to make the products better and cheaper. "In all of human history," he wrote, "less than $200 million has been spent attempting to figure out how to bio-mimic meat with plants," nearly all by two companies that kept what they learned to themselves. Friedrich was appalled by how few of Beyond and Impossible's competitors were laser-focused on science. Eclipse Foods, the ice cream startup with founders from Just and GFI, was building a sophisticated plant-based dairy platform, while Motif FoodWorks, an ingredients firm spun out of the biotech unicorn Gingko BioWorks, engineered a novel way to replicate heme. Otherwise, Friedrich thought most of the new plant-based ventures were marketing plays, trying to cash in on the boom without doing the work. It took Impossible's science team five years to launch a product, so when he saw copycats pitching themselves as the next Impossible and promising products in six months, he thought: *No, dude, you're the next Gardenburger.*

Cultivated meat startups had no existing products to copy, so they took science more seriously. But none of them had even built a pilot plant yet. Impossible had raised seven times more than the entire cultivated meat industry. Friedrich had just read an upcoming scientific paper arguing that cultivated meat would never be economical, and while he still rejected the Cassandra narrative that it was 10 years away and always would be, he didn't buy the emerging Pollyanna narrative that it was inevitable.

"We need to take these findings very seriously," he wrote.

GFI had distributed $1.5 million in scientific grants, making it the world's leading funder of cultivated meat research. But that was chump change compared with the $30 *billion* governments spent every year on agricultural research. Friedrich believed that unless the public sector took a moonshot approach to alternative proteins, his hypothesis that they could replace animal products once they reached taste and cost parity might never be tested. And he feared the uncompetitive offerings already on the market could taint the sector's reputation for years.

He tried to temper the memo's negativity—he titled one section "Optimism (WE CAN DO THIS!)"—and he stuck to happy talk in public. Still, his memo revealed some unhappy thoughts.

"Right now, we are riding a wave. My hope of hopes is that the wave will continue and we will go from success to success to success," he wrote. "However, we should be aware of how fragile this wave is, how uncertain, and how early."

COVID didn't break the wave. Initially, it swelled the wave.

Plant-based sales kept growing, as Impossible's fundraising topped $2 billion; Beyond signed deals with PepsiCo for jerky and McDonald's for a McPlant burger; and Oatly went public at a Beyond-like $10 billion valuation. Just seized 99 percent of the plant-based egg market, then moved again, to sleek corporate offices in Silicon Valley, quite a journey from its original biker garage.

Lesser-known startups thrived, too. Nature's Fynd began selling sausages made of its Yellowstone mycoprotein, which it dubbed "Fy"; Motif FoodWorks released its heme ingredient, "Hemami"; the Chilean firm NotCo unveiled an artificial intelligence platform for plant-based products, "Giuseppe"; and all three closed monster nine-figure Series B rounds. Puris, under former tight end Tyler Lorenzen, kept growing at the speed of Tyreek Hill, and his main concern was that competitors with less neutral flavors would give pea proteins a bad name.

"The market is growing *too* fast," Lorenzen said. "We can't supply it all, and some of the other stuff tastes like shit."

The big dogs were hunting, too. JBS cannonballed into the U.S. with a new plant-based division. General Mills launched a dairy-free cream cheese with Perfect Day whey, declaring "the future of cheese is here." The market pundit Jim Cramer bellowed on CNBC: "You've got to get on the bus or get left behind!"

Just also confounded its critics by meeting Tetrick's 2020 goal for selling cultivated chicken in Singapore, serving a few small plates a week at a posh restaurant. It was an expensive PR gimmick, but it made Churchill's vision of animal-free animal meat real. And after a new film about Upside, *Meat the Future*, portrayed Valeti as the mild-mannered leader of the next agricultural revolution, he raised his own monster nine-figure round, enough to build the first cell-based pilot plant. "Lab-Grown Meat Is Scaling Like the Internet," one analyst wrote.

As COVID tore through supply chains, creating America's first meat shortages since World War II, Josh Tetrick declared meatless meat would soon be as common as streamed music. "We can't go on with the slaughtering and milking of cows," said an executive from Nestlé, which had slaughtered and milked cows since the Franco-Prussian War. The cultivated pork startup New Age Eats vowed to become "the largest and most innovative meat company on Earth"—not cultivated meat company, meat company. The Boston Consulting Group predicted price and taste parity for alternative proteins "in 2023, if not sooner," and a 700 percent market expansion by 2035.

Then the wave broke.

THE TROUGH

GFI did not rent out a stadium in September 2023 for its first conference since COVID, just a cultural center on San Francisco Bay. The sky was again hazy from wildfires, but the euphoria from its last alt-protein-palooza in 2019 had burned out. The time of unlimited growth was over, and fake meat was slumping. The investment boom had busted, and not just the dumb money. The media was throwing dirt on the sector's grave, dismissing it as a fad and a flop.

This time, I didn't have to worry about getting funded on the drinks line. Artificial intelligence was Silicon Valley's new new new thing. Fake meat felt like old old news. Animal protein was enjoying a resurgence in the nutritional world, too, as "meatfluencers" now pushed plant-free "carnivore diets."

"Our industry," Beyond Meat's Ethan Brown dryly observed, "is experiencing turbulence."

Friedrich was again an upbeat host, but he was now as candid onstage as he had been in his confidential 2019 memo, stating matter-of-factly that consumers just weren't that into most meat substitutes: "They don't want to pay more, and they don't want to sacrifice on taste."

Beyond was again the poster child, but not in a good way. Its stock had crashed 95 percent. McDonald's had pulled the plug on the McPlant in the

U.S., Dunkin' took Beyond Sausage off its menu, and Hardee's, its instant love affair with Brown having run its course, stopped selling Beyond Burgers. Almost everything that could have gone wrong had; an ad with "chief taste consultant" Kim Kardashian created an Internet pile-on when she didn't bite into her burger, and Beyond's chief operating officer was arrested for biting someone's nose during a brawl in a parking garage. The company no longer employed, or needed, a chief growth officer.

Brown was trying not to get too low about the low. His public message was that the industry should weather the storm with confidence, because "the dark cloud of misinformation and fear will give way to truth and sunlight." But when we chatted privately, he sounded shaken by the attacks Beyond was facing from journalists, nutritional scolds, and livestock-industry apologists.

"We're getting our *asses* kicked," he said.

So was the sector. Oatly followed up its Beyond-like IPO with a Beyond-like 95 percent drop in valuation, General Mills canceled its "future of cheese" deal with Perfect Day, and Miyoko's Creamery fired its CEO, Miyoko Schinner, an awkward situation when her name remained on its label. Hooray Foods stopped committing its crimes against bacon, and several New Crop ventures failed as well, including Ocean Hugger, the purveyor of tomato-based sushi, and New Wave Foods, the Tyson-backed, women-led shrimp startup; Fora Foods, New Crop's "aquafaba" butter play, also melted away, while Alpha Foods, despite its seasoned executives, collapsed into the arms of a less precarious plant-based firm.

"The transformational outcome did not materialize," said the CEO of the Canadian conglomerate Maple Leaf Foods, "and we no longer believe it will materialize."

Clean-label purists were dumping on fake meat as ultra-processed crap; even the anti-animal-agriculture visionary Frances Moore Lappé told me it was "totally unnecessary," because consumers could eat unprocessed vegan options like lentils that "aren't trying to be something they're not." Meanwhile, right-wing culture warriors were dumping on fake meat as a woke blue-state assault on the American way; when Cracker Barrel added Impossible sausages to its menu, without removing any meat options, it was deluged with Facebook comments like "YOU CAN TAKE MY PORK

SAUSAGE WHEN YOU PRY IT FROM MY COLD DEAD HANDS!" And it wasn't as if ultra-processed-crap lovers or blue-state wokesters were eating much plant-based meat. It was still just 1 percent of the U.S. market. When I ran into MycoTechnology's pugnacious Alan Hahn, I mentioned his 2019 warning that the products needed work.

"I told you they sucked, and they still suck!" he snapped. "The taste, the texture, the price, the nutrition, they nailed zero out of four, and now they're like, 'Why did we get our asses handed to us?'"

Puris had idled one of its factories after pea protein demand crashed; when I reconnected with Tyler Lorenzen, I reminded him about Hahn's trash talk in 2019.

"Maybe Alan was right," the strapping athlete said with a sigh.

Max Elder, a young nose-ringed vegan, had just shut down his pea-based chicken startup, Nowadays, and he now wondered whether he and other founders had been delusional to take on such an entrenched trillion-dollar industry. Was there real evidence that after two million years of ingesting meat in the form of dead animals, humanity was yearning for new delivery devices? Was it so surprising that a bunch of plant-eating idealists had failed to convert carnivores?

"The theory was: If we build it, they will come," Elder said. "Well, we built it. They didn't come."

The industry was inundated with quasi obituaries like "The Vege-bubble Turns to Vege-bust." And "Where's the Beef? Here's Why the Fake Meat Fad Sizzled Out." Former Plant Based Food Association head Michele Simon wrote a scathing *Forbes* essay titled "Plant-Based Fail." She told me she always felt uneasy hyping its early growth from such a tiny base—and now she no longer had to.

"I'm out of the spin zone," Simon said.

Bloomberg Businessweek published the buzziest story, with a cover image of a pile of slop behind the headline "Beyond Impossible: How Fake Meat Became Just Another Food Fad." The thesis was that its customer base had dwindled to a vegetarian core, because only 38 percent of Americans believed it was healthy. The article quoted a mom who quit plant-based meat after her doctor told her: "You are eating processed food." It also quoted the Center for Consumer Freedom's Super Bowl ad featuring a kid at a spelling bee struggling with "propylene glycol," which a judge defined as "a chemical used in antifreeze and synthetic meats." It didn't mention Impossible's video response featuring Pat Brown asking a kid to spell "poop," as in "there's poop in the ground beef we make from cows."

The *Bloomberg* piece, after thousands of words slagging plant-based meat, concluded with a swipe at cultivated meat, noting it had attracted some of the same investors and inspired some of the same save-the-world rhetoric, despite facing "even bigger hurdles." The implication was that it was headed for a similar fall—and it certainly hadn't saved the world yet.

New Age Eats, after vowing to become the world's largest meat company in 2021, folded in 2023. Investment in cultivated meat plunged 80 percent during those two years. A non-peer-reviewed article by UC Davis researchers suggested cultivated meat might be even worse for the climate than animal meat—and while their work was riddled with flaws, including the familiar assumption that land was free, it became chum for another media feeding frenzy. The industry was still nowhere near price parity, and startups like SCiFi Foods, which was using CRISPR to edit bovine cells before cultivating them into beef, were arguing that the original pioneers cultivating unedited cells would never make the economics work.

"There's no point to all this deep tech if you can't make an affordable product," said SCiFi's cofounder and chief technology officer, Kasia Gore.

When U.S. regulators finally approved cultivated chicken from Just and Upside for sale, it only encouraged more Frankenmeat snickering. *The Tonight Show*'s Jimmy Fallon took a typical swipe: "Wow, it's completely safe! Disgusting, but safe." It was barely for sale, anyway. Upside's chicken was only in one San Francisco restaurant that served it once a month. Just's chicken was also available in only one location—a Washington, D.C., restaurant owned by José Andrés, one of its investors—and its bioreactor supplier was suing it over unpaid bills.

The alt-protein sector had clearly overpromised a revolution. The fake-meat slump was not fake news. Still, *Bloomberg*'s fad-and-flop coverage included a sidebar about how "Plant-Based Eggs Are Having a Moment," calling Just Egg's robust sales "undeniable evidence that at least one corner of the plant-based food market can be cost-competitive." Seven years earlier, *Bloomberg* had run that mayo-faced portrait of Tetrick and portrayed his company as a scam. It was a reminder that media pariahs can become media darlings as fast as darlings can become pariahs, that past performance is never a guarantee of future results.

Some Big Food giants still believed. JBS shut down its new U.S. plant-based division, but it forged ahead with the world's largest cultivated meat factory in Spain. ADM actually expanded its alt-protein investments, and its vice president for protein, Greg Dodson, told me it was more bullish than ever on blended products mixing animal meat with plants: "It doesn't have to be all or nothing." Jane Lee, an executive in Cargill's alt-protein division, said it intended to be the largest supplier of meatless meat ingredients.

"We're not concerned about temporary dips in the market," she said. "The boom and bust is to be expected in a young category. We're in it for the long haul."

Another intriguing counterpoint to the fad-and-flop thesis was that Impossible wasn't flopping.

It was nowhere near Pat Brown's goal of doubling production every year, but unlike the larger industry, its sales were still growing. In taste tests, its burgers did almost as well as beef, and its nuggets did better than chicken. Unlike Beyond, which kept getting dumped by fast-food chains, it kept renewing deals with Burger King and Starbucks. It still cost more than animal analogues, but often less than organic versions, and it had plenty of room to grow. Barely 5 percent of Americans had tried it, while only 15 percent had even heard of it.

Impossible hadn't enjoyed a turbulence-free ride, either. The plant-based slump shredded its valuation, inspiring more grumbling about Pat Brown's evangelical leadership. Even as he burned through cash, he kept pushing to expand—into steak and milk, tuna and salmon, China and India. He refused to authorize even modest layoffs to cut costs, pushing for more factories in perennial anticipation of exploding future demand. He also got obsessed with the idea that he and his brother, a translator in Beijing, could persuade Xi Jinping to exempt Impossible from China's GMO restrictions, and he periodically raged at aides for failing to arrange a meeting.

In 2022, Impossible's board ousted Brown for a more conventional food executive, Peter McGuinness, whose background was in marketing, not science. Brown initially stayed on as chief innovation officer, but he went on indefinite leave after the company shelved his expansion plans. When we met at a Palo Alto coffee shop, he complained that the board was far more interested in "short-term money metrics" than his vision of a world without factory farms.

"I was always saying, 'Come on, it's not all about your bank account next year, don't you want to optimize the future for your kids?'" he said. "But my credibility with that group as a business guru was not high."

McGuinness brought an upbeat vibe to Impossible—he punctuates points with "Beautiful!"—and ended the sermonizing about ending animal agriculture. He told his team: We're going to operate like a business, not a church. No more talking like zealots about saving the planet; if our business does well, the planet does well. We can't just be food for liberal coastal elites. We want to be great food a mom in Nebraska puts on her grocery list—oh, and by the way, it's good for the earth. Beautiful!

"The company got drunk on early success, the whole industry did, and who wouldn't?" he told me. "But all the hoopla, the wild valuations, that's over."

He reminded his team that Impossible wasn't a tech company, even if gaga investors had valued it like a tech company. It wasn't a software firm with digital products it could ship worldwide with a click. It was a food company that had to manufacture and distribute perishable products to physical locations. It trafficked in molecules, not electrons.

His back-to-basics plan was: No expansion into China or other new frontiers; first get decent penetration in the U.S. by investing in marketing and sales. No expansion into fish or milk or other new categories, first focus on taking our meat from good to awesome. Impossible even redesigned its packaging, switching from save-the-earth green to this-is-meat red.

"None of this is rocket science," he said. "Don't politicize the food.

Don't insult the customers. Don't expect people to eat something they've never heard of."

McGuinness knew it would be hard to undo perceptions of plant-based meat as ultra-processed Biden-burgers for upscale tree-huggers, but why assume the sector was finished when it was just getting started?

"This industry was positioned wrong and promoted wrong, but that's fixable," he said. "The idea that it's a fad, that's ridiculous."

Fads come and go. They're the next big thing, then they're not a thing. But many transformative innovations—the Internet, virtual reality, electric vehicles—also busted in their adolescent years, following a roller-coaster path known as the Gartner Hype Cycle.

A new technology typically enters the hype section of the hype cycle, the "peak of inflated expectations," after early movers achieve some proof-of-concept milestones, inspiring fawning media coverage and investor exuberance. But a hype-cycle technology at its peak is too immature and expensive to outcompete the incumbent it's supposed to disrupt. It disappoints some early adopters, attracts some bad press, and fails to live up to its inflated expectations.

That's when it enters the "trough of disillusionment," as enthusiasm evaporates and investor focus shifts to its challenges scaling up, cutting costs, and broadening its appeal. Weaker startups fail. The funding squeeze hurts stronger ones, too. Media puffery shifts to savagery.

However, the trough is not always a death sentence. It can be a time of learning. Survivors can make adjustments that lead them up the "slope of enlightenment," as they address shortcomings, to the "plateau of productivity," where they achieve true mainstream adoption. Sometimes, the pessimism of the trough can be as excessive as the exuberance of the peak.

It's certainly possible that fake meat will turn out to be a techno-fad. Maybe humanity will never embrace substitutes for animal carcasses, and we're doomed to eat even more of the earth. But it's also possible that fake meat is on a journey through the hype cycle, from the peak at the dumb-money 2019 GFI conference to the trough at the no-money 2023 GFI conference toward a less dismal future.

The notion that meat substitutes will never catch on because they're highly processed seems especially overwrought, considering how much highly processed food we eat. Kevin Hall, the nutrition scientist credited with exposing the dangers of ultra-processed foods, told me his work has been misinterpreted to "falsely demonize" plant-based meat as a public health menace. It isn't as healthy as kale, but early science suggests it's at least as healthy as animal meat; the American Heart Association approved a "heart-healthy" label for Beyond's new faux steak tips. And it can get healthier, just as locomotives got faster after losing that first race to a horse.

There's also an Occam's razor explanation for fake meat's slump that doesn't assume meat-eaters will only eat clean-ingredient health food: The substitutes weren't yet as cheap or satisfying as the real thing. The early hype around biotech meat made sense, given how much better it was than the old hippie hockey pucks, but it wasn't better than meat, so customers who weren't trying to save animals or the planet didn't keep buying it. That was embarrassing for startups bragging about their sacrifice-free alternatives, but maybe the problem was execution rather than concept.

"It turns out that most people don't want to pay more for products that cost more and don't taste as good—not shocking!" Friedrich said. "That means we need products that cost less and taste better, or else the planet burns to a crisp."

There were still some awful products at GFI 2023. One agribusiness served pea-based milk that tasted so much like paint I asked the server if he was trying to murder me. Beyond tried without much success to give away bags of the jerky it was making with PepsiCo; it tasted like plant-based roadkill, and it was soon discontinued. Giuseppe, the artificial intelligence platform powering NotCo, wasn't intelligent enough to make NotCo's sliders taste like beef.

But I also tried some excellent plant-based products. Beyond's heart-healthy steak tips had a pleasant meaty flavor, perhaps because they were made from fava beans rather than the pea proteins in the burgers and jerky. Climax Foods, another AI startup founded by an astrophysicist who had run data science for Google and Impossible, served a blue cheese that tasted and spread like dairy, while Motif FoodWorks served bioengineered pork meatballs that tasted and chewed like meat. I also met 50/50 Foods CEO Andrew Arentowicz, who had a refreshingly low-tech approach to reducing meat consumption without sacrificing taste: non-vegan "Both Burgers" that were half vegetables, half beef.

"I've got the most advanced food-tech algorithm in history," he said. "Meat divided by two!"

It was also heartening to see an emerging business-to-business ecosystem addressing the Old MacDonald problem, so alt-protein startups wouldn't have to invent everything themselves. I met founders like Omeat's Ali Khademhosseini, a former Harvard professor trying to cut cultivated meat's media costs 90 percent by using plasma from living cows, and Ark Biotech's Yossi Quint, a former McKinsey analyst trying to cut cultivated meat's production costs 90 percent by building smarter bioreactors. Suppliers had set up booths hawking extruders, dryers, synthetic proteins, and anything else a startup might need.

Alt-meat had come a long way in a decade—and that was something, right?

SO MUCH DONE, SO MUCH TO DO

There's a harpoon on the wall at the entrance to the Better Meat Company in Sacramento, above a quote from futurist Buckminster Fuller: "To change something, build a new model that makes the existing model obsolete."

The point is that humans didn't stop hunting whales for oil until they replaced that cruel, inefficient model with fossil fuels, just as they didn't stop exploiting horses for transportation until they replaced that cruel, inefficient model with cars. Paul Shapiro wants to replace the cruel, inefficient model of raising livestock for meat by fermenting fungi called "mycelia" into better meat. The vast networks of mycelia threading our soils are a literal worldwide web beneath our feet, and Shapiro hopes they'll make eating the flesh of a dead animal seem as anachronistic as writing with a quill plucked from a live goose.

Better Meat still blends plants into chicken nuggets for Perdue, but its focus now is growing mycelia into a meat-like superfood packed with more iron than beef, more protein than eggs, and more fiber than oats. Like Nature's Fynd, it uses whole biomass fermentation to produce its mycoprotein, which it calls "Rhiza," but with a common mycelium found in forests instead of an extremophile from Yellowstone, and in brewery-style tanks instead of stacks of trays.

This simplicity is Better Meat's superpower. Unlike cultivated meat startups, it doesn't need complex growth media; its fungi eat grain and food waste. Unlike precision fermentation startups, it doesn't need complex genetic engineering to express proteins; its microbes are naturally packed with protein, and they grow so fast they're ready for harvest after a day in the fermenter. (Their fast growth also helps them outcompete bacteria, so Better Meat can skip some complex sterility measures.) And unlike plant proteins, which have to be farmed, fractionated, dehydrated, extruded, and rehydrated to mimic meat, Rhiza is naturally meaty. Shapiro said that even at craft-brewing scale, it was cheaper to produce than beef.

It was also remarkably versatile. Better Meat was molding Rhiza into hot dogs, fish fillets, foie gras, and just about every other form of meat and seafood. I sampled deli-quality faux turkey slices, compulsively snackable jerky, and grilled faux chicken breast with a smooth mouthfeel. Shapiro was also kind enough to let me try an experimental steak, so I'll be kind enough to say no more about it except that I recommended further experimentation.

Still, it was exciting that Better Meat could make a nutritious form of meat in a day, when after thousands of years of breeding, cattle still took more than a year to become burgers. Fermenting fungi was far more efficient than raising sentient animals that had to breathe, poop, and keep warm—and efficiency will be the key to making factory farms as irrelevant to food as harpoons are to energy.

As an activist, Shapiro had seen that people rarely gave up meat for selfless reasons, even when they were repulsed by the carnage on his FaunaVision. As an author, he had sensed that cultivated meat would take a long time to replace carcasses. So as an entrepreneur, he had concluded that mycoprotein was the most practical solution to the eating-the-earth problem that threatened the climate as well as the eating-animals problem that hurt his heart.

The challenge was scale. The three leading mycoprotein startups—Nature's Fynd, Alan Hahn's MycoTechnology, and another Colorado venture called Meati—had raised a combined $1 billion to ramp up production. (Hahn kept his obnoxious promise to raise more than Puris.) In 2023, Meati opened a splashy $100 million "Mega Ranch" in the Denver suburbs, a factory designed to produce 20,000 annual tons at full build-out. But while that would be a lot for one ranch, it would supply only 0.04 percent of U.S. meat. MycoTechnology is building a similar plant in Oman to make raw material for shawarmas and kebabs, but the world would still need 100 more Mega Ranch–sized operations to supply 1 percent of its current meat diet.

Meati chief supply chain officer Liz Fikes gave me a Mega Ranch tour, showing how spores moved from pipettes to flasks to the four-story-tall feeding tanks where they grew into mycoprotein—Meati called it "MushroomRoot," though it was neither a mushroom nor a root—that got sliced, dried, and finished into cutlets and steaks. She told me the Mega Ranch was nowhere near as efficient as it could be, but her priority was her team's slogan: "Make More Meati." She had just told her chief financial officer she wasn't doing cost math yet, only production math; the CFO had replied that maybe she'd do cost math if she stopped getting paid.

"But that's not our focus now," Fikes said. "We've got to push for scale."

Meati cofounder Tyler Huggins is an outlier in Alt-Protein-Land, a meat-eater who grew up on a Nebraska bison ranch. His initial vision for the firm, then called Emergy, was not even about food; he hoped to use mycelia to create battery storage for renewable energy, the green trend of 2016. He pivoted in 2019, when alt-meat was the trend, and he had raised $300 million to Make More Meati. When I visited Meati's Boulder headquarters, I was struck by its light fixtures shaped like mycelia, rather posh amenities for a startup, but also by its satisfying fungi-based chicken parm, with a better nutritional profile than chicken-based chicken parm.

It still cost more than animal meat, but Huggins believed scaling production from millions to hundreds of millions of pounds would make

Meati much cheaper. While alt-meat founders often compare their ventures to nimble cell phone and digital media startups that disrupted telecoms and newspapers tethered to landlines and printing presses, they'll ultimately need more factories to sell more product.

"It's all about scale. That's how you get profitable. That's how you get impact," Huggins said.

Huggins, the meat-eater from cattle country, and Shapiro, the vegan activist from the D.C. suburbs, had similar spiels about how making more was the key to making a difference, fungi were better than plants at making proteins, and their particular species was the most efficient on Earth. In fact, they were both using the same species. Unfortunately, they were locked in a legal war over that fungus, with Meati claiming a former Better Meat employee stole its tech, and Better Meat claiming Meati had tarnished its reputation and scared off its investors.

The battle was arguably a sign of an industry with a future worth fighting over, but it was diverting a lot of money to lawyers that could have been spent on fermenters. Shapiro had a grand vision of a factory-farm-free future, and paralyzing litigation wasn't going to help make it happen.

"I get out of bed to work on animal welfare and the climate; I want everyone in this space to succeed," he said. "It's insane that we're fighting each other."

In June 2024, a judge ruled in Better Meat's favor, chastising Meati's "shenanigans." The fight was over, but neither company got anything out of it. As this book went to press, Meati was shutting down the Mega-Ranch and flirting with bankruptcy.

Motif FoodWorks was entangled in an even more insane fight.

In 2021, Motif began finalizing deals to sell its Hemami protein to several food companies. But the deals fell apart when Impossible Foods sued Motif for patent infringement, asserting the exclusive right to produce heme for plant-based meat, even though Motif was replicating heme found in cattle rather than soy. Motif was founded as an ingredients play—its name reflected its goal to be a theme throughout the sector—but its customers wouldn't commit to buy its ingredients until the suit was settled, and Impossible refused to even talk about settling. So Motif had to pivot to branded products, which is how its pork meatballs ended up at GFI.

Motif CEO Michael Leonard, who had taken over after the dispute cost the founding CEO his job, envisioned Impossible as a customer, not a rival. Motif's ingredients could have helped the entire plant-based industry with its quality issues. But Impossible's lawsuit kept them off the market, forcing massive layoffs at Motif. It had a similarly chilling effect on several earlier-stage startups that had been exploring heme before Impossible called dibs.

"Impossible showed what technology can do in this space," Leonard said. "But they don't get to own the whole space forever."

It did take gall for Impossible to claim that heme was the essential element in flavorful plant-based meat, and also that nobody else should be allowed to make it—especially when Impossible also claimed it was rooting for the whole industry to succeed. And while its hard-nosed tactics might sound typical for a company that replaces its mission-driven founder, several sources told me its crusade against Motif was launched by the vegan evangelist Pat Brown, the same mission-driven rebel who founded the Public Library of Science to stop monopolists from hoarding information. He seemed convinced that what was good for Impossible was good for the cause of ending factory farming.

"Definitely a 'we alone can fix it' thing," said one insider.

Motif was one of the few plant-based startups whose commitment to science had impressed Friedrich, and it was hard to see how meatless meat could stop the eating of the earth without more Motifs driving more innovation.

"It's heartbreaking," Friedrich said. "We don't have time for this!"

In September 2024, Motif ran out of time, and folded. It finally settled its litigation by letting Impossible take over its heme business.

"Impossible got its message across: Don't even think about trying to innovate in this industry, or we'll sue you to death," a Motif investor told me.

After a decade on the market, plant-based meat still wasn't improving fast enough. Cultivated meat had the opposite problem: It was improving fast, but it hadn't made it to market.

• • •

You don't see slaughterhouses in urban settings like the neighborhood across the Bay Bridge from San Francisco where Upside built its pilot plant. It's surrounded by luxury apartments and restaurants. There's a playground with a dog park around the corner. Its exterior has glass walls that would make Paul McCartney proud, and inside, windows expose the elaborate array of steel tanks and pipes where Upside makes meat. The company plans to offer public tours, which you definitely don't see at slaughterhouses.

"You won't walk out scarred," Uma Valeti greeted me when I visited for a tasting.

I first tried a chicken potsticker, and even though it was a blended product—half cultivated meat, half plants—it tasted exactly like a dead-chicken potsticker. I had tried several plausible plant-based and fungi-based substitutes for ground chicken, which is usually just a vehicle for sauce anyway, but Upside's version was the first to send a "Meat!" signal to my brain, like two million years of evolution saying hi. Upside hoped to start commercial production in 2025 at a cost below $10 a pound, which would be quite a journey from Mark Post's smushed-together $1 million-a-pound burger.

"Everyone said we won't get it to taste like meat, we won't get the cost down. And every month, we get it better and cheaper," said Valeti, who is usually humble by CEO standards, but seemed edgy that morning.

I then tried Upside's chicken fillet, the whole cut that was being served to the public at one restaurant once a month. It was 99 percent cultivated meat, and it tasted better than most chicken, because it was grown from the cells of a heritage chicken that wasn't bred for industrial growth. It tasted like chicken-but-more-so, with the chewiness of real meat that gets stuck between your teeth.

"This is what chicken used to be like," Valeti said. "Pull it apart. Look at the fibers. It's meat! That's what makes all this so frustrating."

Oh, right. *This*. Valeti was in a funk because *Wired* had just posted a splashy investigation accusing Upside of exaggerating its progress, with a quote from an anonymous former employee comparing it to Theranos. The purported scandal was that Upside was hand-cultivating the meat for its fillets in 2-liter flasks called "roller bottles," not brewing it in bulk in its

2,000-liter bioreactors. After raising $600 million, the company was still struggling to grow whole cuts in the bioreactors, and it was often throwing out contaminated batches.

Valeti didn't see why that was scandalous. The goal of the fillets was to give the public a glimpse of the future; he hadn't claimed whole cuts for the masses were imminent. Upside was having problems with fillets, so it would start commercial production with blended products. "That's how innovation works!" he said. "You find the leaks, so you can build something leakproof." *Wired* suggested that by hyping the fillets along with the fermenter, Valeti had left a misleading impression the fillets came from the fermenter. Still, he was making flavorful no-kill meat—in tiny amounts, but the frauds at Theranos never made any accurate blood tests.

"We're still learning to walk," he said. "Of course a heavyweight boxer can come in here and pummel the baby. Do you think that's the story?"

The first solar panels were developed in the 1950s, and solar power remained a rounding error until the 2010s; since the first cultivated meat was developed in the 2010s, it did seem churlish to draw dire conclusions about its immediate failure to disrupt dinner. Only $3 billion had been invested in cultivated meat throughout history, when $239 billion was invested in solar in the first half of 2023. The industry really was a baby. Yet at a time when 10 million animals were slaughtered every hour, and a land mass the size of the Americas was used to make meat, a small group of tissue engineers and biologists were making animal-free meat that required far less land. That did feel newsier than an exposé of the methods Upside was using to produce one pound of meat per month.

On that trip to the Bay Area, I also got to try cultivated fried chicken at Just; a plant-based meatball blended with cultivated pork fat at Mission Barns; cultivated salmon nigiri at Wildtype; and a plant-based burger blended with CRISPR-edited cultivated beef at SCiFi. Aside from the skin on Just's chicken, a prototype that crunched like a potato chip, they were all restaurant-quality dishes. When I asked Just's chef if I was eating breast or thigh, he laughed and said meat, which felt like a Zen koan from the future. I didn't get to try woolly mammoth, but when Vow announced in 2023 that it had grown the first cultivated mammoth, a Belgian rival threatened to sue because it had already grown a mammoth burger. The technology was moving so fast that another industry-funded group linked to Richard Berman launched a new smear campaign against lab-grown meat, claiming its fast-growing cells were "comparable to a tumor."

At GFI, Mark Post told me the 80 scientists he now oversaw at Mosa Meat had replaced almost all their pharmaceutical-grade ingredients. The industry's brainiacs kept overcoming obstacles, just like the brainiacs who worked on early computers the size of a room. "There's been a ridiculous amount of progress with ridiculously small amounts of money," Post said.

But Post still didn't see any way for cultivated meat to get competitive with conventional meat before 2030. He worried about the catch-22 where the industry needed to scale to bring down costs, but wouldn't have enough demand to scale until it could bring down costs. He wished he could lock up his team in the lab to escape all the external noise.

"I do think there's been too much hype, people overselling how quickly this would all happen," Post said. "We're still at the beginning of this battle."

Just's Josh Tetrick admits he was one of the oversellers. He's a congenital optimist and a talented BS artist, so of course he made cultivated meat sound inevitable when he tried to convince his investors it wasn't a detour into lunacy. "I acted like it was definitely going to happen, because the planet is fucked if it doesn't happen," he said. "But who knows?" He didn't see any other killer app with better odds of replacing animal agriculture, but cellular agriculture was a tougher slog than he expected. Just still needed to scale up its bioreactors 30-fold, cut its media costs another order of magnitude, and persuade consumers to eat biotech meat.

"A lot of critics say we'll never solve all these challenges, and they might be right," Tetrick said. "I'm fighting for them not to be right."

SCiFi's leaders were among the harshest critics, insisting cultivated meat would never achieve cost parity without CRISPR. They had compelling technical arguments about the superiority of cells edited to grow without clumping together, and they backed it up by getting the cost of their burger down to $15 at pilot scale. But in June 2024, SCiFi shut down, because it couldn't raise money to build a commercial plant. Cofounder Joshua March thinks its failure is a bad sign not only for cultivated meat but for all alternative proteins; the struggles of Beyond, the only alt-meat startup with a successful IPO, have put a stink on the entire sector, and a capital-intensive industry can't survive without capital.

"Investors don't want anything to do with meat alternatives," he said.

Even the most successful alt-protein ventures have miles to go to make a dent in factory farming. Perfect Day has raised $840 million, partnered with ADM, and ramped up from 1 ton of whey protein in 2019 to 100 tons in 2023. But the world consumes 100 *million* annual tons of whey. "We definitely won't be ruining any dairy farmers anytime soon," said chief scientist Tim Geistlinger. After raising $850 million, Just has dominated the plant-based egg market, displacing 500 million eggs by 2024. Again, though, the world consumes 1.4 *trillion* eggs a year.

Even in that rosy Boston Consulting Group scenario where alt-proteins increase sevenfold by 2035, they'd only seize 11 percent of the market, so meat and dairy consumption would keep increasing, and we'd keep eating more of the earth. In a super-rosy scenario where governments aggressively subsidize alternative proteins and crack down on pollution from animal agriculture, alternatives would get to 22 percent—but the world would still keep eating more animals, just at a slower rate of increase. And those scenarios were modeled before the sector fell into its trough of disillusionment.

Then again, things can change. When we're all carrying handheld devices we can use to video-chat with anyone anywhere, track down any fact, or get anything delivered, who can be sure our big-brained species will never mass-produce desirable animal products without animals? It wasn't long ago that solar seemed hopelessly uncompetitive—and Friedrich calculates that if the world's governments spent as much on bioreactors as they've spent on solar over the last five years, they could build enough to cultivate all the world's current meat and seafood diet.

That was a super-duper-rosy scenario, but governments worldwide were starting to notice the sector. USDA's deputy undersecretary for research, Sanah Baig, said in an effusive speech at GFI that the Biden administration considered biotech substitutes a powerful tool against hunger and climate change.

"This is a watershed moment," Baig concluded. "We will help you harness the power of alternative proteins to create a better future for everyone."

Friedrich was not surprised to hear such a full-throated commitment from such an influential bureaucrat, because Baig was his former chief of staff at GFI. She shared his vision of a moonshot for meat, and it was yet another sign of the sector's progress that she could advance it from the inside.

Searchinger never quite answered Friedrich's last alt-protein question during their Socratic dialogue over donuts. But he later advised the Bezos Earth Fund to make meat substitutes a priority and introduced Friedrich to the foundation's leaders; it soon became a top GFI funder, and has already financed three alt-protein research centers around the world. Searchinger is all in favor of governments funding all-out Manhattan Projects to make alt-proteins better, cheaper, and more scalable.

But he can also see that meat substitutes aren't moving the needle yet, because they still cost more and taste worse than animal meat. He thinks Impossible Burgers are great, but so are beef burgers. Why would meat-eaters who don't devote their careers to studying agricultural emissions spend more for the biotech versions?

He's even more skeptical about cultivated meat, which isn't in the vicinity of the needle yet. Friedrich believes it will start going mainstream with blended products like those Upside pot stickers that mix cultivated meat with plants, but Searchinger doesn't understand the appeal. Wouldn't non-vegans prefer real animal meat blended with plants, like Perdue's Chicken Plus nuggets or 50/50's Both Burgers? That seems like a much easier way to reduce meat consumption, and even that's not happening anywhere near scale.

So Searchinger is still a copper-bullet guy. He thinks alt-proteins are a hugely important item on the sustainable food menu, but he'd also like to see moonshots for better farming, better grazing, and all kinds of eating-the-earth solutions.

So far, there's only one solution starting to attract the kind of public and private funding and support he considers necessary to save the world, which concerns him. And he doesn't consider it a real solution, which terrifies him.

EIGHT
THE SOIL FANTASY

THE UNDERGROUND SOLUTION

Kiss the Ground, like so many documentaries about climate change, opens with apocalyptic images of an overheated Earth: torrential flood, raging wildfire, calving glacier, storm-demolished town. "There's so much bad news about the planet, it's overwhelming," moans the stoner-voiced narrator, the actor Woody Harrelson. "The fear that we're heading for a cliff puts most of us into a state of paralysis." He sounds like a dude slumped across his futon in the fetal position: "The truth is, I've given up. And odds are, so have you."

But then Harrelson pivots from gloom to hope, because the thesis of the award-winning 2020 Netflix film is that one silver-bullet solution can reverse global warming, heal the planet, "and keep our species off the extinction list."

The solution is soil—specifically, regenerative farming and ranching that sequesters carbon in the soil. If we take better care of our literal earth, the argument goes, we can transform agriculture from a climate-killing carbon source to a climate-saving carbon sink.

The world's soils are already vast reservoirs of carbon, holding three times as much as the atmosphere and four times as much as all aboveground vegetation. That's because trees and other plants pull carbon out of the air through photosynthesis, then exude carbon from their roots to feed microbes and other soil organisms. Much of that carbon returns to the air when the microbes respire and the plants die and decompose, but some remains in the soil in the form of organic matter, the dark, spongy muck known as humus. And we know soils could hold even more carbon, because they did before the dawn of agriculture. They've lost more than 100 billion tons since we started clearing and tilling them.

Kiss the Ground makes an impassioned case that regenerative agriculture that rebuilds soils by mimicking nature, as opposed to extractive agriculture that depletes soils by abusing nature, can redistribute enough carbon from the air (where it warms the planet) to the ground (where it helps crops and pasture grasses grow) to offset all our agricultural emissions—and our other emissions, too. The prairie carbon that we plowed up to make food and the fossil carbon that we dug up to make energy can be recaptured and returned to the earth.

"That's why some people are racing to save our soil, in hopes that our soil just might save us," Harrelson says.

It's a lovely theory: By ending a misbegotten era of monoculture fields and factory feedlots, we can harness the power of photosynthesis to make our food not only carbon-neutral but carbon-negative. Even our beef can curb our emissions, if we raise it right; the problem isn't the cow, regenerative ranchers say, it's the *how*. We can restore our degraded dirt by resuscitating subterranean ecosystems of fungi, bacteria, and worms. And we can escape our state of paralysis, because the regenerative playbook[*]—reduced tillage to minimize soil disturbance; offseason cover crops to protect soil from the

---

[*] There's a raging debate about the true meaning of "regenerative agriculture," since unlike its cousin organic agriculture, which has detailed certification rules, it has no official definition. The phrase does get thrown around a lot, and reasonable people can disagree how much of the playbook a farmer has to follow—and whether he needs to produce measurable improvements in soil health—to be considered regenerative. But it's a boring debate. Most practitioners agree on the basic principles and practices, so I'll call those regenerative, without getting into philosophical questions of which farms or farmers qualify as regenerative.

elements; diverse crop rotations; limited chemical inputs; and thoughtfully managed grazing—can repair the damage of the Anthropocene.

Kiss the Ground is mostly talking heads rhapsodizing about how carbon farming can save the climate—and also create healthier food, more profitable farms, and deeper connections to the land. It features advocates from groups like the Rodale Institute, which coined the phrase "regenerative agriculture," and the Savory Institute, the global hub for regenerative grazing; practitioners like North Dakota rancher Gabe Brown, author of the regenerative bible *Dirt to Soil*; and celebrities like Giselle Bündchen, Jason Mraz, and Patricia Arquette. The only action unfolds at the Paris climate summit, as France's agriculture minister launches a "Four Per Thousand" campaign urging all the world's nations to increase the carbon content of their farm soils by 0.4 percent—which, Harrelson says, "would sequester the same amount of carbon that humanity emits."

That's a lot of carbon. And while *Kiss the Ground* offers a maximalist version of the view that kinder and gentler agriculture can cool the planet, that's no longer a fringe view. The *Times* review hailed the film's "persuasive and optimistic plan to counter the climate crisis," and much of the world does seem persuaded.

Four Per Thousand now has 700 partners for its carbon farming crusade, including the FAO, World Bank, and several dozen national governments. The IPCC has estimated that soil carbon sequestration can provide 89 percent of the agricultural emissions reductions the world needs. In his own 2020 film, *A Life on Our Planet*, Sir David Attenborough said regenerative "agroecology" can "rebuild the soil carbon years of continuous farming has removed." Science skeptic Robert F. Kennedy Jr. held a regenerative town hall during his presidential campaign, science evangelist Al Gore hosts an annual regenerative conference at his Tennessee farm, and Searchinger's old nemesis Sara Hessenflow Harper manages an online platform for regenerative farmers. On his popular podcast for anti-establishment bros, Joe Rogan has championed regenerative leaders like Will Harris, a Georgia rancher with a Foghorn Leghorn drawl who supplies grass-fed beef to General Mills. Michael Pollan, who always pushed the regenerative playbook for ecological and agricultural reasons, now pushes it as a climate fix, too.

"We now know how to put carbon back in the soil where it belongs," Pollan said in a video for the regenerative activists at Carbon Underground.

Regenerative principles date back to ancient Mesopotamia, but they got trendy again just as alternative proteins hit their peak of inflated expectations—around the same time Searchinger experienced those Warren Beatty moments when influential officials kept assuming his 2019 report calling for climate-friendly agriculture must be calling for regenerative agriculture. Now that fake meat has stumbled into its trough of disillusionment, carbon farming has emerged as the eating-the-earth solution with the most hype, funding, and political juice.

Environmental groups, which mostly ignored agriculture when Searchinger was lawyering, have embraced regenerative agriculture as "The Secret Weapon in the Fight Against Climate Change," while funders like the Rockefeller Foundation, which once financed Borlaug's work on the Green Revolution, now seem to disgorge cash whenever they hear the word "regenerative." Corporate giants have also boarded the bandwagon, including ADM, Bayer, and Cargill from Big Ag; Nestlé, Mars, and Danone from Big Food; and even Shell and BP from Big Oil. *Wired* opened a 2021 article with a riddle: "What do climate advocates have in common with fossil fuel companies, auto firms, and farm lobbyists?"

"All are joining hands to promote 'carbon farming' as a form of 'regenerative agriculture.'"

The media have promoted it, too, in celebratory features like "Can Dirt Save the Earth?" and "'This Way of Farming Is Really Sexy,'" as well as consumer journalism touting the carbon benefits of regenerative pizza, beer, hot dogs, and milk. The headlines often contrast climate-saving carbon farming with merely climate-friendly fake meat, as in: "Put Down That Veggie Burger. These Farmers Say Their Cows Can Solve the Climate Crisis." The *Times* named "regenivore" its 2023 food word of the year, concluding that "a new generation wants food from companies actively healing the planet through carbon-reducing agriculture." *Rolling Stone* declared: "Far and away, the biggest trend we are seeing in food right now is 'regenerative.'"

It's also a trend in politics. Carbon farming is now the core of the Democratic Party's agricultural agenda, as even urban Democrats now rhapsodize

AGRICULTURE IS EATING THE EARTH

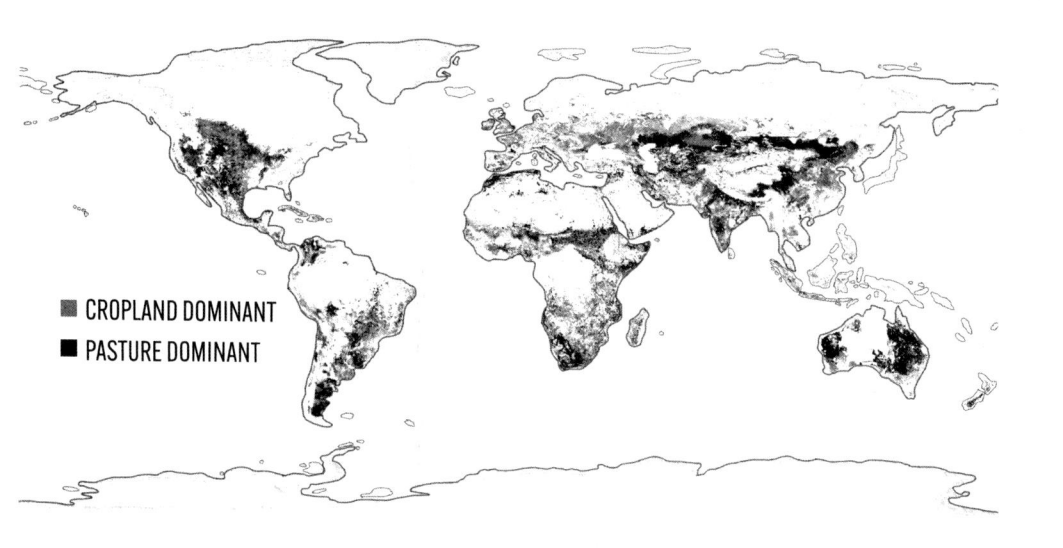

■ CROPLAND DOMINANT
■ PASTURE DOMINANT

Agriculture now covers about two-fifths of the world's land, or about half of the world's land that isn't ice or desert—an area equivalent to all of Asia plus all of Europe. And most of that area is covered by animal agriculture. Two-thirds of all agricultural land is pasture, while only one-third is cropland—and quite a few crops get fed to livestock, too. In other words, the earth is becoming a farm, mostly an animal farm, as forests, wetlands, and other natural landscapes are cleared to create agricultural landscapes.

Tim Searchinger grew up in Manhattan's Upper West Side, where his left-wing parents often took him to anti-war protests (2). He still looked like a kid at Yale Law School, where he's pictured with his future wife, Brigitte (3). He spent the first half of his career as an attorney fighting to protect the environment; that's him canoeing on the Missouri River (4), where he battled the nature-killers of the Army Corps of Engineers, and hiking at Glacier National Park (5).

4

5

Use of U.S. Croplands for Biofuels Increases Greenhouse Gases Through Emissions from Land-Use Change

Timothy Searchinger,[1]* Ralph Heimlich,[2] R. A. Houghton,[3] Fengxia Dong,[4] Amani Elobeid,[4] Jacinto Fabiosa,[4] Simla Tokgoz,[4] Dermot Hayes,[4] Tun-Hsiang Yu[4]

Most prior studies have found that substituting biofuels for gasoline will reduce greenhouse gases because biofuels sequester carbon through the growth of the feedstock. These analyses have failed to count the carbon emissions that occur as farmers worldwide respond to higher prices and convert forest and grassland to new cropland to replace the grain (or cropland) diverted to biofuels. By using a worldwide agricultural model to estimate emissions from land-use change, we found that corn-based ethanol, instead of producing a 20% savings, nearly doubles greenhouse emissions over 30 years and increases greenhouse gases for 167 years. Biofuels from switchgrass, if grown on U.S. corn lands, increase emissions by 50%. This result raises concerns about large biofuel mandates and highlights the value of using waste products.

Most life-cycle studies have found that replacing gasoline with ethanol modestly reduces greenhouse gases (GHGs) if made from corn and substantially if made from cellulose or sugarcane (1–7). These studies compare emissions from the separate steps of growing or mining the feedstocks (such as corn or crude oil), refining them into fuel, and burning the fuel in the vehicle. In these stages alone (Table 1), corn and cellulosic ethanol emissions exceed or match those from fossil fuels and therefore produce no greenhouse benefits. But because growing biofuel feedstocks removes carbon dioxide from the atmosphere, biofuels can in theory reduce GHGs relative to fossil fuels. Studies assign biofuels a credit for this sequestration effect, which we call the feedstock carbon uptake credit. It is typically large enough that overall GHG emissions from biofuels are lower than those from fossil fuels, which do not receive such a credit because they take their carbon from the ground.

For most biofuels, growing the feedstock requires land, so the credit represents the carbon benefit of devoting land to biofuels. Unfortunately, by excluding emissions from land-use change, most previous accountings were one-sided because they counted the carbon benefits of using land for biofuels but not the carbon costs, the carbon storage and sequestration sacrificed by diverting land from its existing uses. Without biofuels, the extent of cropland reflects the demand for food and fiber. To produce biofuels, farmers can directly plow up more forest or grassland, which releases to the atmosphere much of the carbon previously stored in plants and soils through decomposition or fire. The loss of maturing forests and grasslands also foregoes ongoing carbon sequestration as plants grow each year, and this foregone sequestration is the equivalent of additional emissions. Alternatively, farmers can divert existing crops or croplands into biofuels, which causes similar emissions indirectly. The diversion triggers higher crop prices, and farmers around the world respond by clearing more forest and grassland to replace crops for feed and food. Studies have confirmed that higher soybean prices accelerate clearing of Brazilian rainforest (8). Projected corn ethanol in 2016 would use 43% of the U.S. corn land harvested for grain in 2004 (1), overwhelmingly for livestock (9), requiring big land-use changes to replace that grain.

Because existing land uses already provide carbon benefits in storage and sequestration (or, in the case of cropland, carbohydrates, proteins, and fats), dedicating land to biofuels can potentially reduce GHGs only if doing so increases the carbon benefit of land. Proper accountings must reflect the net impact on the carbon benefit of land, not merely count the gross benefit of using land for biofuels. Technically, to generate greenhouse benefits, the carbon generated on land to displace fossil fuels (the carbon uptake credit) must exceed the carbon storage and sequestration given up directly or indirectly by changing land uses (the emissions from land-use change) (Table 1).

Many prior studies have acknowledged but failed to count emissions from land-use change because they are difficult to quantify (1). One prior quantification lacked formal agricultural modeling and other features of our analysis (1, 10). To estimate land-use changes, we used a worldwide model to project increases in cropland in all major temperate and sugar crops by country or region (as well as changes in dairy and livestock production) in response to a possible increase in U.S. corn ethanol of 56 billion liters above projected levels for 2016 (11, 12). The model's historical supply and demand elasticities were updated to reflect the higher price regime of the past 3 years and to capture expected long-run equilibrium behavior (1). The analysis identifies key factors that determine the change in cropland.

1) New crops do not have to replace all corn diverted to ethanol because the ethanol by-product, dry distillers' grains, replaces roughly one-third of the animal feed otherwise diverted.

2) As fuel demand for corn increases and soybean and wheat lands switch to corn, prices increase by 40%, 20%, and 17% for corn, soybeans, and wheat, respectively. These increases modestly depress demand for meat and other grain products beside ethanol, so a small percentage of diverted grain is never replaced.

3) As more American croplands support ethanol, U.S. agricultural exports decline sharply (compared to what they would otherwise be at the time) corn by 62%, wheat by 31%, soybeans by 28%, pork by 18%, and chicken by 12%).

4) When other countries replace U.S. exports, farmers must generally cultivate more land per ton of crop because of lower yields.

Farmers would also try to boost yields through improved irrigation, drainage, and fertilizer (which have their own environmental effects), but reduced crop rotations and greater reliance on marginal lands would depress yields. Our analysis assumes that present growth trends in yields continue but

---

In 2008, Searchinger launched a new career in climate science with a groundbreaking *Science* article documenting that farm-grown biofuels were even worse for the climate than gasoline, because using farmland to grow fuel instead of food would create demand to clear new farmland to grow food. The article provoked a furious backlash, but it ultimately changed the way scientists analyze land-use change and greenhouse gas emissions. It's been cited more than 6,500 times in the scientific literature.

BIOFUELS ARE A CLIMATE DISASTER

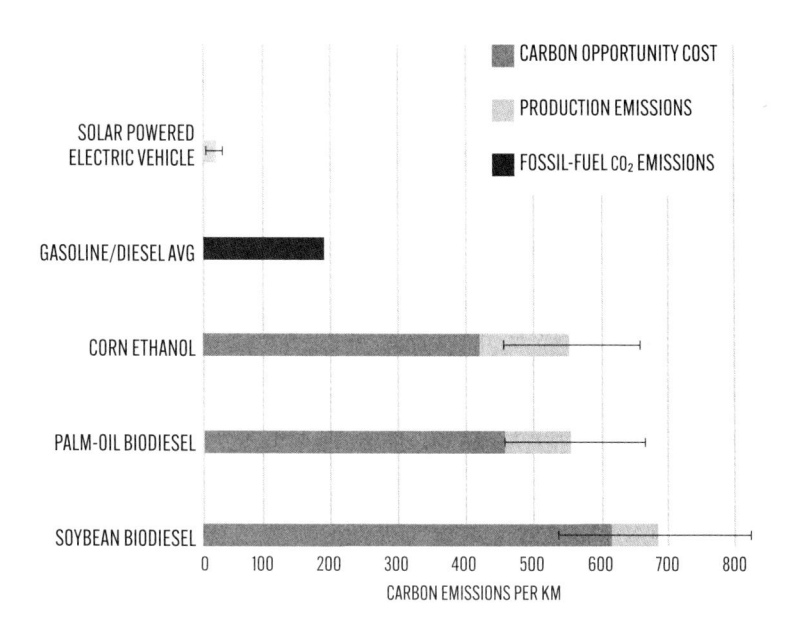

Searchinger's key insight was that earlier studies on biofuels had ignored the "carbon opportunity cost" of using land that could otherwise store carbon in natural vegetation or grow food that would reduce the need to clear other natural vegetation. He also identified that the basic problem with growing energy is its inherent inefficiency. A car fueled by crops generates several times more greenhouse gas emissions per mile than a car fueled by gasoline, and several hundred times more than an electric vehicle powered by the sun (top). Converting every crop, blade of grass, and tree harvested on Earth into energy would meet just one-fifth of our 2050 energy needs (bottom).

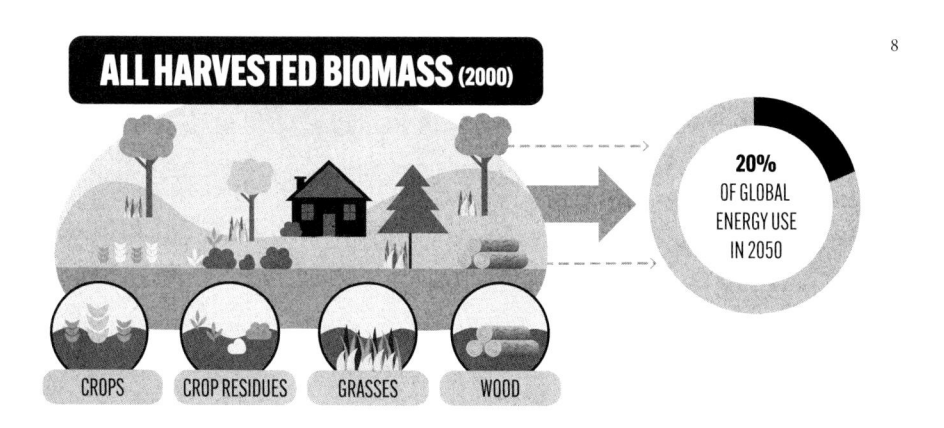

THE WORLD NEEDS TO MAKE MORE FOOD...

CROP PRODUCTION (TRILLION CALORIES PER YEAR)

- 2010 (BASE YEAR): 13,100 TRILLION CALORIES
- 2050 (BASELINE): 20,500 TRILLION CALORIES
- 56% FOOD GAP

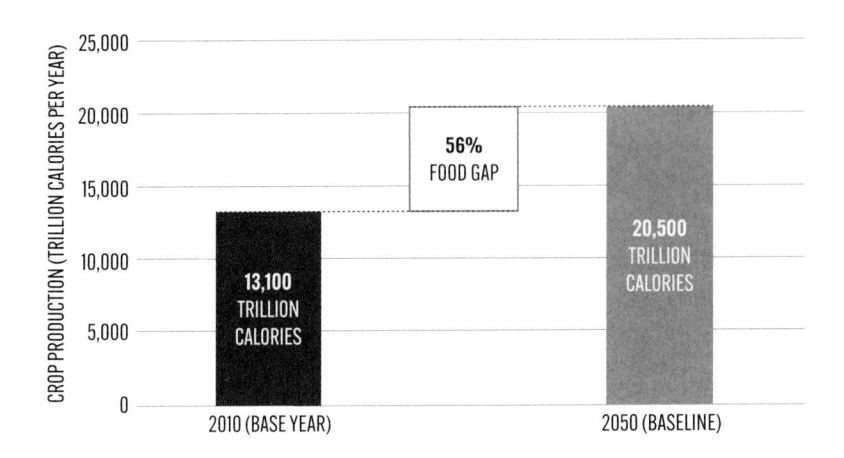

WITH MUCH FEWER EMISSIONS...

GIGATONS OF EMISSIONS PER YEAR

- NONAGRICULTURAL EMISSIONS
- AGRICULTURAL AND LAND-USE CHANGE EMISSIONS

- 2010 (BASE YEAR): 48 (12 agricultural)
- 2050 (BASELINE): 85 (15 agricultural)
- 2050 (TARGET): 21 (4 agricultural)
- 11 GT EMISSIONS GAP

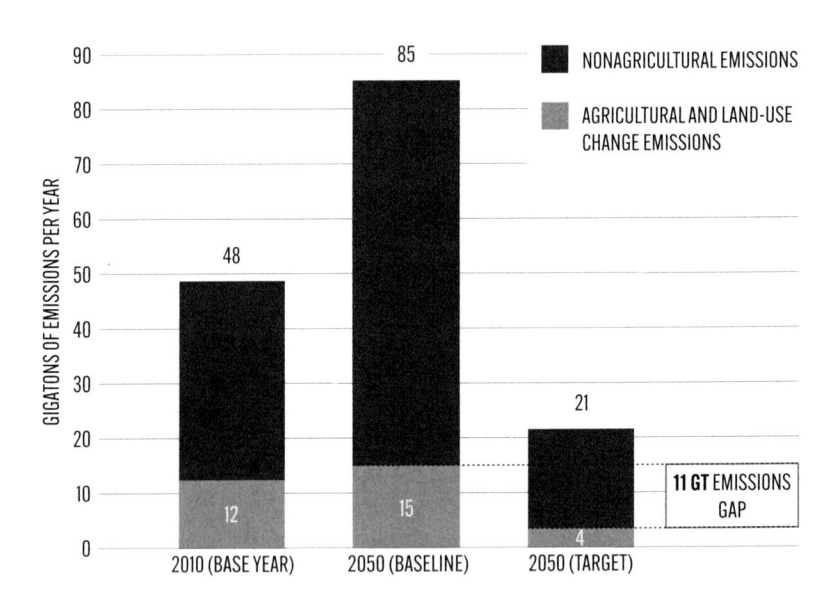

Creating a Sustainable Food Future, Searchinger's 2019 blueprint for how to feed the world without frying it for the World Resources Institute, focused on how humanity could close three global gaps by 2050. Agriculture needed to produce about 50 percent more food (9) and generate about 75 percent fewer emissions (10). And it somehow needed to do that without using any additional land, even though it was on track to clear another two India's worth of forest (11). The report provided a "menu" of 22 different solutions, including using less bioenergy, eating less beef, wasting less food, and increasing crop and livestock yields to make more food with less land—and for the land sector to meet its climate targets, it will need to make dramatic progress in all those areas (12).

...AND WITHOUT EATING THE REST OF THE EARTH

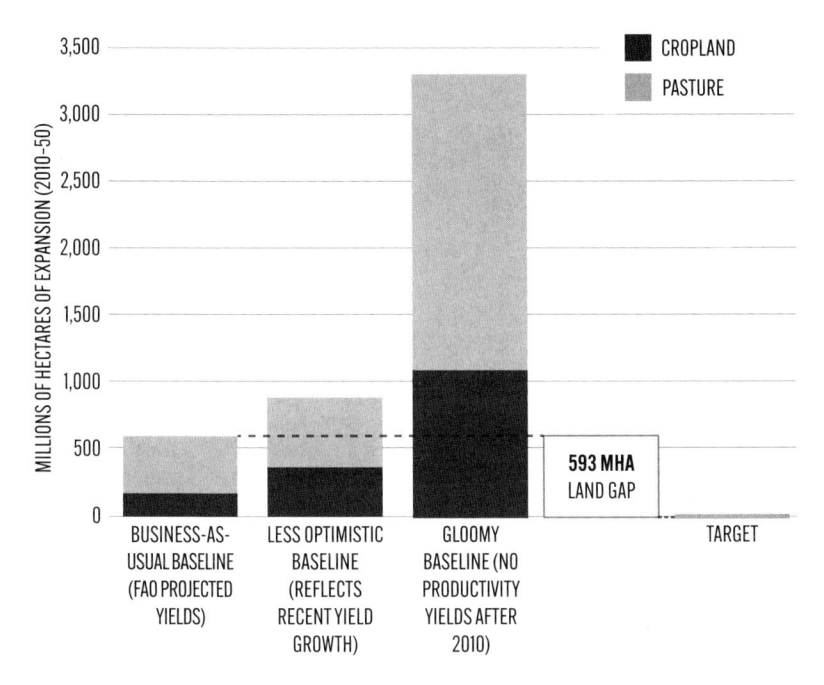

Millions of hectares of expansion (2010-50)

Legend: ■ CROPLAND ■ PASTURE

Scenario	Value
BUSINESS-AS-USUAL BASELINE (FAO PROJECTED YIELDS)	~575
LESS OPTIMISTIC BASELINE (REFLECTS RECENT YIELD GROWTH)	~875
GLOOMY BASELINE (NO PRODUCTIVITY YIELDS AFTER 2010)	~3,300
TARGET	~0

593 MHA LAND GAP

SOLUTION: PRODUCE, PROTECT, REDUCE, AND RESTORE

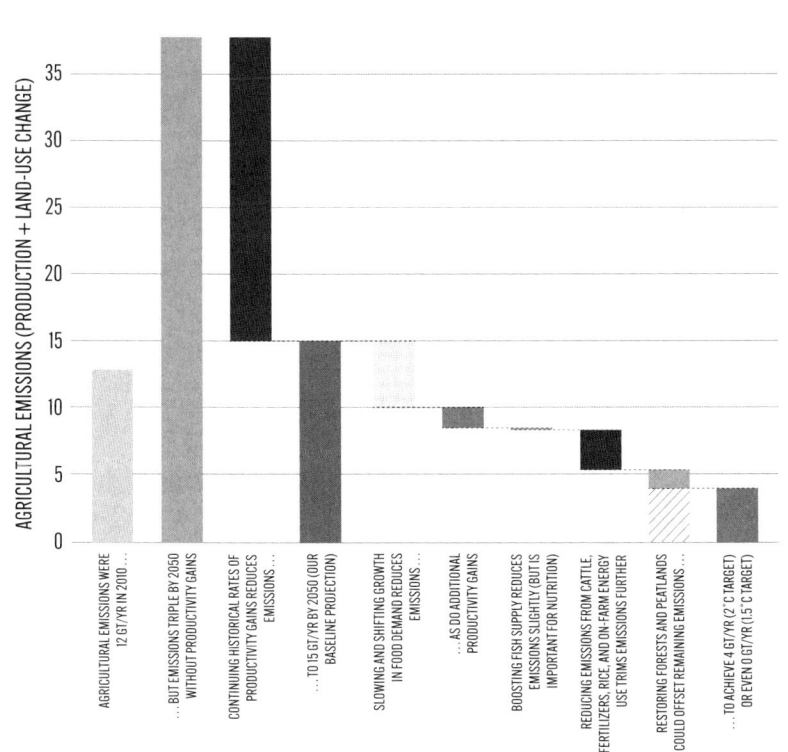

Agricultural emissions (production + land-use change)

- AGRICULTURAL EMISSIONS WERE 12 GT/YR IN 2010...
- ...BUT EMISSIONS TRIPLE BY 2050 WITHOUT PRODUCTIVITY GAINS
- CONTINUING HISTORICAL RATES OF PRODUCTIVITY GAINS REDUCES EMISSIONS...
- ...TO 15 GT/YR BY 2050 (OUR BASELINE PROJECTION)
- SLOWING AND SHIFTING GROWTH IN FOOD DEMAND REDUCES EMISSIONS...
- ...AS DO ADDITIONAL PRODUCTIVITY GAINS
- BOOSTING FISH SUPPLY REDUCES EMISSIONS SLIGHTLY (BUT IS IMPORTANT FOR NUTRITION)
- REDUCING EMISSIONS FROM CATTLE, FERTILIZERS, RICE, AND ON-FARM ENERGY USE TRIMS EMISSIONS FURTHER
- RESTORING FORESTS AND PEATLANDS COULD OFFSET REMAINING EMISSIONS...
- ...TO ACHIEVE 4 GT/YR (2 C TARGET) OR EVEN 0 GT/YR (1.5 C TARGET)

13

WE HAVE A CATTLE PROBLEM

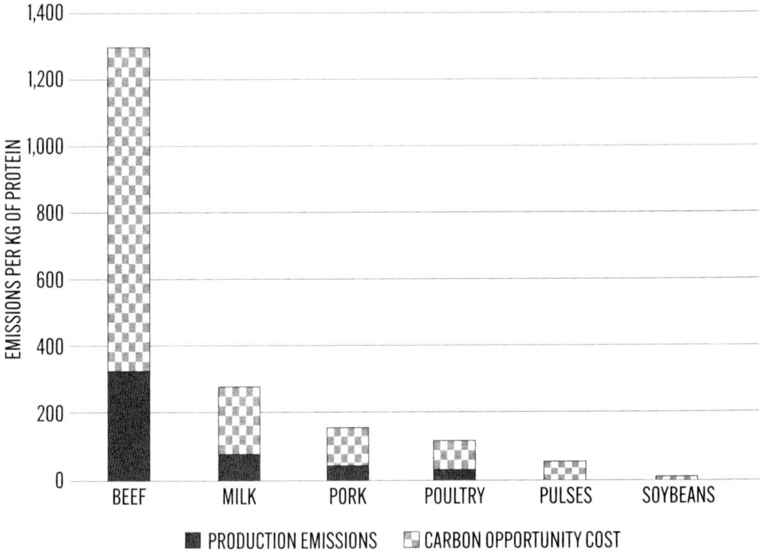
1,400
1,200
1,000
800
600
400
200
0
EMISSIONS PER KG OF PROTEIN
BEEF
MILK
PORK
POULTRY
PULSES
SOYBEANS
PRODUCTION EMISSIONS
CARBON OPPORTUNITY COST

14

EAT
MOR
CHIKIN
Chick-fil-A

When it comes to the climate, all foods are not created equal. While all forms of meat and dairy use more land and generate more emissions than plants do, ruminant meats like beef are much, much worse. Beef accounts for only 3 percent of U.S. calories, but almost half the emissions and land use. The motto of Chick-fil-A's mascot, posing here with Searchinger, is pretty good advice for reducing emissions, and in fact there's been a global shift from beef toward chicken during the last half century, which has prevented even worse climate outcomes.

MORE MEAT AND DAIRY EVERY YEAR

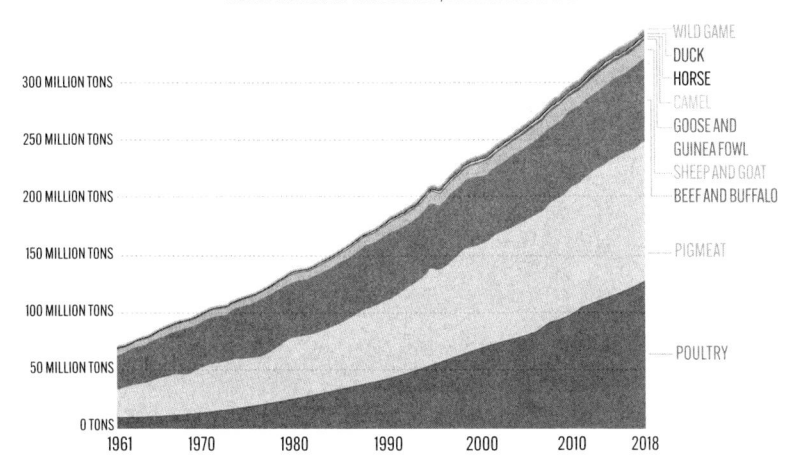

WORLDWIDE MEAT PRODUCTION, BY LIVESTOCK TYPE

- WILD GAME
- DUCK
- HORSE
- CAMEL
- GOOSE AND GUINEA FOWL
- SHEEP AND GOAT
- BEEF AND BUFFALO
- PIGMEAT
- POULTRY

WORLDWIDE MILK PRODUCTION

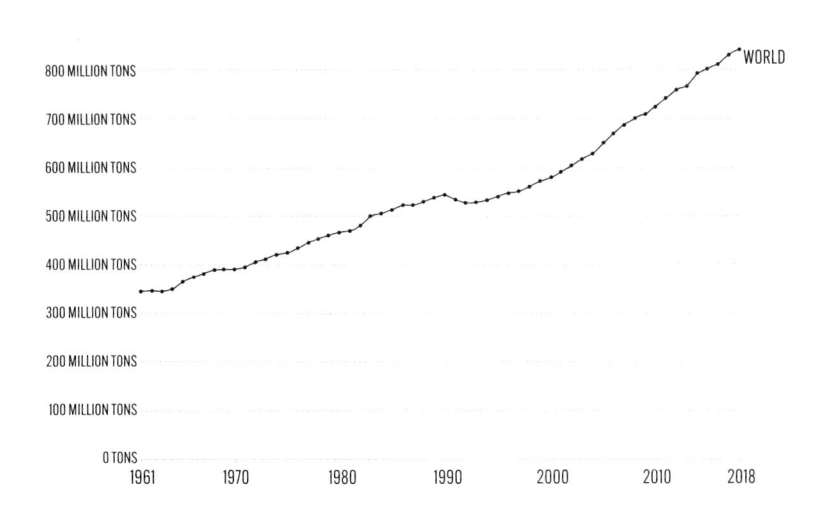

A global shift from meat and dairy toward plant-based diets would be one of the most dramatic ways to reduce emissions from food and agriculture. But the first thing people tend to do when they stop being poor is to start eating meat, and global meat and dairy consumption has been rising steadily for decades.

As a scraggly young PETA activist, Bruce Friedrich tried to reduce meat and dairy consumption by throwing fake blood at fur-wearing models, streaking outside Buckingham Palace, and generally trying to shame carnivores for their diets (16). But it wasn't working, so he shifted his focus to promoting meat and dairy substitutes at the Good Food Institute. He now looks more like a venture capitalist than a rabble-rouser, and in fact he did start a vegan venture fund to try to jump-start alternative proteins (17).

Friedrich now believes the most important achievement of the animal welfare movement will be inspiring the first generation of alternative protein entrepreneurs, including Impossible Foods founder Patrick Brown (20), a renowned biochemist with a messianic streak; Beyond Meat founder Ethan Brown (19), an affable former clean-energy executive; Just cofounder Josh Tetrick (21), a roguish bro with no real record of achievement but a talent for separating rich men from their money; and Upside Foods cofounder Uma Valeti (18), a star cardiologist with a crazy dream of growing cells into meat instead of human organs.

Since beef is such an outsized climate problem, producing it more efficiently is an alluring climate solution. At the Fazenda Tropical ranch and feedlot in the Brazilian Cerrado, Karin Van Den Broek mixes regenerative grazing with conventional industrial practices to generate beef yields far higher than even the most efficient U.S. operations.

At the Nossa Senhora ranch at the edge of the Amazon rainforest, Caaporã CEO Luis Fernando Laranja (center) and manager Laurent Micol (right) show Searchinger how better grasses, better infrastructure, and a bit of fertilizer can help pastures support far more cattle. Caaporã's interventions tripled the ranch's stocking rates—and tripling cattle productivity on degraded ranches in Brazil could allow the reforestation of a land mass twice the size of Texas.

Steve Gabel (24) and Dirk Rice (25) are two of the faces behind super-efficient industrial agriculture. Gabel's Magnum Feedyard can hold more than 30,000 cattle, and efficiency is his North Star; studies show that "grain-finished" cattle who spend their last months in factory feedlots like his generate much fewer emissions and use much less land than purely "grass-fed" cattle. Rice grows 220 bushels of corn per acre, using laser-guided tile plows, genetically modified seeds, chemical fertilizers and pesticides, and a 320-horsepower tractor equipped with precision technology. "Back in my great-great-grandfather's day, men were men and horsepower was horse power," Rice says. "Gotta say, though, we get better yields!"

Two-thirds of the world's agricultural land is grazing land, but increasing yields to produce more crops with less acreage on the other third will be important, too. Aerofarms built a massive vertical farm in downtown Newark to try to grow 400 times as much lettuce per acre as outdoor farms in California, but the economics of making food without sunlight has been tough, and the company slid into bankruptcy. Naveen Sikka of Terviva is having more success with pongamia, a miracle tree he discovered on a trip to India after business school; it can grow seeds much like soybeans with higher yields on lousy land without fertilizer.

Searchinger recently returned to the Everglades, where he saw the manmade filter marshes he fought for as a young lawyer. Not only are they successfully improving water quality by removing nutrient pollution from agricultural runoff, they're inadvertently storing carbon and combating climate change. He had no idea 35 years ago that peatlands restoration was one of the most effective ways to reduce greenhouse gas emissions from the land sector, but sometimes fighting for the earth can have nice unintended consequences.

about zero-emission cattle and carbon-neutral corn. President Biden dropped a plug for regenerative cover crops into his first address to Congress, and with the help of his USDA secretary, Tom Vilsack—previously a fairly conventional aggie who held the same job under Obama, then spent the first Trump term as a dairy lobbyist—launched an unprecedented push to subsidize "climate-smart" regenerative practices. And Kennedy filmed a video in front of USDA headquarters pledging the second Trump administration would end the dominance of industrial agriculture, splicing in clips from *Kiss the Ground* as he vowed to "encourage sustainable regenerative farming that can build soil and replenish aquifers."

Giant agribusinesses and food conglomerates have made pressuring their suppliers to go regenerative the centerpiece of their climate efforts, and climate activists who often howl about corporate greenwashing have mostly praised those efforts. General Mills hired the regenerative rock star Gabe Brown to train its growers, financed a life-cycle analysis that declared Will Harris's regenerative beef carbon-neutral, and partnered with the Kansas-based Land Institute to create a "Climate Smart Cereal" from a new regenerative grain called Kernza.

"We've got religion," General Mills chief sustainability officer Mary Jane Melendez told me.

Corporate America has gotten more religion as carbon markets have made regenerative practices eligible for carbon credits, enabling firms like IBM, Shopify, and JPMorgan Chase to buy regenerative offsets for their own emissions—while enabling farmers to make extra cash. Boston-based Indigo Ag, the leading issuer of farm-based credits in the private carbon markets, has called soil sequestration "a $15 trillion opportunity to fight climate change."

As carbon has become a cash crop, a new ecosystem of facilitators, verifiers, and tech startups like Climate Robotics (which uses AI to optimize soil carbon) and Loam (which coats seeds with microbes to stop soil carbon from decomposing) has emerged to help grow it. Meanwhile, traditional farm lobbyists have hustled to steer the new income into traditional pockets. Democratic Congresswoman Chellie Pingree of Maine, an organic farmer who often fights conventional aggies, found them eager to help her craft legislation supporting agricultural carbon credits.

"I've met with Cargill, McDonald's, the corn growers—it feels like the Twilight Zone," she told me. "These groups used to say I wanted to take the wheels off their tractors, and now they're in my office talking soil carbon."

Carbon farming, clearly, has become a thing—in rural America, in corporate America, and in Washington, which is making it an even bigger thing. The $15 trillion question is: How big a thing should it be?

"Oh, look! That's beautiful!"

Beauty is in the eye of the beholder, and Mark Biaggi, manager of the billionaire environmentalist Tom Steyer's regenerative TomKat Ranch, was beholding a plump and slightly trampled cowpie. He saw beauty in the manure's consistency, which indicated a healthy bovine diet, and the way the cattle mushed it into the ground, so microbes could break down its organic material and enrich the soil. He moved TomKat's herd in patterns that mimicked the natural migrations of wild buffalo, but he saw his real job as growing grass for the herd, which meant feeding the soil microbes that helped grass grow. Good poop made him happy.

"Microbes don't have wings or legs, so their food needs to be down where they can get it," Biaggi said. "That's when the magic happens."

TomKat, a rugged landscape of coastal scrub and grassy hills an hour south of Steyer's San Francisco home, is the passion project of his very California wife, Kat Taylor, the Kat in TomKat, who pushed him to buy it from an Austrian countess after it fell into disrepair. She initially planned to protect it as wilderness, but decided it could do more good as working lands, making all-natural grass-fed beef while drawing down carbon from the air. She believed the world needed a green alternative to industrial meat, so even though she also believed cows were soulful creatures, she made a New Age peace with the idea that death is part of life.

"Our cows have two beautiful years here, then one bad day," she told me during a brisk walk up one of TomKat's hills in the fall of 2019.

Taylor was an in-your-face dynamo with Stanford and Harvard degrees, six tattoos, a day job running the couple's justice-oriented bank, and, according to her bio, "four grown children, each pursuing their one wild and precious life." She was, in other words, a formidable change agent, and also a bit of a weirdo. At a recent climate summit, she had delivered a fiery speech accusing Big Ag of torching our planet and impoverishing rural communities, then suddenly burst into a song:

> *Give me your heart, give me your song, sing it with all your might.*
> *Come to the soils, and you can be satisfied*
> *There's a peace, there's a love, you can get lost inside.*
> *Come to the soils, and let me hear you testify.*

It was Biaggi's job to translate that woo-woo regenerative philosophy into rotational grazing patterns, which he plotted in detail on a wall chart like a general planning a war. Instead of the conventional "Columbus method"—let cattle out to graze in the spring, go discover them in the fall—Biaggi used portable electric fencing to move TomKat's cattle into a fresh paddock almost every day. Then they would devour most of the paddock's grass and crush their manure into its soil, creating a protective mulch to cover its ground for the winter, before Biaggi moved them to the next paddock. It was like a gym workout for the land, intense stress followed by extended recovery, and it clumped the cattle together like the wild herbivores who once ate in packs to protect themselves from predators on the plains.

This anti-Columbus method, or "adaptive multi-paddock grazing," was invented by Allan Savory, an eccentric former game warden in colonial Rhodesia who once ordered 40,000 elephants killed to protect a national park. It didn't work, and he eventually decided well-managed herbivores were actually the planet's only hope for environmental salvation. His TED Talk introducing his theory that bunching up cattle and keeping them on the move can revive grasslands and reduce atmospheric carbon to preindustrial levels has nine million views on YouTube. His Savory Institute claims his method has transformed 75 million acres of land.

The intensive Savory style is much costlier than the Columbus method, and when I met Steyer during an event for his ill-fated 2020 presidential campaign, he told me with an exaggerated sigh that ranching was his wife's most expensive hobby. But then he got serious and said he didn't care how much they lost on TomKat. They had hired the scientific nonprofit Point Blue Conservation

Science to monitor the ranch as a regenerative laboratory, and they hoped the lessons they learned could teach the world a better way to make food.

"If we can show scientifically that this stuff really works and agriculture doesn't have to be destructive, that would be priceless," he said.

Seven years into TomKat's new regime, the ranch was regenerating. Its soil no longer cascaded down its hills during storms, and Point Blue had documented more water retention as well as less soil erosion. Its plant diversity had also improved, with native perennials like needlegrass and oat grass that had almost vanished from the ranch back in every meadow, providing enough extra forage to allow Biaggi to buy less supplemental hay. He said the cattle were healthier than ever, and even after three months without rain, there were green patches throughout the ranch.

"That means we're doing something right," he said.

There was just one problem. Point Blue had collected samples at 42 sites across TomKat, and its soils were *losing* carbon. "Carbon is sexy," Biaggi said, "but honestly, I'm not sure how to manage for it." Point Blue director of working lands Elizabeth Porzig speculated that soil carbon might be a lagging indicator that would take a few more years to show improvement.

"But yeah, those results are a bummer," she said.

Savory has complained that concepts like "empirical data" and "the scientific method" can't quantify the benefits of his holistic approach, but it seems relevant that Point Blue scientists published a study in 2023 revealing that TomKat has continued to lose soil carbon. It doesn't look like a lagging indicator. And experiments elsewhere have confirmed that removing cattle from grasslands generally sequesters far more soil carbon than grazing them either regeneratively or conventionally.

When you also consider cattle's voracious land use—plus their methane burps, and the way the corn and alfalfa fields that help feed them increase fertilizer pollution and water shortages—they don't look anything like environmental salvation. The how matters, but the problem is mostly the cow. Regenerative brands like TruBeef Organic, "the first carbon-neutral meat company in North America," and the milk company Neutral, "carbon neutral from the grass to the glass," can buy offsets for their emissions, but they can't eliminate their emissions. USDA has allowed Tyson to label its Brazen

Beef brand "climate-friendly," which is like labeling a cigarette brand "lung-friendly," but it's refused to release data backing up that claim.

"Regenerative agriculture" is now described as the climate-friendly alternative to industrial farming as routinely as "renewable energy" is described as the climate-friendly alternative to fossil fuels, and the soil-will-save-us message behind *Kiss the Ground* is often portrayed as uncontroversial common sense. The UN has called for a "durable transition to regenerative agriculture." It feels morally and environmentally obvious that farmers and ranchers who take better care of their soils and their animals must take better care of the climate.

But as Searchinger wrote in his report, there's little evidence of that. And even some regenerative diehards are skeptical of the carbon-farming craze.

Ferd Hoefner was into regenerative agriculture before it was cool, before anyone claimed it could help the climate, before it was even called "regenerative." He called it sustainable agriculture, and for three decades, he was the National Sustainable Agriculture Coalition's Washington lobbyist, fighting for natural alternatives to industrial farming.

He didn't have a carbon angle. He just wanted to help family farmers conserve soil, reduce pollution, and resist exploitation by seed, chemical, and grain cartels. He was concerned about a new Dust Bowl, not a hotter planet. But most of the practices he pushed were the carbon-farming practices of today, including no-till to keep soils intact, cover crops to make sure the ground is never bare, and minimizing fertilizers and pesticides.

In 1988, Hoefner helped secure the first $3.9 million grant for a Low-Input Sustainable Agriculture program, which Big Ag lobbyists derided as LISA to distinguish it from macho-man farming that relied on plows and pesticides as God intended. After Congress renamed it Sustainable Agriculture Research and Education, or SARE, some aggies pronounced it "Sara," continuing to mock it as girly farming. Over the next 35 years, Hoefner helped steer $400 million into the program, but that was just 0.1 percent of federal farm spending. Washington mostly focused on subsidizing big growers while protecting them from bad weather and bad markets. Even its farm conservation programs tended to finance fences, drainage, manure lagoons, and other farming expenses with little connection to sustainability.

USDA did offer modest bonuses to farmers who planted cover crops (few did) or reduced tillage (many did), but to Hoefner, they always felt like the token handouts for bike paths in massive highway bills. They didn't disrupt the system. And most growers who did less tilling ended up using more herbicide for weed control. They didn't end up singing regenerative folk songs at climate summits.

But the Biden administration supersized USDA's regenerative investments. It first created a $3 billion Climate-Smart Commodities program that mostly funded regenerative projects, several led by regenerative stalwarts like the Rodale Institute and the Soil Health Institute. Biden's Inflation Reduction Act blasted another $20 billion into "climate-smart agriculture," emphasizing cover crops, no-till, and other regenerative practices. The strategy was all carrots, no sticks; Secretary Vilsack thought it was futile to even try to use climate sticks in farm country, where global warming was often dismissed as a liberal assault on rural values. Instead, he hoped to use federal dollars to incentivize row-crop farmers who wear John Deere caps, drive Ford F-150s, and vote Republican to become carbon farmers, too.

Hoefner had waited his entire career for investments like that. Yet he couldn't shake a sense that the emerging carbon-industrial complex felt a lot like the existing agro-industrial complex, with climate-smart grants to agribusiness giants like ADM, JBS, Cargill, and Tyson; food giants like PepsiCo, McDonald's, and Campbell's; and the trade groups representing soybeans, dairy, and corn. He was especially suspicious of Vilsack's efforts to help develop voluntary carbon markets, which had already been tarnished by scandals over dubious forest carbon offsets that achieved nothing for the climate. He suspected that corporate polluters would greenwash their reputations by buying even more dubious soil carbon offsets, exploiting the regenerative frenzy to avoid having to shrink their own carbon footprints.

"Too much of this money is going to giant agribusinesses, and way too much is going down the rathole of private carbon markets," he said. "Those markets are so far out over their skis, they're tumbling over."

This concern that soil carbon science is lagging far behind soil carbon markets is surprisingly common in regenerative circles. The Land Institute has spent half a century trying to make farms less like factories and more like native prairies, but president Rachel Stroer worries that "carbon exuberance" could discredit the entire natural farming movement. "Some of the claims are just ridiculous," she said. She pointed me to a gung-ho Rodale report suggesting that regenerative agriculture could sequester 55 gigatons of soil carbon every year, enough to offset all human emissions. Rodale came up with that figure by extrapolating data from "exemplary cases"—including its own test plots in Kutztown, Pennsylvania—to billions of totally dissimilar acres around the globe.

"They make it sound like all we have to do is farm differently, and the world can go on with business as usual," Stroer said. "It's fake news."

Gore is worried, too. He converted his own farm to a regenerative operation to try to prove the business case for carbon farming, creating a menagerie of grass-fed pigs, chickens, and cattle foraging alongside organic vegetables, chestnut trees, and watchful scientists. But at his 2022 regenerative conference, he warned that sketchy soil carbon offsets were undermining that business case, rattling confidence in carbon markets while exaggerating agriculture's climate progress.

"People jumped the gun and offered payments without reliability or accuracy, and that's cruising for a bruising," he said. "If you have money changing hands on the pretense that it's sequestering carbon, we could end up thinking we're solving the climate crisis, and then time goes by and, oh my God, we haven't!"

Carbon markets are inherently controversial—partly because they let oil companies and other emitters buy offsets instead of reducing their own emissions, like medieval Catholics buying indulgences for their sins, and partly because it's so hard to make sure the offsets actually reduce emissions. There are additionality questions, since it doesn't help the climate to pay a farmer to preserve some trees or stop tilling a field if he would've done those things without getting paid. There are also permanence questions, since the farmer might cut down those trees or till that field in the future, and leakage questions, since another farmer might just cut down different trees or till a different field right away to meet the demand for food.

So far, though, the carbon-farming debate has mostly focused on the simpler question of whether accurately measuring soil sequestration is even possible. Quantifying carbon is a costly, unwieldy, often haphazard process, reliant on soil samples whose results can vary drastically from region to region, farm to farm, or even within a few feet on the same farm. And they're at best a blurry snapshot of a moment in time. They can't predict how long the carbon will stay underground.

"It's a mess," says Chris Tolles, CEO of a soil carbon measurement startup called Yard Stick. "You take a steel container, hammer it into the ground until you hate your life, bag the soil and mail it to a lab—it's error-prone, it's expensive, it's not scalable. And this is how we're supposed to ensure the integrity of carbon markets that are basically in the Stone Age?"

Yard Stick hopes to replace that mess with handheld devices that use infrared spectrometry to give farmers instant soil carbon readings. Scientists on Gore's farm have developed lower-tech measurement kits they mail to farmers, and they're using the results to build a national soil carbon database. But precision at reasonable price points will take time.

"We've got to be able to count this stuff if we want anyone to believe what we're smoking," said Skidmore College ecologist Kris Covey, who runs Gore's Soil Inventory Project. "Not an easy task."

When even many advocates are skeptical of regenerative agriculture's carbon benefits, less sympathetic observers are predictably scathing. Pat Brown of Impossible Foods dismisses regenerative grazing as "the clean coal of meat," a propaganda effort to greenwash an inherently dirty business; he described TomKat to me as "a bullshit hobby ranch." On the other end of the agro-ideological spectrum, some conventional farmers and ranchers reject regenerative ag as obsolete pastoralism that can never scale up to feed a growing population. And Searchinger's skepticism has only deepened as his criticisms of carbon-farming science have been ignored.

"I thought pointing out the nonsense would kill it," he said. "But it's a hundred times bigger than it was when we started pointing out the nonsense."

In the spring of 2019, Indigo Ag invited Searchinger to Boston to talk to the scientists designing its regenerative offsets. They had embraced the Four Per Thousand idea that tiny carbon increases across gigantic swaths of farm soils could solve climate change. He basically told them he didn't believe what they were smoking.

He explained that estimates of soil carbon potential, like Rodale's 55 gigatons or even the IPCC's less exuberant 3 gigatons, didn't guarantee that potential would be fulfilled, just as a vault's capacity to hold lots of money didn't mean lots of money would end up there. The earth had lots of soil, so Four Per Thousand's goal of increasing the carbon content of all farm soils by 0.4 percent would indeed store lots of carbon. We just didn't know how to achieve that goal. He outlined the lack of evidence that no-till, the only regenerative practice with widespread adoption in the U.S., sequestered any carbon. He left Boston thinking that Indigo's team understood that the data couldn't justify soil carbon offsets.

Three weeks later, Indigo unveiled a Terraton Initiative that read like a sequel to *Kiss the Ground*. It pledged to pay farmers to absorb a trillion tons of soil carbon, as much as humanity had emitted since the Industrial Revolution, through regenerative practices like no-till. Indigo also announced that the regenerative cheerleaders at the Rodale Institute would be its partners.

Did they hear a word I said? Searchinger thought.

Indigo's leaders had strong incentives not to hear. They hoped to build and dominate a multi-trillion-dollar soil carbon market—and by 2020, Indigo's $3.5 billion valuation was the highest in ag tech. They also honestly believed they had the best hoses for a world on fire. Indigo's CEO at the time, David Perry, told me his moral calculus was simple: Photosynthesis was the only scalable way to draw down carbon, so it made sense for the climate as well as his company to start rewarding farmers for regenerative practices, even if the precise soil carbon impact wasn't clear yet.

"We can't afford to wait," he said.

Indigo pledged to validate its credits with soil sampling, modeling, and third-party verification. But sampling was a mess, and Indigo only committed to sample 10 percent of its fields. Soil modeling was also riddled with uncertainty, and a 2023 exposé in *Science* found Indigo's models were widely derided as biased guesswork. And one of Indigo's third-party verifiers, Verra, was embroiled in the scandals over phony forest offsets. Indigo's valuation has plunged 95 percent in five years, and its new CEO, Dean Banks, told me it's still "eating humble pie" about the Terraton Initiative's inflated rhetoric and dubious science.

"There's not a day that we don't look back and cringe," said Indigo's head of sustainability, Ewan Lamont.

The Biden team did recognize that verification would be a problem. David Hayes, a soil carbon skeptic who oversaw ag and climate issues in the White House, pressured USDA to invest in more credible science, and Vilsack, a soil carbon enthusiast, to his credit agreed. The department steered several climate-smart grants to measurement outfits like Yard Stick and Gore's Soil Inventory Project, and created a $300 million measurement and monitoring fund. Hayes assumes the verification efforts will eventually reveal that soil carbon has been oversold, but whatever it reveals, he thinks scientific oversight will be a revolution for USDA.

Searchinger is also glad USDA has promised more rigorous analyses of soil carbon, just as he was glad when Congress required more rigorous analyses of biofuels. But he suspects carbon farming will turn out to be another politically irresistible ag giveaway in a climate costume. He believes the world is putting way too many emissions reduction eggs in the regenerative basket, diverting energy and money from better solutions. As climate-smart cash streams into carbon farming, the official rhetoric about verification hasn't calmed his fears.

"It's all bullshit," he vented.

I wasn't sure what he meant. Soil carbon measurement? Soil carbon modeling?

"No!" he shouted. Then he reconsidered.

"Well, yes, they're bullshit, too. But I meant: Soil carbon is bullshit."

The debate over carbon farming, like Searchinger's dialogue with Friedrich about alternative proteins, has often focused on whether it's the one indispensable solution to agricultural emissions, or just one of many solutions—a 55-gigaton grand slam or a 3-gigaton base hit. But Searchinger spent weeks trying to track down the IPCC's source for 3 gigatons, a figure repeated in scores of scientific papers, and discovered even that relatively conservative estimate was essentially plucked from thin air. Now that he's done the reading, his answer to the $15 trillion carbon-farming question is that it's a zero-gigaton strikeout.

IS CARBON FARMING EVEN REAL?

Even if soil carbon could be reliably measured, modeled, and monitored, Searchinger saw three fatal flaws in the case for carbon farming. His first critique, an on-brand one, was that land is not free.

Most studies concluding that regenerative agriculture could help the climate by sequestering carbon in farmland ignored the opportunity cost of not using that farmland for more intensive agriculture that could produce more food. Even if regenerative practices did increase soil carbon, they could still hurt the climate if they decreased yields, because they'd need more land and induce more deforestation to make the same amount of food. Their carbon opportunity costs could far exceed their carbon sequestration benefits.

For Searchinger, the déjà vu was exasperating. Carbon farming, like biofuels, only looked climate-friendly when everyone ignored the land-use change elephant in the room. The General Mills–funded study that deemed Will Harris's regenerative beef carbon-neutral mentioned in passing that it used more than twice as much land as regular beef, then didn't bother to assess the climate cost of its larger footprint. It felt like *Moonstruck* again, if inefficient land use had been the sin that Cher tried to slip past her priest at confession. If all ranches used land that inefficiently, the U.S. would require nine more Californias' worth of pasture to make the same amount of beef. How could that be carbon-neutral? And speaking of California, TomKat had an even more profligate stocking rate of 17 acres per cow, when conventional ranches had rates as low as one or two acres per cow, so it used much more land per pound of beef.

Yet again, the world was overlooking the impacts of farms beyond their gates. A lengthy *Washington Post* puff piece about the Land Institute's regenerative Kernza grain, headlined "A Recipe for Fighting Climate Change and Feeding the World," waited until the 37th paragraph to point out, *Moonstruck*-style, that "under ideal conditions, it can provide as much as 30 percent of the yield of traditional wheat." *Come on*, Searchinger thought. A wheat substitute that under ideal conditions used more than three times as much land as wheat was not a recipe for fighting climate change or feeding the world! Replacing all wheat with Kernza at those abysmal yields would expand global agriculture by an area at least three times the size of all U.S. cropland. What kind of recipe was that?

Not all regenerative practices hurt yields—more about that soon—but studies have found that cover crops and no-till can hurt a little, and limiting chemical inputs can hurt a lot. The yield penalty for organic row crops grown without them usually ranges between 15 and 40 percent. In general, nature is much better at storing carbon than making food, which is why humans began clearing nature to farm in the first place. It shouldn't be surprising that less intensive farms designed to mimic nature tend to make less food. Yet groups like the Nature Conservancy kept flacking for regenerative agriculture, when its lower yields would mean less nature conserved.

The Land Institute, which calls its regenerative approach "Natural Systems Agriculture," is refreshingly honest about this trade-off. Kernza is the first perennial grain crop domesticated from a wild grass in thousands of years, sinking roots much deeper into soils than annual crops that have to be replanted every spring. Since natural prairies are all perennials, while grain crops before Kernza were all annuals, it's as close to natural as farming gets. Still, Kernza pays a yield price for sending so much energy to its roots. After decades of breeding, its seeds are a third the size of wheat seeds. The Land Institute believes yields shouldn't be the only measure of a crop's value—its Twitter handle is @natureasmeasure—but research director Tim Crews acknowledges that until Kernza's yields improve dramatically, "it won't be ready for prime time."

For Searchinger, the point was that it was irresponsible to hype the benefits of storing more carbon without even assessing the costs of using more land. Indigo claimed regenerative practices could draw down three tons of soil carbon per acre per year, while the renowned soil scientist Rattan Lal, a Gore adviser who won the 2020 World Food Prize for his carbon farming research and advocacy, estimated half a ton; either way, shouldn't they at least analyze the possibility that lower yields could imperil forests that store as much as 20 tons of carbon per acre? In an article in *Nature*, Searchinger used Swedish data to show that when the analysis included the carbon opportunity cost of land, organic wheat generated nearly 70 percent more greenhouse gas emissions than conventional wheat.

In other words, when you eat low-yield organic bread, you're eating more of the earth. Even if yields aren't the only metric that matters, they matter.

Searchinger's second critique of the science suggesting big soil sequestration opportunities from carbon farming was even more Captain Obvious: Most of those opportunities weren't really from farming. Incredibly, the bulk of the soil carbon benefits in many of the studies touted by regenerative advocates came from practices that *removed* land from agricultural production, like leaving natural buffer strips along rivers or converting marginal fields and pastures into forests. The advocates were not only making the familiar error of ignoring the indirect land-use effects of taking farmland out of production, they were actually using the carbon gains from ceasing agriculture as evidence of the potential of regenerative agriculture.

That bait-and-switch sounds almost too *duh* to be true. But Searchinger and David Powlson, a 76-year-old British soil scientist, found numerous studies that lumped together environmental practices that produce big carbon gains, such as restoring drained peatlands, with agricultural practices that at best produce minuscule carbon gains, such as no-till and cover crops—and then got cited as proof that the agricultural practices contribute to big climate gains. They were reminiscent of Cleveland Cavaliers scrub Robin Lopez's observation, after a 71-point outburst by his star teammate Donovan Mitchell, that he and Mitchell had just combined to score 72—except basketball analysts didn't claim Lopez was the key to victory.

"The soil science world wants a nice story to tell about how soil carbon can solve the climate crisis. And everybody wants farmers to get paid. So nobody wants to say, 'Wait, those climate benefits aren't from farming, they're from *not* farming,'" Powlson said. "But I'm a miserable old git, so I say it. Tim says it, too."

Searchinger thought rewilding unproductive farmland, rewetting drained peatlands, and restoring natural buffers were all excellent ways to store carbon. But they all had opportunity costs when they took farmland out of production, and ignoring those costs was another accounting error. He felt like he was playing more Whac-A-Mole, pointing out glaring mistakes that felt like rationalizations for predetermined conclusions, only to see similar mistakes and rationalizations pop up elsewhere. The bullshit asymmetry principle suggested there were better ways to spend his time. These fact checks would never make news, and he doubted he'd win a World Food Prize for debunking myths the food world wanted to believe.

Still, billions of public and private climate dollars were starting to cascade into carbon farming, justified by science that exaggerated its benefits and ignored its costs. And after reading some science out of rural Australia, he came around to his third critique of carbon farming: The climate benefits weren't just exaggerated. They were probably almost nonexistent.

Crops, like us, are carbon-based life-forms. That's why John Kirkegaard, an agronomist in Australia's science agency, always assumed that regenerative no-till farmers who left crop residues like stalks and leaves in the field after harvest would return some carbon to their soils. He wasn't sure exactly how, because the inner workings of soils were still as mysterious as the inner workings of oceans. But he figured the billions of microbes and other organisms in soils must break down the carbon-rich plant material into carbon-rich humus.

He tested that sensible-sounding theory for three decades in the New South Wales farm town of Harden—and it turned out to be completely wrong. His test plots where wheat stubble was left on the ground ended up with no more soil carbon than his plots where the stubble was burned. Often, they ended up with less.

"A real puzzle," recalled Kirkegaard, who called from Darwin during a break from his annual jaunt through the outback.

One of Kirkegaard's colleagues, Clive Kirkby, was such a staunch advocate of leaving residues on fields that he initially questioned the Harden results. But the data kept conflicting with his priors, so he started questioning his priors instead. He became so obsessed with the puzzle that he enrolled in graduate school to study it in his sixties. There he came up with a new theory: Maybe the microbes that ate carbon-rich stubble needed a more balanced diet. One thing that was known about the inner workings of oceans was their ratio of carbon to nitrogen, sulfur, and phosphorus. Kirkby hypothesized that soils might also have a fixed "stoichiometric" ratio—and therefore needed more of the other nutrients to absorb more carbon.

That turned out to be correct. He first demonstrated it in the lab by leaving wheat straw in two pans of soil. He added nitrogen fertilizer to one of them, and its soil became a dark humus that looked like compost. Microbes devoured the straw. He left the other pan unfertilized, and its soil remained a lifeless beige, with bits of uneaten straw. The same thing happened in the field: Fertilized plots consumed stubble, developed richer soils, and sequestered two tons of extra carbon per acre. Unfertilized plots retained stubble, looked like sterile moonscapes, and lost a ton of carbon per acre. Kirkby kept working with Kirkegaard after finishing his PhD—he may have been the world's oldest postdoc—and when they stopped adding fertilizer, the plots started losing carbon again.

The inner workings of soils were still mysterious, but one thing the Australians made clear in obscure journals was that you couldn't boost soil carbon without adding the other nutrients in proper stoichiometric ratios. Those ratios turned out to be fairly consistent everywhere; to sequester a ton of carbon, you had to add 150 pounds of nitrogen. No more N meant no more C. In fact, adding carbon in the form of wheat straw without adding nitrogen ended up cannibalizing nitrogen and releasing carbon from organic matter already in the soil.

In ag circles, the Harden studies were greeted as a tweak to the regenerative playbook. *Grist* wrote them up with the headline "The Secret to Richer, Carbon-Capturing Soil? Treat Your Microbes Well." To Kirkegaard, though, the results undermined the entire concept of regenerative carbon farming. They suggested the key to sequestering carbon was not keeping soils covered, or minimizing soil disturbance, and certainly not using less fertilizer, even though carbon markets were starting to reward those regenerative practices. The key was using *more* fertilizer, which was a problem, because fertilizer already generated more greenhouse emissions than India. The world needed to use less of it, or at least get more of it into crops—and the Harden work showed it only helped sequester carbon when it *didn't* get into crops.

"There's this belief in magic, where you can sequester lots of carbon without using any nutrients people don't like," Kirkegaard said. "Sorry, you can't go against basic chemistry."

Even though Searchinger was already a carbon-farming skeptic, the Harden papers shocked him—partly because they revealed there was no way to sequester more soil carbon without more nitrogen, partly because he had never read that before. He had wondered why African farmers struggled to build soil carbon: *duh*, because they don't have enough fertilizer. Chinese farmers had more soil carbon, because they used more fertilizer. Why hadn't this made any top journals?

"I'm this guy in the outback waving my hand and saying 'Wait, none of this regenerative propaganda makes sense!" Kirkegaard says. "Tim had that environmental background, so he got the nitrogen stuff right away."

Searchinger had worried about nitrogen ever since his first case as a full-time environmental lawyer, when he learned that nitrogen from Manhattan sewage plants was fueling algal blooms in Long Island Sound. He later watched nitrogen from chicken manure pollute Chesapeake Bay and nitrogen from fertilizer in the Mississippi River basin create the dead zone in the Gulf. Now that he worked on climate, he saw the Harden papers as a red-alert warning: Farmers who wanted to get paid for sequestering more carbon could just slather more nitrogen on their fields, the opposite of what they needed to do to curb emissions or protect water bodies.

When Searchinger wrote about these problems with carbon farming, a group of scientists in the regenerative camp—including the soil celebrity Rattan Lal and the head of the IPCC's land-sector work—responded that since farmers already applied excessive fertilizer, they wouldn't need to apply more. They could just use the excess nitrogen that was already in their soils instead of letting half of it escape to the environment—"a positive benefit!"

But farmers overapplied nitrogen because they expected some to escape; if they knew how to prevent that, they wouldn't waste money on insurance N in the first place. Anyway, when Searchinger did the math, he calculated that even if the world's farmers somehow did divert half the excess nitrogen in their soils into organic matter, they'd only be one-three-hundredth of the way to Four Per Thousand's carbon targets. Yet again, he felt like critics were throwing spaghetti against his wall, trying to make a clear-cut issue look complicated.

There was one way farmers could add nitrogen to their fields without adding fertilizer: adding manure. And "integrated crop-livestock systems" that bring livestock into fields to graze and poop were part of the regenerative playbook. Will Harris and Gabe Brown reported dramatic increases in soil carbon after they phased out chemical fertilizers, because they both got more nitrogen from manure.

But manure wasn't a real climate solution, either—not only because it released nitrous oxide and methane, but because the amount of manure on earth was mostly fixed, so if one farmer increased his soil carbon by using more manure, that usually meant less manure and less soil carbon for another farmer. Searchinger called it robbing Peter to pay Paul—except Paul got paid for building carbon, while Peter didn't have to pay for losing carbon. It was another accounting error, creating incentives to move livestock, manure, and nitrogen around the planet without changing the overall carbon content of the planet's soils or atmosphere.

More bullshit, so to speak.

"There's no free lunch," Kirkegaard said. "Anyone who's worked with soil carbon and measured the bloody stuff knows it can't make miracles happen."

It wasn't total bullshit.

As hard as Searchinger worked to debunk outlandish carbon farming claims, he saw value in some regenerative practices. For example, most cover crops helped stabilize soils and reduce runoff, while leguminous cover crops like clovers that fixed nitrogen from the air might even help sequester a bit of soil carbon. And while he thought "agro-forestry" was often a euphemism used to greenwash conventional oil palm, coffee, and cacao plantations that replaced native forests, he saw value in agroforestry practices that integrated trees and shrubs into fields and pastures, storing carbon aboveground while improving soil health. He was a fan of the "Great Green Wall of the Sahel," a multination African resilience initiative to hold back the Sahara desert by planting millions of acres of vegetation.

So he tried to emphasize regenerative ag's benefits for the soil even as he questioned its benefits for the climate. If Tom Steyer wanted to spend his money raising cattle in a way that reduced erosion and improved biodiversity on his ranch, Searchinger didn't see much harm, even if it failed to soak up carbon.

What was harmful was the emerging consensus that carbon farming could solve the eating-the-earth problem, when he considered it the least promising of his 22 menu items. The world had limited bandwidth for climate action, and the attention and cash that governments and corporations were pouring into regenerative agriculture could help a lot more elsewhere.

Even if all you cared about was the amount of carbon in the world's soils, trying to sequester new carbon in farm soils wasn't nearly as effective as preserving the carbon already sequestered in nature. As one paper put it, leaving "grasslands unplowed, wetlands undrained, and forests uncut" kept much more carbon from entering the atmosphere than regenerative agriculture on former grasslands, wetlands, and forests removed from the atmosphere. Protecting nature was much climate-friendlier than farming in harmony with nature.

This was another roundabout case for sustainable intensification. Less intensive agriculture sounded greener, and more in tune with our cultural nostalgia for preindustrial scenes of farmers in overalls feeding pigs by hand or baling hay, but anything that decreased food production increased land destruction. Cultural nostalgia for preagricultural scenes of forests, wetlands, and savannas would drive much better climate outcomes.

Searchinger always assigned his classes *The Omnivore's Dilemma*, Michael Pollan's elegy to the small pastoral farmsteads that covered most of the Midwest before the Green Revolution converted them into monoculture mega-fields. But he also assigned *A Sand County Almanac*, the 1949 classic Aldo Leopold wrote on a quaint preindustrial farm in Wisconsin, because it's an elegy for the biodiverse grassy prairies wiped out by those small pastoral farmsteads. Today's Midwestern farms have even less biodiversity than the farmscape Leopold described, but Searchinger wanted his students to see that almost all the Midwest's habitat and carbon losses happened when Pollan-style agriculture replaced nature, not when intensive agriculture replaced small-scale agriculture. That's when most of its birds and wildflowers and oaks disappeared. Restoring the Midwest of Leopold's day wouldn't restore the lush wild ecosystems that Leopold mourned, and by reducing global food production, it would endanger tropical ecosystems that are still semi-intact.

Anything that reduced productivity would make the global land squeeze worse. Still, as long as the regenerative train seemed to be leaving the station, Searchinger hoped there might be a few truly climate-smart solutions that could climb on board.

SAME YIELDS, LESS MESS

The agro-wizard Norman Borlaug used to visit Iowa farms with his British friend Ted Cocking, a University of Nottingham plant biologist. They'd marvel at the productivity the Green Revolution had wrought, the green-and-gold expanses of densely packed cornfields providing food and feed for the world. But they knew that runoff laced with nitrogen from those fields was fueling algal blooms in Iowa's lakes and rivers, then traveling down the Mississippi to suffocate the Gulf. And they knew the source of that nitrogen.

"Fertilizer," Borlaug told Cocking, "is the Achilles' heel of the Green Revolution."

Nitrogen is all around us. It's 78 percent of our air. But it's not in the form plants need to grow, which is why we've always searched for alternative forms of nitrogen for our soils. In the 19th century, two Guano Wars erupted over tiny South American islands covered in nitrogen-rich bird poop. In the 20th century, no invention was more consequential—not television, air-conditioning, or the atomic bomb—than the "Haber-Bosch" industrial process two German chemists devised to alchemize nitrogen from the air into nitrogen fertilizer. Scientists believe that without Haber-Bosch fertilizer supersizing our harvests, nearly half of the eight billion of us on Earth wouldn't be here. In class discussions of *The Omnivore's Dilemma*'s comprehensive case against industrial agrochemicals, Searchinger always asks if anyone noticed a *Moonstruck*-style passage where Pollan briefly acknowledged that "without synthetic fertilizer, billions of people would never have been born." It's a pretty important point in fertilizer's favor.

Still, fertilizer is a problem. It's usually the most expensive farm input other than land, and since it's made of natural gas, its prices fluctuate when petro-thugs like Vladimir Putin create geopolitical disruptions. Half of it gets lost after it's applied, reappearing in the form of haze, acid rain, and "eutrophication" that chokes water bodies like the Great Lakes with algae. Even before scientists knew fertilizer warmed the planet, through its fossil-fueled manufacturing process and the nitrous oxide it expels into the atmosphere after it's applied, they knew it fouled our estuaries and aquifers with nitrate pollution. And when yield-conscious farmers see downpours carrying off their fertilizer, they often get back on their tractors to reapply, continuing the cycle of waste and pollution.

More than half the world's fertilizer is applied to corn, rice, and wheat, while legumes like soybeans and alfalfa that can fix their own nitrogen don't need as much. That's why Cocking embarked on a multidecade quest to train the major cereal crops to fix their own nitrogen, to help maintain Green Revolution yields without Green Revolution chemicals. In an interview before his death at 92, he told me Borlaug urged him not to give up on self-fertilizing grain.

"Borlaug always said: 'Keep going, Ted,'" Cocking recalled. "He'd tell me: We can't sacrifice yields, so you need to get nitrogen into crops without making such a mess."

Cocking eventually did develop a natural strain of nitrogen-fixing bacteria that could invade and colonize the cells of corn plants, a step toward Borlaug's vision of nutrients without mess. His discovery led to the creation of Azotic Technologies, a startup that sells those microbes as a biological fertilizer designed to help farmers replace chemical fertilizer without sacrificing yields.

"It's finally happening!" Cocking exulted.

Biotech alternatives to fertilizers and pesticides, like biotech alternatives to meat and dairy, offer intriguing possibilities for limiting environmental damage. Yields ought to be nonnegotiable, not only because land is not free, but because farmers lose money when they lose yield, and not even the cleverest agricultural innovation can fix the eating-the-earth problem if farmers won't use it. But solutions that can help them use fewer chemicals without making less food would be truly regenerative solutions, even if they came from wizards like Cocking rather than the eco-prophets of *Kiss the Ground*. And they could help close the emissions gap without widening the food or land gaps.

"It could be huge, if it works," Searchinger says. "We need to find out if it works."

Born half a century after Cocking, Karsten Temme shared his dream. Temme started trying to create self-fertilizing crops while getting his doctorate in bioengineering at Berkeley, hoping to repair the Green Revolution's Achilles' heel. He kept trying after founding a startup called Pivot Bio and raising seed money from Bill Gates. The crops, however, did not cooperate.

"We were chasing the holy grail in agriculture," Temme said. "We wanted to do it. Bill wanted us to do it. We just couldn't get it to work."

Eventually, Pivot Bio discovered a different grail. After failing to engineer crops to fix nitrogen, its scientists figured out how to reengineer nitrogen-fixing soil microbes to spoon-feed crops throughout the growing season. The microbes already did this naturally, but they had gone dormant in fertilized fields. Pivot edited their genes so they could no longer sense when there was abundant nitrogen in soils, so they would no longer get lazy and stop fetching their own nitrogen from the air. It then packaged them into a biofertilizer called Proven that could replace hundreds of pounds of volatile chemicals with a tablespoon of microbial powder, swapping 20th-century chemistry for 21st-century biology.

I first visited Pivot's headquarters in Berkeley before the peak-hype Good Food Institute conference in 2019, and it had the same high-energy, mission-driven feel as the fake meat startups. A whiteboard displayed microbe jokes. ("What do you call a mushroom at a party? Fungi.") Posters offered inspirational messages. ("We MUST Take Care of the Soil for Future Generations.") Scientists were studying dirt under microscopes, dissecting roots with scalpels, sequencing a microbe's genome on a machine the size of a copier, and tending corn plants in a grow room with club-vibe magenta lighting.

That spring, historic floods had whisked tons of conventional fertilizer off Midwestern farms, and Temme showed me photos of stunted Iowa cornstalks alongside healthier plants treated with Proven. James Backman, a 26-year-old Minnesota farmer and Proven sales agent, told me rainfall that washed away its chemical competitors was his best sales tool: "Then growers start telling you, 'Hey, I can see where my Pivot's at!'" It felt like the right product at the right time. The Netherlands was paralyzed by a *"stikstofcrisis,"* or nitrogen crisis, and thousands of Dutch farmers had driven their tractors to The Hague to protest fertilizer restrictions. Really, all of global agriculture faced a *stikstofcrisis.*

But Proven was only applied on 10,000 of America's 90 million acres of corn that year. Its initial formulation was only designed to replace one-tenth of a farm's fertilizer, so it probably reduced the earth's fertilizer load by about 0.000005 percent. All Temme could think about was expanding its scale.

"I'm wrestling with this every night," he said. "People are clamoring for this!"

That sounded like typical startup bluster. I was skeptical that farmers, who tend to stick to their routines, in the Midwest, where farm runoff isn't regulated much, were clamoring to put their yields at risk in order to clean up their runoff with untested "bugs in a jug."

But when I returned to Pivot in 2023, before the funereal post-hype GFI conference, its microbes were on five million acres, an area the size of New Jersey, a 500-fold increase in four years. Its new formulation could replace twice as much fertilizer, and farmers seemed to like the results; Pivot's revenues from its early customers had tripled. Temme had raised $600 million at a unicorn valuation, mostly from big agribusinesses like Bayer, Bunge, and Continental Grain, and it was starting to feel like a big agribusiness itself, chasing the kind of gigaton scale that could narrow the global emissions gap.

One possible path to scale was tapping into the rivers of climate finance that governments and carbon markets were starting to pour into regenerative farming. Much of that money was flowing into soil carbon credits that used questionable sampling to measure the benefits, or rewarded farmers for regenerative practices without even measuring the benefits. Pivot's benefits were much easier to verify, because the emissions from the fertilizer it replaced were much easier to calculate. Scott Faber, still lobbying on land issues three decades after hearing Searchinger rant about the Mississippi River in his tighty-whities, believed fertilizer emissions were agriculture's lowest-hanging climate fruit, a maddeningly unexploited opportunity.

"It's like if I were bleeding to death, and you were standing next to me with a tourniquet, but nobody bothered to pay you to put it on my wound," Faber said.

Pivot was starting to broker private deals to compensate farmers for reducing nitrous oxide from fertilizer—not by selling offsets to companies looking to buy indirect forgiveness for their sins, but through simpler "insets," where food and ag companies like ADM reduced their sins directly by paying for emissions reductions in their supply chains. One bourbon company bought Proven for four Kentucky corn farmers to make climate-friendlier mash; the distiller got to count their reduced emissions toward its climate goals, while the farmers got free nitrogen. Pivot's prices were already competitive, especially after Putin's war in Ukraine ratcheted up fertilizer costs, but free was better.

Still, Pivot would need another 20-fold expansion to hit 1 percent of the global fertilizer market. And despite its commercial success, Proven's agricultural success was not yet scientifically proven. Farmers aren't idiots, and Searchinger considered the fact that so many of them kept buying Proven the best evidence it performed as advertised. But he was concerned that several field trials of Proven, as well as Azotic and other alt-fertilizers, had found little or no yield benefit. Even scientists who remain confident about Pivot agree more research is needed.

"The growth has been astounding, both the scale and the speed," said New York University environmental studies professor David Kanter, a nitrogen expert and Pivot adviser. "But I keep telling them, you need to address these concerns."

Soil carbon skeptic David Hayes argued inside the Biden White House that since most on-farm emissions come from nitrous oxide or methane, at least half the climate-smart funding should focus on them, too. But he lost that argument. "There's just too much excitement about soil carbon," he told me. Soil carbon is agriculture's shiny new object, as influencers like Mark Wahlberg, Jane Goodall, and the Dalai Lama hail its potential to reshape our relationship to our land. By contrast, technological solutions like Pivot seem like workmanlike tweaks to a broken system. *Common Ground*, the actual sequel to *Kiss the Ground*, features glamorous celebrities like Jason Momoa and Laura Dern; *The Nitrogen Dilemma*, a new documentary about the fertilizer problem, features NYU's Kanter and me.

It's easy to see the allure of alternatives to high-intensity agriculture's chemical pollution and soil degradation and animal imprisonment at industrial scale. But since carbon farming couldn't fix the eating-the-earth problem, and alternative proteins weren't ready to help much with the demand side of the problem, Searchinger truly believed his supply-side mantra: *We need incredible yield growth, or we're incredibly screwed.*

NINE
MORE BEEF, LESS LAND

FEEDING THE WORLD WHILE SAVING THE FOREST

Fazenda Tropical, a corn, soy, and beef plantation the size of Disneyland amid the woody savannas of the Brazilian Cerrado, produces incredible yields. And it produces them with many of the carbon-farming practices promoted in *Kiss the Ground*. It's five square miles of proof that some soil-pampering regenerative approaches are at least compatible with off-the-charts productivity, even if they don't sequester much carbon.

In early 2023, I followed Searchinger to the rural outskirts of Goiás state, the "Heartland of Brazil," to learn the high-yield secrets of Fazenda Tropical. He was most interested in its high-yield beef, for the same reason Willie Sutton was most interested in banks; it was where the emissions were, and the best opportunities for emissions reductions. Brazil's pastures covered an area three times the size of Texas, and many barely had any cows, so tripling grazing productivity on those pastures could free up a couple Texases' worth of land.

Fazenda Tropical's owner, Mario Van Den Broek, was a no-nonsense businessman with a stern jaw and sturdy build. But even though he did not sing soil-health folk songs, he checked quite a few regenerative boxes. He hadn't tilled his soils in three decades. He planted a diverse mix of regenerative cover crops that nourished an even more diverse mix of underground organisms. In one of his luxuriant cornfields, he scooped up humus that looked like brown cottage cheese; two earthworms, the engineers and signifiers of healthy soils, were wriggling in the muck on his callused palm.

Van Den Broek's cover crops also nourished organisms aboveground: his black Angus and white Nelore cattle, which his 26-year-old daughter Karin moved from his pastures to his croplands to graze in the dry season. Fazenda Tropical is one of those "integrated crop-livestock systems" regenerative advocates love, where crops help feed the livestock and livestock help fertilize the crops. When the cattle were back in their pastures, Karin rotated them to fresh paddocks every three weeks—not the TomKat Ranch's daily rotations, but not the Columbus method, either. The Van Den Broeks even composted their extra manure into an organic biofertilizer.

Ultimately, though, what set Fazenda Tropical apart was its productivity. Unlike typical U.S. fields that rotate between soybeans one year and corn the next, or "integrated" fields that rotate between crops one year and grass the next, its fields were a triple threat, producing harvests of soy and corn as well as extra forage for cows every year. Its soy yields were above the U.S. average, its corn yields were near the U.S. average, and its dry-season cover crops contributed to astronomical beef yields. Fazenda Tropical's pastures supported seven head of cattle per acre, an unheard-of stocking rate. It's not a perfect apples-to-apples comparison, but that's 10 times the standard rate in the most efficient U.S. ranching regions, or 100 times TomKat's rate. And the more cattle Fazenda Tropical can support, the less pressure there will be to clear more Cerrado for new pastures.

Karin Van Den Broek, a petite blond with her father's intimidating jawline, studied livestock management at the University of Nevada, and she's disgusted by online vegans who portray animal agriculture as irredeemably evil and destructive. Her family produces beef because consumers demand beef. Producing as much as possible on every acre helps preserve natural landscapes.

"We make the most of this land," she said. "We make meat for people, and we return energy and organic material to the soil."

As Brazil has become an agricultural superpower, its farmers and ranchers have developed a villainous eco-reputation, fueling boycotts of Brazilian beef and soy linked to deforestation. Fazenda Tropical arguably counts, since Mario carved it out of the wilderness in 1985, when the Cerrado was a mostly pristine savanna. But the American Midwest was once wilderness, too, and its growers don't get blamed for what it used to be; nobody links Indiana soybeans to deforestation, even though 85 percent of Indiana was once forested. The Brazilian agriculture community is still frosted about a 2010 report out of the U.S. advocating "Farms Here, Forests There." Why should only rich nations that deforested their breadbaskets long ago be allowed to make the most of their land?

Anyway, no one can stop Brazil from having farms there, so boosting the output of its existing farms can help it retain forests there. Charlton Locks, whose company certifies sustainable Brazilian farms and ranches, took us to Fazenda Tropical to show us sustainable intensification at its most intense.

"This is how we can feed the world *and* save nature," Locks said.

So what's the secret behind this regenerative cornucopia in south-central Brazil? There are two, and the first is, well, Brazil.

Triple-cropping corn, soy, and beef on the same land in the same year is a windfall possible only in the tropics, with year-round warmth and abundant rain. The Cerrado's soils were once so infertile that even Borlaug called the region unsuited for agriculture, but its climate was so perfect that in 1973, Brazil created the research agency Embrapa to figure out how to bring the Green Revolution there. To make a long agronomic story short, it worked. The Cerrado now produces a majority of Brazil's agricultural output. This has been a disaster for its flora and fauna, as Searchinger and I saw back in 2008 when we met that shopkeeper who risked divorce to save a slice of Cerrado. But as the soybean king Blairo Maggi said, Brazilian pioneers only followed the lead of 19th-century Americans, and some farm even more efficiently than 21st-century Americans.

The second secret behind Fazenda Tropical is that it's as much a factory farm as a regenerative farm.

The Van Den Broeks ran a modern industrial operation with

top-of-the-line John Deere tractors and the latest vaccines and antibiotics. They were completing a $3.5 million irrigation project to make sure they could keep triple-cropping during droughts. They applied chemical fertilizer, not just manure, and on pastures as well as crops. Fazenda Tropical didn't even produce purely grass-fed beef. Even though its cattle, like all cattle, ate mostly grass, they also got supplemental grain from an on-site feed mill every day in its pastures, then spent their last two months gorging on grain in a crowded on-site feedlot. Where did you think it got the extra manure for its biofertilizer?

The point is, Fazenda Tropical isn't some twee organic farm growing artisanal Kernza, and the Van Den Broeks aren't back-to-the-land progressive farmers outraged by Big Ag. They *are* Big Ag. When the subject turned to Brazil's left-wing then-former President Luiz Inácio Lula da Silva, who was talking up sustainable agriculture as he campaigned to get his old job back, Mario grabbed a hoe and scoffed: "Lula wouldn't know what to do with this!" Like most Brazilian farmers, he liked right-wing President Jair Bolsonaro, the "Trump of the Tropics," a climate denier who declared open season on the Cerrado and the Amazon. Mario borrowed from the regenerative playbook because he thought better soil created better profits, not because he had drunk the *Kiss the Ground* Kool-Aid.

Modern agriculture has been polarized into two camps, Team Regenerative's nature-minded prophets and Team Industrial's efficiency-minded wizards. At conferences, on Twitter, even in journal articles, Team Regenerative slams Team Industrial as rapacious destroyers of underground ecosystems, rural families, and pastoral traditions, while Team Industrial dismisses Team Regenerative as cultists incapable of feeding the world. So it was interesting to see Fazenda Tropical deploying ideas from both sides of the divide, nurturing vibrant soils in the service of kick-ass yields. It suggested the two camps could learn from one another.

But Searchinger wasn't focused on farm ideologies. He just wanted practical details about what worked and how it could scale. He grilled the Van Den Broeks about their fertilization costs, feed conversion rates, and the percentage of soybean cake in their finishing rations. When Karin called up real-time answers on her phone, he cross-examined her about the ambiguities of calculating beef yields in an integrated crop-livestock system.

This much was unambiguous: If all producers were as efficient as the Van Den Broeks, the eating-the-earth problem would be much less of a problem. Unfortunately, not all farmers can shell out $3.5 million for drought-insurance irrigation. Not all ranchers know how to optimize pasture grasses and feedlot rations to maximize weight gain and cattle health. As Karin put it: "Being sustainable costs money and requires knowledge."

Thirty-nine local ranchers pay the Van Den Broeks to finish cattle in Fazenda Tropical's feedlot, and Karin said only two had adopted any sustainable practices. The others used the Columbus method, and when we drove past their pastures, we saw bare ground, invasive shrubs, termite mounds the size of refrigerators, and bony cows that looked like they were on hunger strikes. Most of Brazil's pastures are classified as degraded, while only 9 percent are integrated systems; Embrapa scientist Roberto Guimarães Jr. said that with the right incentives, almost all its pastures could integrate crops, and almost all its croplands could integrate cattle.

"It's like adding a farm to your farm," he explained. "But it's a little complicated, so landowners don't invest the time and resources."

That was frustrating, because Searchinger believed the easiest way to start closing the global food and land gaps would be to ratchet up yields on degraded pastures in Brazil. In the last half century, a Nigeria-sized swath of the Amazon and a France-sized swath of the Cerrado had been cleared for pastures. If some of them could be improved to support more cows, others could be rewilded.

The online vegans were right that if we all quit beef, we wouldn't need cattle pastures. But how likely was that? I quit beef while researching this book, to be less of a climate hypocrite, but I relapsed in Brazil, because, well, Brazil. And nobody has created steaks out of plants, cells, or fungi as exquisite as the carcasses I devoured medium rare on that trip. That's why Searchinger was on the prowl for solutions that could help ranchers generate more beef with less land.

Fazenda Tropical showed it could happen. The question was how to make it happen at scale.

• • •

We saw one answer a thousand miles north at the Nossa Senhora ranch on the Amazon frontier, the contentious borderlands between agriculture and rainforest. Laurent Micol, a former management consultant who switched careers to help save the Amazon, took us to a lush pasture where a few dozen chunky and healthy-looking cattle were resting in the shade of a Brazil nut tree. They were making a languid noise that honestly sounded less like "moo" than the "uhh" middle-aged dads release when we get up from our comfy chairs.

"They're almost ready to be steaks," Micol said.

He didn't mean to sound callous, but his goal wasn't saving cattle; it was saving the wildlife that cattle displace when they expand into the jungle. When we eat burgers and steaks, we're eating not only cows, but jaguars, capybaras, and other Amazon critters that are losing their habitats to cows. We're eating marmosets, cockatoos, and the rest of the cast of *Rio*.

The company Micol works for, Caaporã, is named for an indigenous god believed to protect the Amazon, because by upgrading degraded pastures to support more cattle, it reduces the pressure to deforest more Amazon for more pastures. It's a sustainable intensification play. Caaporã makes deals with absentee landowners—Nossa Senhora's ran a shoe company in São Paolo—to improve their ranches, manage their herds, and split the profits from the increased productivity.

Nossa Senhora wasn't a spectacularly optimized food factory like Fazenda Tropical, but it was way more productive than it used to be. In its degraded state, it supported 0.5 cows per acre and took four years to turn them into steaks. Now it supported 1.5 cows per acre, and they reached slaughter weight in two years. Tripling yields while halving the time cows spend burping, farting, and pooping pollutants was exactly the kind of grazing efficiency revolution Searchinger had called for in his sustainable food report. If pasture was the Rodney Dangerfield of land use, Nossa Senhora showed what was possible when it got a little respect.

The actual improvements were less exciting than the results. Micol's team ripped out invasive weeds; replaced the grass with a fast-growing variety called Mombaça; installed fences to divide the ranch into paddocks;

and built a reservoir and other water infrastructure so cattle no longer had to waste energy trekking to a river to drink. The new managers also began fertilizing pastures; supplementing the cattle's diets with grain and a super-fast-growing elephant grass called Capiaçu; and periodically rotating them to new paddocks instead of letting them munch where they pleased. The per-acre cost of the entire renovation was less than a third of Fazenda Tropical's luxury irrigation project, so it seemed quite scalable.

"It's not rocket science," said Caaporã's energetic CEO, Luis Fernando Laranja. "Better grass! Fertilize the fucking pasture! Feed the fucking cattle!"

Two-thirds of Nossa Senhora was still overgrown jungle, thanks to Brazil's limits on deforestation on existing farms, so we trudged into it to get a feel for what the Amazon was once like, and what more efficient agriculture could save. I lost my bearings a dozen steps into the brush. We saw palm trees as tall as 10-story buildings, including some açaí, which I had only seen in bowls. We saw dense clumps of hardwoods, including some ipê, which I had only seen in decks. We saw two red macaws and three blue-and-yellow macaws glide overhead, iridescent reminders that shrinking agriculture's footprint can preserve nature's.

The Caaporã mantra was a nice cheat sheet for higher-yielding beef: better grass, fertilize the pasture, feed the cattle a little more than grass. Better breeding and veterinary care was crucial, too, to avoid wasting land and feed on cattle that die before becoming beef. Integrating crops with livestock required more work, but Embrapa found that ranchers who started farming some of their pastures increased their profits as well as productivity. Planting trees in pastures looked promising, too. Not only did the trees provide shade for cattle, timber for sale, and measurable carbon storage, dairy cows in silvopasture systems pumped out more 11 percent more milk and had almost twice as many viable pregnancies.

But the pasture-revival business was not booming. Caaporã was only upgrading four ranches nationwide. The margins were tight, and it was tough to justify the up-front investments without back-end rewards for the climate benefits. For owners of degraded ranches, it was usually more profitable to extract new pastures from the Amazon, often by buying land deforested by slash-and-burn scofflaws.

"It's amazing how irrational our species can be," Laranja said. "There's no need to cut down 400-year-old trees for shitty pastures! But people will keep doing it until they get paid more not to."

This was the kind of problem carbon markets were supposed to solve, by leveraging cash from big emitters to reward practices that reduce emissions. Nossa Senhora was reducing overall emissions by reducing the deforestation that was converting the Amazon from carbon sink to carbon source. But that was not a service the carbon markets were rewarding. In fact, from the perspective of the markets, Nossa Senhora had a worse climate impact *after* its improvements: It had more methane-emitting cattle, and it now used nitrous-oxide-emitting fertilizer. It didn't get credit for avoiding land-use emissions beyond its property line. Its owner might have had more luck attracting climate dollars if he had removed its extra cows, ripped out its new infrastructure, and let the ranch degrade again.

In other words, the markets treated land as if it were free.

To Searchinger, this was an accounting travesty as perverse as Medicare's reimbursement rates. He had just read in Nate Silver's *The Signal and the Noise* how bad accounting drove the 2008 financial meltdown, creating incentives to package high-risk mortgages into supposedly low-risk securities that eventually went kablooey. The lesson, once again, was that bad incentives could drive bad behavior on a grand scale. If the world didn't want the climate to go kablooey, it needed to encourage more intensive rather than extensive animal agriculture. Tripling yields on Brazil's degraded ranches was clearly doable, but it wouldn't happen unless it made financial sense for ranchers and meat packers.

"Instead of rewarding inefficiency," Searchinger suggested in his everything-is-so-stupid tone, "*maybe* we should try rewarding *efficiency!*"

Our tour guide Charlton Locks was trying to make that happen in the private sector, with limited success. He had brokered a deal for Unilever to buy sustainable soy oil from some efficient growers, including Fazenda Tropical. But Unilever's premium was only 1 percent above the regular soy oil price, and the Van Den Broeks didn't think that justified the compliance headaches. Locks had spoken with several foreign institutions about financing similar efforts, and they all sounded enthusiastic until the discussion turned to actual financing.

"The main input for a sustainable farm isn't fertilizer or seeds. It's money," he said. "Everybody talks about sustainability, and nobody wants to pay for it."

We heard the same thing from Embrapa researchers, who were struggling to disperse the sustainability hacks they had discovered in their test fields. One soil scientist, Robert Boddey, had found that a purple-flowering clover called *Desmodium* helped increase weight gain in cattle by 60 percent. But only 1,000 acres of *Desmodium* had been planted nationwide, about 0.0002 percent of Brazil's grazing land.

"It works like a dream," Boddey said. "We just need to get the bloody seeds into people's hands."

Searchinger had a similar reaction whenever we saw promising vegetation like Capiaçu, the super-fast-growing elephant grass at Nossa Senhora; or *Leucaena*, the Colombian super-shrub that Embrapa was testing in Brazilian pastures; or *Moringa*, another nitrogen-fixing silvopasture tree with spicy leaves that cattle loved. When Laranja showed us a row of *Moringa* that boosted productivity on his former mother-in-law's ranch, Searchinger was captivated by its potential to produce high-protein feed while fertilizing nearby grasses. But he was also frustrated: Billions of dollars were flowing into soil carbon schemes that didn't store carbon and often reduced crop yields, yet no one could spare a few bucks to plant trees that stored carbon in plain sight and might increase pasture yields?

"This is what climate finance should finance!" he shouted.

We'll get back to climate finance and policy. But it's clear there are plenty of agricultural practices that can increase yields and decrease emissions, even if it's less clear how to spread them around the planet.

BIG DOESN'T MEAN BAD

Steve Gabel knows beef has a bad reputation, which is why an "I ♥ Beef" sign greets visitors to his Magnum Feedyard in northeast Colorado. He also knows factory feedlots where confined cattle get stuffed full of grain before getting shipped off to slaughter have an even worse reputation, which is why he wanted to show me the reality of industrial beef production. The first thing he did was drive me to the middle of a black-and-brown sea of ear-tagged heifers and steers, and roll down the windows of his mud-splattered Chevy Silverado.

"You hear that?" Gabel asked. He was a gruff prairie lifer with a white goatee, chewing a toothpick and staring me down. I wasn't going to lie.

"Uh, I don't hear anything," I said.

"Exactly!" he shot back. "You're surrounded by 25,000 animals. You think any of 'em were mistreated today? You hear 'em bellering?"

If that was a cross between bellowing and hollering, they weren't doing it. They were eating, drinking, and milling around their pens with blank expressions, like guests at a dull cocktail party with dirt floors. They weren't free to roam the open range, but I had just visited a hog farm where each pig was squeezed into eight-square-foot stalls, and these cattle had much more room. Unlike the pigs, they didn't get to winter in a climate-controlled barn, but as Gabel reminded me, they did get to wear leather coats.*

"People say we're cruel, we're abusive—well, I'm not ashamed of anything we do," he said. "Our industry has a good story to tell."

The story he wanted to tell was about efficiency. Activists portray the "concentrated animal feeding operations" that feed 99 percent of U.S. livestock as filthy torture chambers out of *The Jungle*, but mistreating cattle in a CAFO, Gabel said, is inefficient. Stressing out cattle: also inefficient. Even inconveniencing cattle by making them walk through mud or manure is inefficient, which is why he was spending millions on a new drainage system to keep his pens dry, and why he had tractors tricked out like snowplows to scrape poop into easily avoidable piles. He also sent riders on horseback into every pen every day, to make sure all his "beef animals" were healthy and comfortable.

"If I don't create the friendliest possible environment, they might gain

---

* I had a similar conversation with the manager of that Illinois hog farm, who told me he treats his pigs as well as he treats his kids, and probably keeps his pigs even more comfortable. "Except that part at the end," I said. "Yeah, we omit that with our kids," he agreed.

4.1 pounds a day instead of 4.5. That's money out of my damn pocket!" he said. "We maximize our efficiency, so they can maximize their genetic potential."

Magnum is an outdoor assembly line, manufacturing protein from raw materials that happen to be hefty ruminants. Live animals do add uncertainty to Gabel's business—he's used a wheelchair since he broke his neck handling an ornery bull—but he minimizes financial risk with *Moneyball*-style analytics. He told me without checking notes that he uses 10.23 gallons of water per head per day, his finishing feed is 72.5 percent corn, and his mill converts kernels into flakes to increase their digestibility from 82 to 95 percent. He seemed to know everything happening on his lot, down to the latest cases of foot rot. He sounded like the busybody mayor of a town whose residents all had hooves, hides, and upcoming execution dates at a nearby JBS slaughterhouse.

Gabel likes to contrast his old-school family business to Big Ag conglomerates run by suits with sprawling org charts. His wife of 43 years keeps the books. His son buys the cattle. His daughter manages the pharmaceuticals. He exasperates his lawyers by closing seven-figure feed deals without putting anything in writing. But Magnum is Big Ag, too. It's in the top 1 percent of U.S. cattle operations, and that top 1 percent feeds half of U.S. cattle. The day before my visit, federal agents stopped by to investigate whether the Gabels had been improperly squeezed while liquidating hedges on the Chicago Mercantile Exchange.

"A guy chasing a few cows don't gotta worry about that," he chuckled.

Gabel's larger point was that economies of scale unlock efficiencies. When he bought Magnum in 1994, it held 3,000 cattle, not enough to justify an on-site mill, expensive drainage projects, or manure lagoons with advanced pollution controls. Now that it's 12 times larger, Gabel can afford veterinary, nutritional, and environmental consultants; a hospital with electronic medical records for every animal; hormone implants and feed additives like "ionophores" and "beta-agonists" to help his cattle gain weight; and 10,000 acres of corn and alfalfa fields that reduce his feed costs and provide a destination for his manure.

"Being big doesn't make us evil," Gabel said. "It makes us efficient."

Many critics of industrial agriculture see efficiency as a pitiful excuse for boiling the planet with burping cows. Gabel is a Trump Republican who sees global warming as a pitiful excuse for city slickers to gripe about the heartland families who put food on their plates. But no matter what he or his critics believe, the inefficiency of beef makes it worse for the climate than other food, while the efficiency of Magnum makes its beef better for the climate than other beef. Feedlot-finished beef have lower emissions than purely grass-fed beef; it doesn't take as long to fatten them for slaughter, so they spend less time emitting methane, and they use much less land, even accounting for the fields that grow their feed. Gabel has reduced the feed he needs to grow a pound of beef by a third over three decades, so his cattle use even less land.

Overall, the U.S. makes 18 percent of the world's beef with just 6 percent of the world's cattle. Beef from sub-Saharan Africa or the Indian subcontinent can have carbon footprints 50 times larger than U.S. industrial beef. To Searchinger, those statistics illustrate the importance of making even more beef in efficient nations—and also exporting their efficiencies to the rest of the world.

He's often asked: If beef is so awful, what should American beef producers do? His answer: Produce beef! The idea that the Magnum Feedyard is something to make more of rather than get rid of is a heretical environmental idea. But as long as humanity keeps eating beef, it's less awful for the climate when it's produced with yields as high as Magnum's.

The most aggressive advocate of that idea is Frank Mitloehner, a German-born professor in the UC Davis animal science department who runs a research center funded by the livestock industry. He's built like a retired linebacker in one of those old light beer ads. He has an accent that, for historical reasons that aren't his fault, sounds creepy when he speaks emphatically, which is usually how he speaks. He's especially emphatic when he discusses the problems he has with people who have problems with animal agriculture.

Mitloehner has devoted his career to understanding and mitigating the impacts of livestock. He did his graduate work studying the environmental effects of cattle management in Amazon pastures and Texas feedlots, and his current work focuses on reducing emissions and other agricultural pollution in California. But he is undeniably, as the *Times* described him in a front-page hit piece about his industry ties, "An Outspoken Defender of Meat." He's notorious on Twitter, where he goes by @GHGGuru to highlight his greenhouse gas expertise, for trolling vegans with snarky #yes2meat posts and gloating about bad news for fake meat. He has a habit of using accurate facts (as in, methane doesn't linger in the atmosphere as long as carbon dioxide; fossil fuel operations also emit methane; wild ruminants used to emit lots of methane, too) to push dubious pro-meat conclusions (as in, methane from modern cattle is overhyped).

But just because he's an animal-protein partisan doesn't mean he can't do science. It was Mitloehner who discovered the methodological error in *Livestock's Long Shadow*, which shouldn't have discredited the entire report but was undeniably an error. He thinks his pro-livestock advocacy gives him credibility with climate-skeptical cattlemen and dairymen when he tries to persuade them to reduce their impacts. He just happens to believe that more efficient meat and milk, like more fuel-efficient cars and more energy-efficient buildings, is the best way to reduce livestock's long shadow. He considers higher-yield animal proteins as indispensable an eating-the-earth solution as Friedrich considers alternative proteins.

They're certainly better than lower-yield animal proteins. Mitloehner points out that since World War II, the U.S. dairy herd has shrunk by two-thirds, yet produces two-thirds more milk. That means less land, methane, and manure behind every carton. The average California cow produces 23 times as much milk as the average Indian cow. In other words, to match a California dairy's output, an Indian dairy needs enough feed to support 23 times as many lungs, tails, and other bovine body parts that breathe, swat flies, and do other non-milk-producing things.

"I get allergic when people say, 'Oh, American farmers are destroying the planet,'" he said. "Think of the difference between 23 people all driving their own cars or 23 people getting on one bus. That's a win for the climate!"

To illustrate the power of productivity, Mitloehner drove me in his Tesla—he often points out that petroleum is more destructive than livestock—through the Central Valley, the top milk-producing region in what is, contrary to popular belief and Wisconsin license plates, America's top milk-producing state. We stopped at the DeSnayer Dairy, another productive industrial operation whose 1,900 Holsteins made 37 times as much milk as Indian cows.

The DeSnayer story, like the Magnum story, was about efficiency through scale. Like Magnum, it had expanded 12-fold since it opened; it was now part of the top 10 percent of U.S. dairies, which produced 70 percent of U.S. milk. Its bulky machines milked more than 100 pounds out of every cow every day. Its chemically treated lagoons smelled quite mild even though they captured more than 100 pounds of manure from every cow every day.

DeSnayer manager Paul Van Puijenbroek believed healthy and happy cows were the best producers. He was up at four every morning taking their temperatures. He personally drew their blood every week and made sure his workers wiped them down with clean towels after every milking. I had no idea if they were comfortable—their udders bulged like overstuffed shopping bags—but they weren't bellering.

Van Puijenbroek was annoyed that afternoon because for the first time in three months, one of his cows needed an IV. He tried to avoid giving them antibiotics—because he was then required to throw out their milk, not because he was worried about the global antibiotic-resistance crisis—but when they did get sick, he did whatever he could to get them well. He noticed that organic dairymen who bragged about their zero-antibiotics policies often sold him their sick cows.

"I'm the bad guy, because I treat them?" he groused.

Mitloehner is concerned that cultural revulsion for this kind of industrial food production has become so pervasive that even high-efficiency Big Ag operations sentimentalize Old MacDonald approaches. He fears they're losing confidence in the results-oriented factory mentality that's helped them manufacture more food with fewer resources.

"Every time I see a creamery truck with a picture of a 1950s dairy and a red barn and happy cows prancing around a pasture, I want to scream," he said. "Those cows were tied to a pole and milked by stooped labor. The milk sat in buckets for hours. The manure went straight into the creek. Why romanticize it?"

Every year, he polls his students whether they'd want a classic 1950s car or a modern vehicle—and 90 percent vote retro. He then reminds them the antiques had no airbags, no seat belts, and lousy fuel efficiency—and most switch their votes to modern. He'd like to see agribusinesses lean into their technological superiority, instead of running cutesy ads with anthropomorphic livestock.

"These animals are not our buddies," he said. "They're our food!"

Environmentalists and other activists rarely have anything nice to say about "large-scale," "efficiency-obsessed" industrial agriculture. They trash factory farms for "exacerbating deforestation and water scarcity," "polluting rivers, lakes, and tap water," or "using approaches better suited to making jet fighters and refrigerators than working with living systems." And global institutions are often just as critical. "Industrialized farming," the UN Environment Programme reported, "costs the environment about $3 trillion every year."

Nevertheless, industrial operations produce a lot of the world's food, and their share continues to increase as the world continues to move away from Old MacDonald toward bigger, more productive farms. More than 90 percent of America's dairies have vanished since 1970, while the average herd size increased 15-fold. Three-quarters of America's hog farms have also disappeared, and the smaller half of the ones left produce less than 1 percent of U.S. pork, while the 5 percent with at least 5,000 pigs produce nearly three-fourths of the pork.

Mitloehner has an appointment at a university in China, which has started building automated 26-story pork skyscrapers that can house over a million pigs a year. A *Times* story compared one to an iPhone factory, which wasn't intended as a compliment, but accurately reflected its efficiency. Mitloehner said he's watched China's Communist leaders aggressively promote U.S.-style consolidation toward larger and more-efficient farms, while pressuring inefficient rural smallholders to quit farming and move to cities. China wants to be a rich nation, and in rich nations, most citizens are no longer yeoman farmers.

"A lot of China's agriculture is like ours was 200 years ago, and they know it's unacceptable," Mitloehner said. "They have 400 million pigs dying every year before weaning! Of course they want bigger modern farms."

Animal agriculture in much of the world is so spectacularly inefficient that it could enjoy major productivity gains even without pork high-rises, ionophores, or other industrial innovations. Many pastoralists could improve their yields just by timing births before wet seasons, when forage is most abundant. At ranches like Nossa Senhora, modest upgrades like planting better grasses and supplementing them with a bit of hay or grain can help cattle grow faster.

But for livestock yields to increase in the future even faster than they have in the past, some of the technologies that make factory farms so efficient will have to spread. Poor countries need better genetics, nutrition, and other Big Ag innovations that help make more meat and dairy—preferably without more antibiotic overuse, anti-environmental lobbying, and other downsides of many Big Ag operations.

New innovations could help, too.

"*EW*, GROSS, YOU'RE USING *TECHNOLOGY*"

"Those are some big boys," said Ermias Kebreab, a colleague of Mitloehner in the UC Davis animal science department. "And they *eat*."

On that sunny morning in the Central Valley, the boys were nine months old and weighed around 800 pounds, a healthy size for Angus steers. They spent their days in cramped pens at the department's ramshackle feedlot, so they didn't have much to do except eat. When Kebreab showed me their barn, they were all munching on breakfast: enough corn and alfalfa to fill a steamer trunk, plus an addition that might help make cattle less of a climate problem someday.

Kebreab, a six-foot, three-inch African immigrant whose blue-striped Oxford, charcoal slacks, and black dress shoes seemed to gleam through the dusty barn, was trying to reduce the boys' methane emissions. He had laced their feed with a red seaweed imported from Australia that suppressed methane-producing microbes in their stomachs, and the early results were astounding: Steers with traces of seaweed in their diets were burping 80 percent less methane. They also seemed to be converting their feed into weight gain more efficiently, raising the tantalizing possibility that they could make more beef with less land.

For Kebreab, feeding the world is more than a farm-lobby talking point. He grew up in war-torn, drought-crushed Ethiopia, as the famine that horrified young Bruce Friedrich was killing a million of his countrymen. He was a professor's son, so he never went hungry, and spent summers in a rural village tending cattle and goats for his relatives. But he's never seen livestock yields as mere numbers on a spreadsheet. He's working on another project to help Ethiopian pastoralists produce more milk, by developing Amharic-language software to help them optimize their feed with local ingredients.

"That was my dream as a kid, to make agriculture more productive so there could be enough for everyone," he said. "If we can help reduce emissions, too, that's even better."

Kebreab is a mild-mannered academic, not a provocateur like Mitloehner, but he gets similarly irked when well-fed Americans dismiss animal agriculture as a climate crime. He cares about the climate, but he considers animal agriculture a noble calling that can provide wholesome nutrition to hungry people. He attributes his unusual height for a famine-generation kid to the unusual amounts of meat and milk in his diet.

"Ethiopia still has way too many kids with stunted growth because they get *no* animal protein," he said.

Kebreab then began riffing about the first time he decapitated a chicken, and how the 98 percent of Americans who don't farm don't understand where their food comes from, before he stopped to point out a spry-looking middle-aged woman with a mop of blond hair striding purposefully across the feedlot. If you're interested in the future of food, he said, you need to meet her.

He waved, and the woman shouted at us in a crikey-mate Australian accent: "Come on over, you need to see this! We've waited five years for this moment!"

Alison Van Eenennaam, a biotechnology professor who ran a UC Davis genetics lab, was about to oversee the key procedure in an experiment with as much potential to transform animal agriculture as Kebreab's seaweed. She led us to an enclosure where a one-armed man appeared to be wiping a cow's rear end. The body part being wiped, it soon became clear, was not the rear end; with a mischievous grin, Van Eenennaam identified the wiping implement as a "vulva squeegee." It also became clear that the man wielding the squeegee had two arms. The other wasn't visible because it was up the cow's actual rear end.

We were witnessing the intrusive prep work for the first-ever implantation of an embryo genetically edited to produce male offspring. Van Eenennaam's team had harvested eggs from dead cows at a slaughterhouse, then used CRISPR to tweak their DNA in her lab, creating nine potentially viable embryos. She confessed to nerves as we watched the embryos inserted with pipettes into nine potential moms—not because she felt uneasy about creating Frankencalves, but because she was so eager for a pregnancy.

"If this works," she said, "we can change the world."

Van Eenennaam understood the sci-fi implications of editing genes to create sex-selected designer babies. She wryly dubbed her experiment the Boys Only Project. But she wasn't doing it to perpetuate the ruminant patriarchy. She was doing it to make more food with less land. Steers convert grass and grain into beef more efficiently than cows, because they don't expend energy developing ovaries, udders, and other maternal equipment; that's why many ranchers use "sexed semen," which increases the odds of male offspring but reduces fertility rates. Boys Only could more reliably produce the ranching equivalent of a fuel-efficient Prius instead of a gas-guzzling Escalade, while Girls Only edits could one day assure dairy and poultry farmers milk-producing cows and egg-laying hens.

The genomic possibilities are almost limitless. Brazilian scientists have edited Angus cattle to better tolerate tropical heat, while a biotech firm has created CRISPR pigs immune to a virus that stunts their growth. Van Eenennaam was also working to engineer hornless cattle, so ranches and dairies don't have to burn horns off calves to reduce their injury risks. And Kebreab was teaming up with the Nobel laureate gene-editing pioneer Jennifer Doudna to try to create cattle that emit less methane without expensive imported seaweed. Fake-meat entrepreneurs are right that the cow is a relatively mature technology, but it could still be improved.

Van Eenennaam had spicy opinions about that's-not-natural technophobes, vegan radicals, niggling bureaucrats, and anyone excessively skittish about harnessing the genomic revolution. She blamed much of the fear-mongering about CRISPR and GMOs on organic farming interests trying to protect their overpriced products by dragging agriculture back to its primitive roots.

"More efficient means more sustainable. That's fundamental," Van Eenennaam said. "There are always going to be people who want to turn their backs on progress—*ew*, gross, you're using *technology*—but inefficient systems will always have high greenhouse gas outputs, and technology can help."

A few weeks later, Van Eenennaam emailed me that Cow #3113, named Elle by nerds in her lab who translated 3113 into upside-down calculator-ese, was pregnant. She attached a sonogram, adding: "IT'S A BOY!!!!!"

Obviously.

It was just happenstance that I saw two techno-fixes for livestock yields in the same lot, concocted by yin-and-yang scientists from opposite ends of the earth. But all kinds of fixes are in the works—tunnel-ventilated barns with Pentagon-level biosecurity, Fitbit-style health monitors for animals, virtual fencing for pastures, AI-driven robots that can vaccinate chickens and recognize sounds of distress, a hog-farm-monitoring software system called PigBrother, even "in ovo sexing" that helps egg farmers remove male chicks before they hatch instead of tossing them into meat grinders. Historically, the human race has been much better at inventing new stuff than changing how it does old stuff. It's especially good at inventing efficient ways to produce food, because higher yields usually mean higher profits for farmers, ranchers, and agribusinesses.

"The stars ought to be aligned, because the big guys like Cargill and JBS believe in efficiency," Searchinger said. "If they have an ideology, that's it!"

Still, innovations don't always change the world as quickly as innovators want.

Kebreab's seaweed has not yet been approved as a feed additive, and while dozens of countries have approved a chemical additive called Bovaer that can reduce methane by a third, it was fed to only 0.01 percent of the world's cattle in 2023. And Van Eenennaam's CRISPR cattle seem likely to remain at 0.00 percent for a while. Elle gave birth to a bull named Cosmo, and his sperm was used to create more Boys Only embryos, but U.S. regulators wouldn't let Cosmo enter the food supply without years of safety data, so Van Eenennaam had to have him incinerated.

The sad 35-year saga of AquaBounty salmon, the first genetically modified animals approved for human consumption, is more evidence of the obstacles reengineered animal protein will face before it can move the climate needle. In 1989, researchers successfully modified an Atlantic salmon to make it grow faster, and its descendants reached harvest weight twice as fast as regular salmon with one-fourth less feed. But the FDA didn't approve them for sale until 2015. And even though they had grown for decades in secure tanks on an inland farm in Indiana, anti-GMO groups persuaded a judge to order another multiyear review of their danger to wild salmon if they somehow escaped. AquaBounty eventually got its permits, but it never got much traction, as anti-Frankenfish activists continued to denounce its salmon as biotech poison. In December 2024, the company ran out of cash and shut down its operations.

Still, aquaculture seems ripe for disruption with or without genetic hijinks. The global supply of farmed fish has soared while the wild catch has flatlined—there are only so many fish in the sea—and environmental regulators are cracking down on near-shore "net pens" that pollute coastal waters. As seafood demand rises, indoor farms are emerging as the best hope to supply it. And since most fish are more efficient than land animals at converting feed into food—they don't burn much energy keeping warm (because they're cold-blooded) or growing bones (because the water supports their weight)—getting more protein from salmon and tilapia rather than beef and lamb could help us eat much less of the earth. We currently eat about one-third more meat than seafood, but if that could be reversed without emptying the oceans, it could ease a lot of pressure on forests.

The Norwegian firm Atlantic Sapphire has launched the most ambitious effort to attack the eating-the-earth problem with fish, a "Bluehouse" designed to grow half the current U.S. salmon diet on just 160 acres of former farmland. That's a lot of lox. And it's not where you'd expect it.

• • •

In the early '90s in the Norwegian harbor city of Ålesund, 15-year-old Johan Andreassen and his cousin started fishing the local fjords for wrasse, which they sold to Norway's aquaculture industry to use instead of pesticides to get rid of parasites like sea lice. They gradually built their two-teens-and-a-boat hustle into an organic salmon farming operation near the Arctic Circle, becoming the exclusive supplier to Whole Foods. But they always knew that even pesticide-free fish farms fouled pristine estuaries. And it always felt environmentally absurd to grow perishable food in frigid bays at the ends of the earth, then fly it to the United States to truck it "fresh" to market.

"We were trying to do it as sustainably as we could, but it didn't make sense," Andreassen said. "Nobody puts beef or pork or chicken on a plane."

So they sold the company and began scouting the U.S. for locations to build a more efficient farm closer to American customers. They decided to re-create the salmon's entire life cycle in indoor tanks, not only the freshwater early stage but the saltwater "grow-out" stage. They envisioned an artificial environment with digitally controlled temperatures, water chemistries, and currents the fish could swim against. Then they could churn out omega-3-rich protein without antibiotics or pesticides; without exposure to diseases, parasites, or microplastics; without damage to oceans or estuaries; and without climate-killing jet travel.

The question was where. Andreassen investigated a dozen states before learning from a random YouTube video that the limestone underneath South Florida was like an ancient layer cake with everything the Bluehouse needed—a freshwater aquifer near the surface, a salty aquifer below, and a boulder zone where it could inject wastewater further down. Atlantic salmon are cold-water fish that had never been found anywhere near Florida, but Andreassen decided to build the world's largest inland fish farm on a tomato field at the edge of the Everglades.

"Who would've thought hot and humid Florida would be the perfect place to grow salmon?" said Andreassen, who is now a tall blond in his mid-forties, with the rugged cheekbones of a Nordic movie star.

I went to see the Bluehouse during the COVID spring of 2020, when its first building was almost done and the first few million Florida-born salmon were swimming in its tanks. Its footprint was already the size of the Miami Heat's arena, and the company expected it to grow 20 times larger. Eventually, it was supposed to produce one billion annual salmon fillets.

"You think of potatoes from Idaho, lobsters from Maine, now you'll think of fresh salmon from Florida!" an Atlantic Sapphire executive gushed when I arrived.

I couldn't decide if the plan sounded awesome or bonkers. It called for a $2 billion complex the size of the Mall of America, including an oxygen plant, a breeding facility, a factory to recycle bones and guts into supplements and pet food, and four million square feet of cylindrical tanks connected by hundreds of miles of pipes. When I visited, the Bluehouse was a frenetic construction zone. Some workers were scrambling to finish the roof above the Olympic-pool-sized grow-out tanks, where the first batch of adult salmon had just arrived. Others were just starting to build a fully automated processing facility—nothing like the blood-and-guts operation where Searchinger worked in Alaska—where full-grown salmon would soon be stunned, gutted, and filleted.

There's no Columbus method for this kind of farming. It requires intensive 24/7 oversight, from a control room that looks like a nuclear reactor's. Thue Holm, a grizzly-bearded Dane who was Atlantic Sapphire's chief technology officer, explained that if you've ever owned an aquarium, you know how hard it is to keep a few fish alive indoors. You've got to maintain the filter, clean the tank, and feed the fish properly—and you still sometimes find them floating upside down.

"These are the same kinds of challenges," Holm said, "just harder."

A lot harder. Every drop of Bluehouse water must be recycled every half hour. Precise levels of oxygen, carbon dioxide, and other gases must be maintained at all times. Andreassen compared it to running a giant cruise ship with millions of passengers, where "every system has to run perfectly or else the passengers die." He wasn't exaggerating. Two months earlier, excessive nitrogen in a tank at the company's pilot plant in Denmark had wiped out 200,000 salmon overnight.

"That's what keeps me awake at night, trying to make sure we don't fail the fish," he said.

Still, Andreassen assured me the first batch of Florida-born salmon that had grown from tadpole-like hatchlings to silvery adults were thriving: "The fish have ways of talking to us, and these are happy fish." I couldn't evaluate their moods, and some were crammed even closer together than factory-farmed hogs as they swam in circles against the artificial current. A few two-foot-long adults leaped out of the grow-out tanks, as if making a break for freedom, into nets that bounced them back into the water. But in public filings, Atlantic Sapphire was reporting low mortality rates, and the fish did seem to be eating well. I watched hundreds of juveniles that looked like anchovies line up below a horizontal bar that dispensed their feed, then swarm to the surface the instant it hit the water. Needless to say, they weren't bellering.

Investors didn't think Atlantic Sapphire was bonkers. Its valuation was nearly $1 billion on the Oslo exchange. Its confident founders who looked like Vikings and talked like tech bros had a nice slogan—"Blue Is the New Green"—and a nice eco-story about saving an industry that was running out of wild fish and clean estuaries. They claimed the Bluehouse was already producing the world's most efficient protein, using as little as 1.05 pounds of feed per pound of fillet, making about 24 times as many tons of food per acre as the tomato fields it replaced. And they expected to keep improving yields, because they wouldn't have to breed for disease resistance indoors, just rapid growth and even better feed conversion.

"We can give the fish ideal conditions, and then we can breed the future generations for those conditions," Holm said. "They're only going to get better."

Except they didn't.

A month after my visit, Atlantic Sapphire had to harvest 200,000 ailing Bluehouse fish who got stressed by all the construction commotion. Then the Bluehouse lost another 500,000 salmon to filtration problems. There were more die-offs at the Danish pilot plant, then a total die-off when it mysteriously burned down. And once the Bluehouse's filtration issues were resolved, problems chilling Florida water to Arctic temperatures dragged its production down to 700 tons in 2023, just 7 percent of its current capacity, a mere 0.3 percent of its eventual goal.

Not only did Atlantic Sapphire fail its fish, it burned through $650 million from investors. Its stock price declined 98 percent, and Andreassen was ousted at the end of 2023. One industry insider compared him to a mechanic who convinced the world he could build the most advanced Formula One race car in history while he was driving it.

"It all looked wonderful on Excel," the source said. "But then they had to make food out of live animals. That's really hard!"

Reconfiguring nature to make animal protein more efficiently really is hard. Turbocharging crop yields might be even harder.

TEN
MORE CROPS, LESS LAND

BIGGER TRACTORS, BIGGER FIELDS, BIGGER YIELDS

The Grand Prairie of east-central Illinois was once a panoramic green expanse of marshy grasslands, bedazzled with a colorful profusion of wildflowers that shimmied in the breeze. Many of the 19th-century pioneers who encountered it on their way west to tame the American wilderness were awestruck. "Life, in all its fresh and beautiful forms, was leaping forth in wild and sportive luxuriance at my feet. But all was vast, measureless, Titanic; and the loveliness of the picture was lost in its grandeur," one wrote. Others who slogged through it were awed in a different way, denouncing the region as a waterlogged wasteland unfit for human settlement. Some became delirious from swamp fevers, or lost so much blood to mosquitoes they passed out.

But the settlers who put down roots in the Grand Prairie after the Civil War knew how to get rid of water. They transformed its wetlands into fertile farmland, driving tile drains into the mud by hand, tilling its black soils with moldboard plows. They destroyed an ecosystem and created a breadbasket. They replaced the wild prairie celebrated by Aldo Leopold with the pastoral farms celebrated by Michael Pollan.

Now those are gone, too. Today, the Grand Prairie is still home to some of the American Corn Belt's most productive land, but those bog-tough homesteaders who wrestled it away from Mother Nature would barely recognize it.

Dirk Rice still grows corn and soybeans on the 200 acres of the Illinois prairie his great-great-grandfather settled in 1881. But it's now less than one-tenth of his land; the only farmers he knows with only 200 acres work full-time jobs in town. He installs drains with a laser-guided tile plow. He works his fields with a 320-horsepower tractor equipped with precision technology to make sure he only applies chemical fertilizer and pesticides where they're needed.

"Back in my great-great-grandfather's day, men were men and horsepower was horse power," Rice said. "Gotta say, though, we get better yields!"

The story of the Corn Belt, and of croplands in much of the world, is another story of steadily increasing scale and efficiency toward megayielding mega-farms. It's another sustainable intensification story of better technology, better breeding, and big farms getting bigger by swallowing less productive ones. Since the Green Revolution began, America has lost half its farms and a quarter of its farmland, but its output has tripled, with 78 percent of its farm sales from the largest 6 percent of its farms. Globally, crops now cover an area the size of South America, but without the land-sparing yield increases of the Green Revolution, they'd cover three South Americas.

In one of his barns, Rice showed me his father's tricycle-red Farmall 400 tractor, a technological marvel from before the revolution. It was the size of a Kia Soul. Then he showed me his John Deere combine, which weighed as much as 10 Kia Souls. It looked like a Zamboni on steroids, with a touch-screen yield monitor and a grain cart that held more corn than a semitruck. Its tires were six feet tall.

"My grandfather ruined his shoulder shucking corn," Rice said. "This thing picks corn, strips it, sorts it, weighs it, and measures the moisture content of its kernels. And this is probably the smallest one John Deere makes."

Rice now grows 36,000 cornstalks per acre, twice as much as his dad did in 1958. That's partly because his plants are bred to grow closer together, partly because his planter drops seeds in exactly the right place; regular GPS is only precise within a few inches, but John Deere's "real-time kinematic GPS," or RTK, boasts sub-inch accuracy. Rice's corn is also genetically modified to kill rootworm beetles that required chemical warfare when he was young.

"We had to dump pounds and pounds of stuff with skulls and crossbones that killed everything around," he said. "Man, you didn't want that stuff on you."

Rice makes a conscious effort to limit his environmental impact and enhance soil health. He's mostly no-till, which is common in the Midwest, and plants rye as a cover crop to stabilize his soil, which is less common. He even integrates cattle that eat the rye and fertilize his cash crops, which is extremely uncommon. His precision tech has improved his nitrogen-use efficiency by a third, reducing his contributions to the dead zone in the Gulf. He's also started rotating in a little winter wheat, which practically counts as regenerative crop diversity in a region where planting decisions are usually either corn-then-soy or soy-then-corn.

But Rice's top priorities are yield, yield, and yield. He's up to 220 bushels of corn per acre, five times what his grandfather reaped with a Farmall 400 and a bum shoulder. His farm is as sophisticated a manufacturing operation as the Magnum Feedyard or DeSnayer Dairy—and like the Van Der Broeks at Fazenda Tropical, he takes pride in making the most of his land.

"We feel a responsibility to feed the world," he said. "You can't do it on a little organic farm with no herbicides."

Even Big Ag's yields will need to rise dramatically to feed the world of 2050, which won't be easy for efficient growers like Rice. He can't just use more fertilizer, since he already uses more than his crops can absorb, or spray more pesticides and herbicides, since pests and weeds are already developing resistance to his chemicals. He's also afraid climate change will hurt his yields, though it's a touchy subject in local diners. Rainstorms that used to be one-in-five-year events now flood his fields almost every planting season.

"We can argue how much man is contributing, but it's happening," he said.

In much of the world, climate-driven heat waves and droughts are an even worse threat to yields—not only because they kill crops, but because they dry out already overtapped sources of irrigation water like the Colorado River and the aquifers underneath California. By 2040, groundwater depletion is expected to fallow half a million acres of farmland in the southern Central Valley alone, which definitely won't help feed the world. In large swaths of sub-Saharan Africa where parched soils already produce meager yields, more intense drought and desertification could lead to large-scale abandonment of agricultural lands.

Rice's corn yields are double the global average, and six times the average in Africa, where most farmers don't even have fertilizer, much less RTK tractors that pinpoint where to apply fertilizer. The most obvious way to boost crop yields, like livestock yields, would be to distribute Green Revolution technologies to less-efficient growers abroad. Searchinger and his team calculated that sub-Saharan Africa is now on track to convert every acre of its non-agricultural land to agriculture by 2050; its land-use emissions alone would push the world past 2 degrees warming, even if all other emissions stop everywhere else on Earth. But if it can triple its yields, it has a chance to stabilize its agricultural footprint—and for poor smallholders, a little irrigation water, a "microdose" of fertilizer, or even access to weather data on a phone can spell the difference between starvation and subsistence.

Rice could also nudge his own yields even higher, as his bigger competitors already have. I drove an hour from his spread to visit Doug Schroeder, who farms twice as much ground and averages 25 percent higher yields. His machines were all a step up in size, horsepower, and analytical power. He couldn't explain exactly why he shelled out half a million bucks for a 36-row planter and a 500-horsepower RTK combine, but his corn was as high as the proverbial elephant's eye.

"I feel stupid paying for all that stuff—and then I see the difference in the field," Schroeder said. "When you're sitting in your combine, with the yield monitor ringing like a Vegas slot machine, you're glad you wrote the check."

Farming is an unpredictable business, but as weather gets funkier, modern technology that harnesses Big Data and artificial intelligence can give farmers more information they can use to manage risk. I got a glimpse of the future of precision ag far from the Midwest, in a Tel Aviv skyscraper.

On the 16th floor, I visited a crop analytics outfit called Prospera Technologies, whose engineers called themselves the Data Wizards in Muddy Boots. It looked like a typical tech startup—pool table, twentysomethings in T-shirts, desks with rainbow flags—except it had soil monitors and water gauges lying among its extra printers. When I asked one of the data wizards about a poster in a futuristic font that read: "If It Grows, We Watch It," he showed me an incoming text from a sensor attached to a pivot irrigation system on a Nebraska farm. It was a photo of an aphid, one-sixteenth of an inch long, chewing a soybean leaf, with GPS coordinates showing the farmer on the other side of the world where to spray to avoid an infestation. Often, Prospera's algorithm decides to spray without consulting the farmer.

"Humans have all kinds of biases," said Prospera founder Daniel Koppel, a data scientist who served in an Israeli army cyberintelligence unit renowned for spawning tech entrepreneurs. "The machines can make better decisions about what a particular crop needs at a particular time, and they're only going to get smarter."

The agribusiness Valmont bought Prospera for $300 million in the ag-tech boom of 2021, and dozens of other precision ag startups hope to follow its path, touting proprietary data and AI platforms to help farmers do more with less. The scientists of CGIAR, the global consortium for agricultural research, also created the nonprofit Platform for Big Data in Agriculture, leveraging satellite imagery and other available data to help farmers who can't afford venture-funded products make better decisions. In modern agriculture, CGIAR says, information is power: "power to predict, prescribe, and produce more food more sustainably."

But information is not magic, and data is a competitive business. In 2023, Valmont wrote off half its investment in Prospera after disappointing sales.

For most of the last decade, the queen bee of the crop analytics world was Gro Intelligence founder and CEO Sara Menker, a charismatic, TED Talking, Ethiopian-born former commodities trader who left Wall Street to create the world's largest agricultural data platform. Menker pitched Gro's

AI-driven models as indispensable tools for corporations and governments that could boost yields and avert the kind of famines that plagued her native land. She raised $125 million at a near-unicorn valuation; opened offices in New York, Nairobi, and Singapore; packed her board with bigwigs like former Goldman Sachs president and Trump adviser Gary Cohn; and made the Time 100 list of influential people. Then in February 2024, Gro suddenly failed to make payroll, and soon collapsed. One burned investor told me it wasn't Theranos, because it had a useful product, but it rhymed; Menker generated worshipful media coverage as she hobnobbed around the world charming Davos types, but apparently forgot to generate revenues.

"My God, all the engineers she hired, we could've made more money if she had them sell Girl Scout cookies," the investor said.

The truth is, for rich-world farmers who already manufacture grain at industrial scale, the march of precision technology only provides incremental yield benefits. Their operations were already intelligent and data driven without Gro Intelligence data. And when droughts, floods, and other disasters strike, they can still lose entire harvests just like their low-tech counterparts, though U.S. crop insurance eases the financial blow.

This is the problem of farming in the great outdoors, and it's getting more problematic. That's why a new wave of entrepreneurs are trying to solve that problem, and the eating-the-earth problem, by moving farming inside.

The industrial ruins of downtown Newark were a strange location for an agricultural revolution. But the ag-tech startup AeroFarms built the world's largest vertical farm there in 2016, on the site of an abandoned steel mill. It looked like a warehouse the size of the White House designed by a closet-organizing company, lined with neat rows of rectangular "grow tables" stacked three stories high. Each table was covered in high-end leafy greens that had never seen the sun. The goal, CEO David Rosenberg told me, was to create a new agricultural paradigm, liberating farming from the inconveniences of weather, winter, and nighttime.

AeroFarms grew kale, arugula, and other greens without pesticides or herbicides, using 99 percent less land and 95 percent less water than its outdoor competitors in California and Arizona. It released no runoff into the environment. It used no soil or sunlight, misting its plants with nutrients and bathing them in LED light all day and night. Its greens weren't exposed to bird droppings or other contaminants, so consumers didn't even have to wash them. And they grew so fast they were ready for harvest every two weeks, instead of a few times a year.

This was the ultimate factory farm, substituting robots and automation for backbreaking labor. It deployed AI and thousands of surveillance sensors to create algorithmically optimized nutrition, temperature, and lighting, eliminating risks farmers had agonized about for 12,000 years. Just as high-rise buildings rein in urban sprawl, high-rise farms could rein in agricultural sprawl, freeing up conventional farmland for rewilding. By going up rather than out, AeroFarms could produce 400 times as many greens per acre as a conventional farm.

"The future," Rosenberg said, "is happening a lot faster than we expected."

When we spoke in 2021, Rosenberg was about to take AeroFarms public at a $1.2 billion valuation. That was a lot of lettuce for a firm that had just started to sell salad. But with fake meat entering its trough of disillusionment, indoor farming looked like a better bet on a future of agricultural land constraints and weird weather. Vertical-farming unicorns like San Francisco–based Plenty, Berlin-based Infarm, and Manhattan-based Bowery Farming were attracting investors like Walmart, Google, and Bayer as well as Natalie Portman and Justin Timberlake.

Cash was also pouring into multiacre horizontal greenhouses outside cities that used sunlight as well as the new indoor-farming tech, using more land than vertical farms but way less than outdoor farms. The leader was AppHarvest, a Kentucky startup that supplanted Indigo Ag as the ag-tech valuation king at $3.7 billion. It was heralded as the economic engine that would replace coal in Appalachia, with a board that included *Hillbilly Elegy* author and future Vice President J. D. Vance as well as Martha Stewart. Its slogan was "Building a Movement from the Ground Up."

These "controlled environment agriculture" startups were all proving their concepts with high-end salad greens. Some were also experimenting with tomatoes and berries. But their elevator pitches were more about tech than food or ag.

"The phone in your pocket has more computing power than the entire Apollo mission," Plenty said on its website. "Isn't it time farming caught up?"

Turns out, growing crops indoors, like growing fish indoors, or for that matter growing just about anything just about anywhere, is also really hard.

AeroFarms never did go public, and in 2023, it went bankrupt. The entire vertical farm boom busted; Bowery failed after raising nearly $500 million, Infarm failed after raising more than $600 million, and Plenty filed for bankruptcy after raising nearly $1 billion. "Investors went wild for the utopian promise of tech-driven vertical farming," one analyst wrote. "Now the industry is on the brink of collapse." The greenhouse business also staggered into a trough. AppHarvest, the unicorn that was supposed to build a movement and replace coal in Appalachia, went bankrupt, too.

Vertical farming's most obvious problem is that it's a ludicrous energy hog. While LEDs have gotten much cheaper, they're still much costlier than sunlight. I did some back-of-the-envelope math with a disillusioned investor, and we calculated that farms like his would require every megawatt of America's renewable electricity to grow 5 percent of America's tomatoes.

Greenhouses use less energy, at least when the sun is shining. But not even zero-energy indoor salad-green farms could solve the problems of global cropland, because they're not salad-green problems. They're not tomato, berry, or marijuana problems, either. Almost all the world's cropland grows grain, vegetable oil, and other staples. Less than 3 percent grows fruits or vegetables. The high-priced greens in indoor farms have about as much to do with feeding the world as the $3 million Lamborghini Sián has to do with transporting the world.

Ted Caplow cofounded BrightFarms, an entrepreneurial success story that raised $200 million and built six multiacre greenhouses, but he was never satisfied with its limited ambitions. "We were solving the lettuce problem. I thought we might solve the strawberry problem," said Caplow, who was ousted well before the firm was sold in 2023. "But we were never going to solve the food problem."

At the peak of the hype cycle, David Rosenberg told me he agreed that indoor farming remained on the fringes of the food system, and that it might do more short-term good helping to improve outdoor farming. His Newark operation was collecting more data about how plants grew than any other farm in history, and he had a deal with Cargill to deploy what it learned to help cacao farmers in the developing world improve their yields. "We're unlocking the mysteries of plant biology," he said. "We image every plant every day, so we can see when they're spotting and ripping and curling, and we can figure out why." In other words, AeroFarms wasn't just running a farm. It was also running a lab.

Its problem was, it was still running a farm. Aerofarms has emerged from bankruptcy under a new CEO, but it had to shut down its unprofitable Newark operation. Like the alt-protein startups that pitched themselves as tech plays, it was dealing with molecules, not electrons, making products that were physical, not digital. They had to be manufactured and kept fresh and transported to market. It was neat that AeroFarms could tweak its algorithms to make its arugula spicier, but it still hadn't figured out how to make its arugula profitable. It hadn't even solved the lettuce problem.

In fairness, solving the lettuce problem would be an admirable achievement. It could free up 300,000 acres of U.S. farmland, an area the size of New York City. It could ease California's water shortages and end the wasteful cross-country trucking of waterlogged greens. Solving the strawberry problem could save another 50,000 acres, an area the size of Washington, D.C. As Searchinger says, no acre should be left behind.

Still, Caplow was right: The food problem won't be solved indoors. "Controlled environment agriculture" sounds like the perfect antidote to the uncontrolled environments where farmers ply their trade, especially as land gets scarce, aquifers dry up, pests and crop diseases expand their range, and governments get serious about regulating farm emissions and pollution. For the foreseeable future, though, the overwhelming majority of our food will be grown on the outdoor farms that cover two-fifths of our land, use nearly three-fourths of our fresh water, emit one-fourth of our greenhouse gases, and are the leading threat to terrestrial ecosystems, water bodies, and endangered species.

We've just got to find ways to grow it better. In the 20th century, that meant better chemistry. In the 21st century, it often means synthetic biology.

A GREENER REVOLUTION

Colorado potato beetles are eating machines that look like orange-and-black jelly beans. They're among the world's most voracious pests. When they invade a field, they often wipe out the entire crop, then burrow underground until spring, when they rise again like gluttonous phoenixes to destroy the next crop. They're very hard to kill and very prolific baby makers, which helps them develop rapid resistance to the chemicals farmers spray on them. One scientist told me they combine the resilience of an M1 tank with the appetite of Cookie Monster.

In the spring of 2021, I watched thousands of these ravenous bugs defoliating several rows of potato plants at a University of Wisconsin research farm, leaving little behind but sad-looking stems. An ecologist named Mark Singleton explained to me that they would soon lay eggs, and their larvae would be even hungrier. Local potato farmers were fighting the beetle with a cocktail of a half dozen chemical pesticides, but it kept returning for annual feasts.

"It's the Terminator," said Singleton, chief commercial officer for the Boston-area startup GreenLight Biosciences. "It won't stop until you stop it, and even then it just keeps coming."

But several other rows in the field were lined with normal potato plants, bushy and leafy with white flowers, sending energy to their tubers underground. That's because they were sprayed with a GreenLight biopesticide that sounded like it had been sent from the future to terminate the Terminator. It did its killing with RNA cellular messaging technology similar to the active ingredient in the COVID vaccines, except its messages blocked the genetic impulses that prompted the beetles to move their bowels. So they couldn't poop, then they couldn't eat, then they stewed in their own toxins until they dropped dead.

Constipation is an unusual murder weapon, but GreenLight's beetle juice, Calantha, was as effective at protecting yields as conventional pesticides—and it didn't kill aphid-eating ladybugs, soil-enriching earthworms, or other beneficial organisms. It was like the difference between firebombing a city and sending in an assassin who only takes out the bad guys. A spoonful of Calantha can protect a football field's worth of potatoes, and unlike chemical bug killers that can ravage soil organisms, decimate bird populations, and poison farmworkers, no trace of it remains in the environment a few days after it's applied.

GreenLight executives told me they expected EPA approval for Calantha by the end of the year. Since U.S. potato farmers used an area larger than Rhode Island, and globally, potatoes covered an area larger than Florida, it looked like a nice fix for another little slice of the eating-the-earth problem.

But that wasn't why Greenlight had raised $400 million, or why it was going public at a unicorn valuation. Calantha was, well, small potatoes. It was proof of concept for a biomanufacturing platform designed to mass-produce RNA for a dollar per gram, when Big Pharma paid thousands per gram. Even though GreenLight manufactured molecules, like fake meat and indoor farming startups, it was a more legitimate tech company; RNA is like software carrying information and instructions to cells, so it really was trying to create a biological Intel Inside. I knew it was playing for high stakes when it got the Washington power broker and A-list hostess Juleanna Glover, a former Elon Musk and Dick Cheney adviser renowned for connecting the tech and political worlds, to schlep to Wisconsin with me to watch beetles. It felt like hiring Beyoncé to sing at my daughter's birthday.

The special sauce that made GreenLight's manufacturing so much cheaper than Pharma's involved scavenging the building blocks of RNA from dead cells instead of manipulating living cells into making them. Its pilot plant in a former Kodak factory in Rochester could already produce as much RNA in a few weeks as Pfizer and Moderna had produced for hundreds of millions of vaccines in a year.

GreenLight founder and CEO Andrey Zarur, a biomedical engineer trained at MIT and Harvard, joined Pharma in the 1980s, when its focus began shifting from chemical drugs to genetically engineered "biologics" that were more targeted and had fewer side effects. But he always hated their exorbitant prices, and he suspected synthetic biology could save more lives with cheaper agricultural products.

"We lose enough crops to pests and disease to feed a billion people.

We spray too many chemicals and take too much land away from Mother Nature," he said. "I figured, why not work on that?"

GreenLight was working on RNA solutions for all kinds of pests and pathogens. It had also developed hacks targeting the *Varroa destructor* mite, a pinhead-sized parasite that was wreaking havoc on the honeybees that pollinated many of the world's crops; a mildew that damaged vineyards; a fungus that destroyed wheat kernels; and *Botrytis*, the ubiquitous gray mold that beards berries and tomatoes. In field tests, the biopesticides were all taking out their targets without collateral damage to other species or the environment, a tantalizing upgrade on the blunt force of chemistry. GreenLight was applying its RNA tech to human health, too, running clinical trials for a COVID vaccine in Africa and studying a new therapy for sickle cell anemia.

"They're like planes on the runway, waiting to take off," Zarur told me. "The potato beetle solution is just waiting for the tower to say go. The others are right behind. Once we get them in the air, we start changing the planet."

But the tower didn't say go that year. Or the next year. Or almost all of the year after that.

By the time EPA approved Calantha, just before Christmas in 2023, GreenLight had run out of runway. After going public at a $1.2 billion valuation, its stock price plunged 98 percent, so its investors took it private with a $45 million bailout. It had to shed two-thirds of its staff and halt its work on human health. "We got clobbered," said GreenLight marketing chief Catie Lee. She blamed Calantha's holdup on backlogs in EPA's biopesticide division, which had more new products to review than the office for conventional pesticides—an encouraging sign of the times, but a potential source of additional delays for GreenLight's other products.

Synthetic biology solutions like RNA pesticides and Pivot Bio's microbial fertilizer feel like the future of crop protection and crop amendments, but it will take time to develop and scale them. The genetic revolution could also drive the future of the crops themselves—maybe sooner, maybe not.

Consumers everywhere fear genetically modified organisms. In global surveys, half say GMO food is dangerous, while only one-sixth agree with the scientific consensus that it's safe. That's why U.S. supermarkets sell "non-GMO" oranges and salt, even though there's no such thing as GMO oranges or salt, and why GMO crops are banned in most of Europe, even though they're imported to feed European livestock. Jimmy Kimmel did a fun segment asking shoppers at a farmers market what they thought of GMOs (definitely bad) and whether they knew what GMOs were (definitely not). Reason does not drive our dietary choices.

That's also why Golden Rice, genetically modified to address the vitamin A deficiencies rampant in the Global South, remains commercially unavailable more than two decades after *Time* announced on its cover: "This Rice Could Save a Million Kids a Year." It's been stymied by a political and regulatory logjam even worse than the one that held up AquaBounty's GMO salmon; anti-GMO activists who considered the scourge of agricultural technology worse than the scourge of childhood disease once destroyed a test field of Golden Rice in the Philippines.

Even GMOs that have made it to market haven't always produced the wondrous results their supporters expected. But there are high hopes that next-gen GMO crops like blight-resistant potatoes, insect-resistant cowpeas, and drought-tolerant grains can protect yields as the climate changes, chemicals lose punch, and irrigation water gets scarce. There's also a new wave of "molecular farming" ventures modifying crops to grow animal proteins without animal agriculture—alt-alt-proteins, perhaps. Nobell Foods, the woman-run long shot seeded by New Crop Capital in 2015, has engineered soybean plants to grow casein, the dairy protein that makes cheese stretch and melt. It doesn't have regulatory approval yet, though, and no one knows how pizza lovers will react to modified mozzarella.

Anything with a GMO label will always face complaints that it's unnatural. But as UC Davis plant geneticist Pam Ronald points out, almost everything we eat has been genetically altered in some way. Ronald helped develop a flood-tolerant rice variety that's now being planted on millions of acres in South Asia, where poor farmers lose a quarter of their harvests to flooding. Her question is: Why *wouldn't* we use the most promising genetic techniques to grow more food in tough conditions?

"The challenge for agriculture is to feed more people with less damage to the environment," Ronald says. "This can help!"

The most promising new technique is gene editing, which also gets more regulatory leeway because it doesn't transfer genetic material between species. (Flood-tolerant rice isn't "transgenic," either, which smoothed its path to the field.) Agriculture's future may depend on startups like Yield10 Bioscience, which is using CRISPR to boost the yields of a new oilseed crop called camelina. CEO Oliver Peoples, a synthetic biology pioneer from MIT, believes grain farmers will be able to plant fast-growing Yield10 camelina as a winter cover crop that also functions as a cash crop, preventing erosion and enhancing biodiversity while also producing food, feed, and fuel on land that previously lay fallow six months a year.

"It's a free lunch," Peoples said. "This could be like bringing 100 million acres of new land into production without harming nature. Plus it's got great benefits for soil health and sustainability. It's win-win-win!"

Yield10 has not yet done so much winning that anyone is tired of winning. In 2023, U.S. farmers grew just 1,000 acres of its high-yield camelina, while Chilean farmers grew another 50 acres the company genetically modified to produce omega-3 oils for fish feed. That's not 100 million.

Still, it took less than two decades for GMO crops to go from zero to 90 percent of America's corn, soy, and cotton. And there are a slew of CRISPR approaches in development, promoting not only yield but resistance (to pests, fungi, or diseases), tolerance (of extreme temperatures, flood, or chemical inputs), and a variety of other traits, from gluten-free to low-allergen to long shelf life.

The most mind-blowing effort to engineer high-yield crops is aiming to edit out the inefficiencies of a process that's been underperforming for two and a half billion years. Yes, for more than a decade, scientists at the University of Illinois and affiliates around the world have been trying to hack photosynthesis, even though it's done a pretty solid job of sustaining life on Earth since the Archean Eon.

We wouldn't be here if photosynthesis didn't transform sunlight, water, and carbon dioxide into the sugars plants need to grow and the oxygen we need to breathe. But as Searchinger says, the algae that invented it were not great thinkers, which is why plants convert such a tiny portion of the sunlight that reaches them into energy. The key enzyme in photosynthesis, Rubisco, is hilariously inept at capturing carbon dioxide. It often inadvertently grabs oxygen instead, producing a toxic compound that mucks up the process. Photosynthesis also sputters whenever clouds pass overhead, and it's slow to get back in gear when the sun reappears.

The Gates Foundation and other funders have poured $120 million into the RIPE program, short for Realizing Increased Photosynthetic Efficiency, to try to make photosynthesis better at what it does. It's the brainchild of director Steve Long, the British plant biologist whose confident ignorance about biofuels and deforestation once gave Searchinger the impression he was a moron. He wasn't. He was the first scientist to recognize that the twin revolutions in supercomputing and genetic engineering could be used to troubleshoot a process that had coasted on evolutionary autopilot since life began.

His team began by using a supercomputer to create a digital simulation of all 170 elements of the photosynthetic process, an almost impossibly complex model that required millions of calculations just to simulate a leaf rustling in a breeze. (A NASA contractor who helped debug it told Long the work was actually harder than rocket science.) Then the team used its super-model to flag bottlenecks in the process and specific genes that could be tweaked to relieve them. Then it tested the tweaks on tobacco plants, the easy-to-manipulate lab rats of agronomy, and eventually on food crops in greenhouses. The final step was to transplant them to test fields in Champaign, where I got to see the world's first potato plants engineered to photosynthesize more efficiently.

"We didn't know if any of it would work," Long said. "We just did the science, and yes, some of it works!"

For example, RIPE's model identified three proteins that helped plants adjust to cloud cover and other abrupt fluctuations in light, the way your eyes adjust when you come in and out of the sun. When the team inserted extra copies of those proteins into soybeans, yields increased 20 percent. The model also suggested that another enzyme called SBPase ought to be more abundant during photosynthesis, and adding extra genes coding for that enzyme accelerated plant growth as well. At first, Long and his colleagues were puzzled why nature had undersupplied SBPase, until they figured out it was because plants had evolved in an atmosphere with less carbon dioxide. They now suspect crops will need even more of the enzyme to maximize their growth as atmospheric carbon levels keep rising.

The haplessness of Rubisco, the world's most abundant protein, is still the most glaring glitch in photosynthesis, and RIPE's scientists have not yet trained it to stop grabbing oxygen by mistake. But they did devise a workaround that streamlines how plants recover from the mistake, and the reengineered potatoes I saw ended up with another 20 percent yield gain. Overall, RIPE's first three tweaks to photosynthesis could conceivably raise global yields as much as 50 percent, if they're demonstrated to work in other crops and eventually distributed worldwide.

"Conceivably" and "eventually" are uncomfortable words for a planet in crisis. Science doesn't march forward in a straight line, and neither does politics. RIPE plant biologist Amanda Cavanagh, a Rubisco expert at the University of Essex, told me there's nothing scientists can do about that except keep working.

"We can't expect a miracle every day," Cavanagh said.

Sometimes, though, miracles happen.

THE SUPER-TREE

Naveen Sikka's father was an engineer. His mother was a doctor. They lived in the New Jersey suburbs, where he finished second in his high school class and was named the state's Mr. Future Business Leader. His "super-classic child-of-Indian-immigrants upbringing," as he described it, would not have been complete without trips to New Delhi every summer to visit family. He always felt like he was stepping back in time, to an era before air-conditioning and reliable electricity. India opened his eyes to the world outside his bubble.

"I was living this great American life, but India kept me grounded," he said.

He majored in French and political science at Columbia—he didn't want to be the super-classic child-of-Indian-immigrants math-and-science guy—and spent a formative summer interning at the U.S. embassy in Togo. The poverty he saw among Africans living in mud huts and farming tiny patches of dry dirt was even worse than India's. It got him thinking about inequality, agriculture, and using his Mr. Future Business Leader skills to build a more sustainable world. He went to business school at Berkeley, where he ran the energy club at that time when BP was launching its research institute and biofuels fever was sweeping the campus.

"There was so much fucking money, and people were following the money," he recalled.

After graduation, he got a call from former Democratic National Committee chairman Joe Andrew, who had met him on the Obama campaign, and remembered his interest in biofuels. Andrew asked him to take a quick trip to rural India to check out an oilseed tree that Goldman Sachs was looking at for a biofuels deal. It was called pongamia, and when Sikka first saw it growing in the arid badlands eight hours outside Hyderabad, he was blown away.

It looked like an ordinary orchard tree, with bulging seedpods and pink-and-white flowers, except it was thriving in a rocky, barren, quasi-desert landscape. In photos from that trip, Sikka looks amusingly out of place taking notes in a blue dress shirt and khakis, like a management consultant transported into a scene from *Dune*, but not as out of place as the pongamia sprouting out of the dust. "I couldn't believe it could grow on such shitty land," he said. He was aware of Searchinger's ILUC critique that was causing such angst in the industry, the idea that growing fuel instead of food on good farmland accelerates global warming as well as global hunger. But growing fuel on land too lousy to grow anything else sounded like a victory for the climate and the hungry as well as Goldman Sachs.

The deal fizzled, but Sikka's obsession did not. He came to see pongamia as a miracle crop, an answered prayer for a planet running out of fertile farmland and fresh water. It was high-yield, low-impact, drought-tolerant, carbon-storing, nitrogen-fixing, regenerative, resilient, and just about everything else agriculture needed. If it could thrive without chemicals or irrigation in the barren lands of India, it wouldn't have to compete for valuable acreage in Iowa or the Cerrado.

It was not a new or rare tree. It had grown wild for millennia in South Asia and Australia, and was cited for its healing properties in India's ancient Ayurvedic texts. Plant guides warned that its beans had a "bitter taste and disagreeable aroma," which was why it had never been domesticated as a crop. But it had spread across the subtropics, including several hundred ornamentals in a park near my Miami home, and its seeds were often used in India for lubricants, lamp oil, and fertilizer. Pongamia oil also provided the active ingredient in antifungal remedies like Heel-Tastic that Americans could buy on Amazon.

Pongamia seeds were a lot like soybeans, containing protein meal and oil, so Sikka realized that if someone could figure out how to get rid of its bitterness, it had the potential to produce food and animal feed as well as fuel on infertile land. Since it could grow much taller than a soybean plant, it also had the potential to use its airspace to produce much higher yields, the eternal agricultural dream. It was like a natural outdoor vertical farm—as Sikka called it, vertical soy.

Productivity was only one of its superpowers. It was a leguminous tree that fixed nitrogen from the air, so it needed little or no fertilizer. It had natural resistance to pests, so it needed little or no pesticide. Not only was it drought-tolerant, needing little or no irrigation, it was flood-tolerant and salt-tolerant. Unlike annual grain crops, which required farmers to buy seeds every year and burn diesel planting them, it was a low-maintenance perennial that only had to be planted once. Its lateral roots helped prevent erosion by stabilizing the soils around them. And it sequestered easily measurable carbon above the ground. It was hard to imagine how it could be any climate-smarter.

Sikka incorporated Terviva in 2010 with a simple mission: to produce truly sustainable fuel and food by planting trees instead of cutting them down. He wanted to create a genetic library of pongamia traits, an arboreal 23andMe, then breed the elite cultivars that could make the most oil and protein on the worst land. He also planned to create a traceable supply chain from geo-located trees, so he could offer buyers a verifiably deforestation-free alternative to dubiously sourced commodities like soy and palm oil. That wouldn't have been possible a few years earlier, but like Upside, Pivot,

GreenLight, and other biotech startups, Terviva could exploit the genomic revolution that had cut the cost of DNA work a thousandfold.

Sikka saw pongamia as the next revolution. His theory of the case was that just as cleaner coal and more-efficient gasoline cars would never fix the fossil fuel problem, cleaner and more-efficient versions of existing crops couldn't fix the eating-the-earth problem. Pongamia could be the disruptive agricultural equivalent of solar power or the electric car. It could be soy without the man-boob reputation, spurring reforestation instead of deforestation.

Still, it's never easy to introduce a new crop, even a high-yielding miracle crop. Terviva would need a few more miracles.

Farmers tend to skew conservative, in business and politics, so Sikka knew they'd be reluctant to devote land to an unproven green idea. Still, he was a persuasive salesman with a passion for his trees, a data-driven pitch, and a soothing voice that sounded more NPR than CEO. And he was pitching yield increases to yield obsessives, so he figured he could recruit a few early adopters.

Then he started making his pitch to actual farmers.

"They looked at me like I was an alien descended from Mars," he said.

Farming was risky enough with traditional crops. Nobody wanted to take on a new crop, especially a new tree crop that wouldn't produce a harvest for four years. Sikka quickly concluded that no matter how wonderful he made pongamia sound, farmers would never take the leap unless they had no other options.

"What we needed," he said, "was some farmers who were totally fucked."

Lo and behold, a miracle arrived, in the form of a bacterial disease that was wiping out Florida's citrus groves, inspiring a few totally fucked farmers to take a flyer on pongamia. They only planted it on 150 acres of their worst fields with the most depleted soils, but they grudgingly agreed to give a new tree crop a chance.

"Our first stroke of dumb luck," Sikka said.

He was also lucky the first grower to embrace pongamia was Ron Edwards, a former Tropicana executive who ran a major citrus and real estate operation in central Florida. Edwards had the country drawl and beefy jowls of a good ol' boy, but he was a shrewd entrepreneur who had cofounded the SoBe energy drink and Blue Buffalo pet food startups that were later sold to PepsiCo and General Mills. He kept telling his fellow growers: Citrus ain't coming back. Not only did he help persuade a few to field-test pongamia, he invested in Terviva and became chairman of its board.

"Folks thought I was P. T. Barnum pushing this thing," Edwards said. "But here was this jungle-tested tree that was sustainable a crop as you'll ever find. Growing without inputs! Hard to beat that."

Another miracle occurred around the same time: the de-bittering of pongamia.

Sikka founded Terviva as a biofuel play during the biofuel boom, but he always hoped to make food and feed as well—partly because the economics would be tenuous without them, partly because he felt queasy about growing fuel, even on bad land, when the global poor needed protein and vegetable oil. He just wasn't sure pongamia could ever be palatable, even to livestock, and he feared making it palatable might require an exotic or toxic recipe. His team experimented with dozens of solvents, many of them nasty chemicals like benzene, before hitting the jackpot with an affordable organic substance that was already regularly consumed by humans: alcohol. The serendipity of the discovery reminded Sikka of Doc Brown in *Back to the Future* slipping on his toilet, banging his head, and accidentally inventing the flux capacitor.

If that wasn't miraculous enough, a compound extracted in the de-bittering process turned out to protect trees from a fungal disease that was ravaging banana plantations. And pongamia oil turned out to be a stunningly pure vegetable oil, like olive oil with a neutral flavor and striking golden color; a top food consultancy found it created "an indulgent mouthfeel" reminiscent of "foods fried in butter." Terviva also won a federal grant to test its pongamia meal as cattle feed, and it performed as well as soybeans. Pongamia protein even showed promise as a plant-based food ingredient. It had all nine essential aminos, as well as a mild nutty flavor that worked nicely in burgers, butters, and flours.

Sikka figured the odds against all those coincidences happening back-to-back had to be a billion to one. And they allowed him to shift his business model. In 2016, a big biofuel firm offered to buy all of Terviva's oil for a decade—a potential nine-figure deal, a no-brainer for a seven-figure startup. But he said no. He didn't want to be part of a biofuel story. He wanted to help feed the world.

He didn't even want to feed animals. He kept thinking about something the CEO of one of India's largest agribusinesses told him: *People need food, too.* "It was humbling. It reminded me of my time in Africa," Sikka said. Why run pongamia through inefficient livestock when it could provide efficient nutrition directly to humans? He canceled a feed trial in Australia and began hiring food executives and scientists to create marketable ingredients. Terviva's new mission statement was "Planting millions of trees to feed billions of people."

One climate investor I knew was so excited about pongamia, he made it his home Wi-Fi password. But the miracles Sikka witnessed in India and his Oakland lab wouldn't matter much without the high yields he was promising in Florida; great farming ideas often fail to thrive on actual farms. The miracle tree sounded a bit too miraculous, so in the spring of 2019, I drove up I-95 to the wreckage of citrus country to see how it was doing in the field.

It was doing great.

I met Sikka in the citrus hub of Fort Pierce, and we drove past miles of abandoned groves to Terviva's five-acre test plot. The company had cloned 30 high-performing trees—he called them "LeBrons"—and we could see their offspring competing in the Florida sun. Most were six years old and around 25 feet tall, even though none had ever received a drop of fertilizer, pesticide, or irrigation water. There were a few runts, but Sikka didn't need all his cultivars to be superstars, just enough to avoid the kind of homogenous gene pool that made Florida citrus so vulnerable to disease. Terviva was even sequencing the pongamia genome, so it could control pongamia genetics and ensure the next generation of super-trees were even more super.

"Some of these are bad hoopers," he said as we walked the field. "But some of them are awesome, and we'll clone them, too."

It was a swelteringly humid day. Annoying pairs of mating lovebugs kept swooping around our faces. But Sikka brightened every time he spotted a LeBron Jr. with branches drooping under the weight of its seedpods. "Look at that bunchy one!" he shouted. "It's loaded!" The bunchiest were producing at rates that would have netted 10 times the average U.S. soy yield over an entire acre.

"I don't even mention those numbers in my pitches," he said. "I'm afraid I'll get laughed out of the room."

He did tell potential investors that Terviva's early harvests were already beating U.S. soy yields and tripling Indian soy yields. Within a decade, he expected an acre of pongamia without inputs to produce as much oil and protein as three or four acres of U.S. soy sprayed with expensive chemicals.

Terviva had also stumbled into another mini-miracle: The harvesting equipment for pistachio trees worked perfectly on pongamia, which made the crop much more economical to grow. And when we visited the old citrus facility where Terviva was starting to crush beans, Sikka's team showed him a machine from the sunflower industry that it had just retrofitted to shell pongamia seeds. We watched it fill a bucket in 15 seconds, hundreds of times faster than shelling by hand, another step toward industrial efficiency. We even saw a four-foot-tall tangelo plant that had shriveled and nearly died a few months earlier, before a Terviva scientist had nursed it back to health with, what else, pongamia oil.

"We really are the clown car that drove into the gold mine," Sikka said.

His lucky-me schtick aside, he wanted to create an actual new paradigm, anchored not only in sustainable intensification but justice for farmers and farmworkers. He took me to see Terviva's nursery, where Mexican and Central American migrants were tending thousands of saplings. They were all earning more than they had in citrus, plus stock options, and they kept telling me Terviva treated them *como familia*, like family. Perhaps they laid it on thick because the boss was nearby. But after Trump's election in 2016, that boss used Google Translate to give an emotional speech promising to stand with immigrants, so perhaps they meant it.

Sikka hoped to do more good in India, where Terviva was paying poor rural villagers to collect wild pongamia seeds. It was a complicated exercise in empowerment, requiring awkward negotiations with male village elders over a female workforce. But a McKinsey study he had commissioned found India had one million tons of pongamia seeds available for harvest every year, a bounty that was mostly being wasted. He saw a chance to pump hundreds of millions of dollars into communities that had never participated in the global economy, while injecting new oil and protein into the food supply. And it wouldn't require any new land, not even bad land. The opportunity was literally there for the picking.

Terviva's main challenge, for impact making and profit making, would be scale. In that sense it was no different than Pivot Bio, Meati, or the Magnum Feedyard. Sikka expected Terviva's processing costs in 2020 to be 10 times what Dreyfus paid per ton, because he expected to process a few hundred tons while Dreyfus handled tens of millions. It was hard to get efficient without getting big.

Sikka liked thinking big. He relentlessly pitched pongamia to the ABCD grain behemoths (ADM, Bunge, Cargill, Dreyfus), other food giants (including Danone, Kellogg's, Nestlé, and Unilever), and less famous ingredients giants (like IFF, Kelco, and Givaudan). He talked about doubling Terviva's acreage every year, using pongamia to de-desertify the Sahel, making burgers better than Beyond's and milks better than Oatly's. He told me his dream was that just as soybeans replaced oats on many U.S. farms, and palm oil replaced rubber in much of Southeast Asia, pongamia could one day replace most soybeans or palm oil.

Then he paused, to think bigger.

"You know, it would be amazing to take out both."

By 2020, Florida growers had planted 1,000 acres of pongamia. "We're looking to sign up a yuuuuuuuge amount of acres by year end," Sikka told me in an email. He was looking to launch his food business, too: "I predict big shit by year end—principally getting a major food co to commit to using pongamia oil instead of palm. It's also possible we will have a deal in place for our protein."

Alas, Terviva didn't sign up a yuuuuuuuge amount of acres that year. The major food company—it was Danone, which was considering pongamia oil for its nondairy creamer—didn't commit. The protein deal didn't happen, either.

After 12 years of miracles, Terviva had only altered 0.00001 percent of the world's agricultural landscape, and it still hadn't sold any food.

"We're the luckiest fuckers on the planet. The universe has smiled at us so often. And we haven't made a dent!" Sikka said. "It shows how hard it is to change agriculture."

Since 2020, the universe hasn't smiled at Terviva as often. Sikka pursued a variety of deals that fell through—5,000 acres in Florida, a $200 million pongamia fund, a big contract with a food services company, a licensing agreement for the banana-protecting biopesticide. Unilever considered pongamia oil for its vegan mayo, and Impossible and Beyond both tested it for their plant-based meats, but nobody pulled the trigger. By 2023, Terviva had only 1,500 acres of trees, when the world had 300 million acres of soy. Searchinger tried to get a pension fund with major land holdings in Brazil interested in funding a large-scale pongamia planting, but financial firms were just as skittish as farmers about unproven crops.

That year, the plant-based brand Aloha finally debuted Terviva's pongamia oil in a sustainable protein bar. But one food customer couldn't cover Terviva's $20 million-a-year cash burn, and the company had several near-death experiences before the Mitsubishi Corporation came to the rescue with a major strategic investment in December 2023. Venture capitalists looking for fast returns were always an awkward fit for the agricultural sector, so Sikka was thrilled to have a patient new Fortune 500 partner.

There was a catch, though: Mitsubishi wanted Terviva's oil for low-carbon biofuels. Sikka had thought he was out of the biofuels business, but like Michael Corleone, he was getting dragged back in. Chevron's biofuel arm soon invested, too. Indulgent mouthfeel just wasn't paying the bills, and even though Sikka wasn't happy to make Terviva's story a biofuel story, even temporarily, it was better than a bankruptcy story. And using pongamia oil grown on bad land for fuel would be way better for humanity and the climate than using corn ethanol.

"It was emotionally difficult to make that pivot, but the market is what it is," Sikka said. "Food tech has cooled. Biofuels are back, big-time."

At least Terviva had sold a little edible vegetable oil. It still hadn't sold any protein. So despite his humbling "people need food, too" experience, Sikka made another grudging pivot to animal feed. India had a lot of dairy cows, Mitsubishi's holdings included a feed business, and with plant protein in its trough of disillusionment, the market was what it was.

Sikka was even scaling back Terviva's efforts to plant trees, shifting its focus from Florida to the existing wild supply chain in India, a faster path to scale. By the end of 2023, Terviva had bought 2,000 tons of beans from rural harvesters. Still, the world produced 350 million tons of soybeans that year alone.

What made these growing pains even more exasperating was that the miracle trees were still doing great, doubling soy yields without inputs and trending higher. Sikka was also finalizing a deal to sell carbon offsets to a tech firm, which would make the economics of carbon-storing trees much more attractive. A half dozen pongamia startups had launched globally, hoping to cash in on a now-proven super-tree, and Sikka was glad to welcome them to the party. Maybe they'd license Terviva's patents—and in any case, pongamia was good for the world.

It was just a shame that the pongamia party was still so tiny, and that it was becoming a biofuels party. The main lesson, as Sikka put it, was that "it takes forever for this shit to scale." Fixing food and agriculture, as Max Weber said of politics, will require a slow boring of hard boards. Six hundred million farmers won't plant pongamia tomorrow—or GMO Golden Rice, or gene-edited camelina, or soybeans engineered for better photosynthesis. They won't all start using RTK combines, Pivot biofertilizers, or GreenLight biopesticides, either. It hopefully won't take forever for solutions that produce more food with less land and fewer emissions to scale, but it will take time. Even with billion-to-one miracles, this shit is hard.

"Can we breed sustainable crops and get them to scale fast enough to meet the world's needs?" Sikka asked. "It's an open question."

The other lesson of Terviva's saga, especially its pivot back to biofuels, was that incentives matter. Biofuels were back because governments had created new mandates and subsidies for biofuels, especially aviation biofuels. Sikka was suddenly fielding inquiries about planting pongamia for fuel all over the world—in Belize, in Indonesia, even in fields littered with spent munitions in Laos, the ultimate degraded land. People were still following the money.

As we'll see, Searchinger was distressed about the land-use implications of people following the money back to biofuels. He wanted to see more money for people to follow into climate-friendly solutions, not only miracle crops like pongamia but alternative proteins that could reduce meat and dairy consumption, alternative fertilizers and pesticides that could reduce emissions without reducing yields, genetic advances that could increase yields, and scientific research to figure out which solutions could qualify. He also hoped to see negative consequences for climate-unfriendly activities. He was heartened by the innovation in the food and agriculture space, but it was clear consumers and farmers wouldn't make green choices out of the goodness of their hearts. The free market, for all its virtues, was not providing the carrots and sticks that could make that happen on its own.

"You don't need tenure at Princeton to understand this," he said. "Let's encourage things that are good, and discourage things that are bad. Simple!"

Well, not exactly simple.

ELEVEN
HOW TO SAVE THE WORLD

THE MODEL NATION

At U.S. congressional hearings, the politicians sit on a dais facing the audience, while the witnesses sit at a table with their backs to the audience. The politicians make partisan speeches and ask gotcha questions their aides can edit into viral videos. When it's not their turn, they usually leave the room or check their phones. It's rare to see anyone listen or anything learned.

When Searchinger testified at a Danish Parliament hearing in early 2020, in the government castle Danes and Netflix fans know as Borgen, the witnesses sat on the dais, because they were the stars of the show. The politicians sat with their backs to the audience and stayed for the entire show. They asked thoughtful questions designed to gather information, not to score points. They seemed genuinely eager to hear from the experts, which, it occurred to me for the first time, must be why these things are called hearings.

The topic of this one was biomass power, which wasn't even the reason for Searchinger's trip to Denmark. But the hearing played out like one of his wonk fantasies, as the pro-biomass politicians seemed thunderstruck by his explanations of why wood-burning was even dirtier than coal-burning.

They kept thanking him for setting them straight, as if he had let them know they had spinach in their teeth, and vowing to change their policies to align with the science.

"Excellent points from Timothy. We should no longer say biomass is carbon-neutral," said the center-left party's committee chairwoman. The center-right party's climate leader agreed: "We all need to do some rethinking!"

A bearded professor with the haughty air of the second husband of a Woody Allen character's first wife tried to push back, claiming biomass plants only burned climate-friendly residues. Searchinger shut him down with industry data confirming they mostly burned whole trees. The professor then claimed biomass was still carbon-neutral when it was sustainably harvested. Searchinger pointed out that the sustainability rules had nothing to do with carbon.

"I must stress, Mr. Searchinger is correct," another expert chimed in. Yes, several Searchinger-pilled politicians agreed, definitely correct. Even a utility executive said he was grateful for "wise people like Timothy who keep us on our toes," and urged Parliament to ban companies like his from burning whole trees.

"We beg you to make sure we only do good biomass," he said.

Searchinger hoped the good-government (and of course pro-Searchinger) vibe was a sign that Denmark was ready to show the world how to quit biomass power. It was a model nation, perennially ranked one of the happiest, healthiest, and best-governed on Earth. It was also a climate leader, with some of the world's strictest legal commitments to emissions reductions. Maybe it could set another global example of how to convert facts and logic into responsible policy—and maybe after ignoring the Massachusetts example, the world would follow.

In fact, biomass was only a small part of his vision of Denmark as a global model for cutting-edge land and climate policy. After the hearing, he met with that center-right climate leader, Tommy Ahlers, who congratulated him for showing Borgen the error of its ways.

"Thanks to you, we will change!" Ahlers said.

"This is a good moment to pivot to agriculture," Searchinger replied.

"That's another opportunity for Denmark to show the rest of the world what to do."

That was the real reason for his trip. The Danish Agriculture and Food Council had hired WRI to draw up a plan to decarbonize the nation's farm sector, and he wanted climate-forward, fact-focused Denmark to become a laboratory for eating-the-earth solutions. With a population smaller than Maryland's, it generated only 0.2 percent of the world's agricultural emissions. But by test-driving the ideas in his food report, maybe it could create a roadmap for addressing the other 99.8 percent.

The problem, he told Ahlers, was that Danish politics seemed to be turning against high-yield production agriculture, when low-yield agriculture was as damaging to forests as burning trees for electricity. Why not show the world how to do even higher-yield agriculture with less environmental impact? Why not become a global model of sustainable intensification?

Ahlers leaped to his feet to give Searchinger a high five.

"Perfect, I love it!" Ahlers gushed. "Now we've got to do it!"

Denmark had twice as many pigs as people. It devoted more of its land to agriculture than any other country except Bangladesh. And it was decarbonizing the rest of its economy so quickly that its farm sector was on track to generate half its emissions by 2030. When Searchinger visited the ag council to share his recommendations, policy director Ebsen Nielsen told him Danish farmers were counting on him, because they were under intense pressure to go green. Environmentalists were clamoring to get rid of half their livestock.

"People used to take pride in our farming. Now we're the bad guys, the whining polluters on the dole," Nielsen said. "Too much manure! Too many pigs!"

Searchinger came bearing welcome news: Shrinking the Danish livestock industry would be "unbelievably stupid for the climate." His team had calculated that Denmark's large-scale pig farms were the most efficient on Earth. Its dairies were a close third behind the U.S. and the Netherlands. Fewer pigs and cows in Denmark would just mean more pigs and cows in less-efficient countries where they would make more of a mess.

He then treated Nielsen to one of his snarkathons: "Why don't you just shoot all the pigs in Denmark? Then your emissions from pork production would be zero! Maybe everyone should shoot their pigs. Then the world would have zero emissions from pork! . . . Just shut down global agriculture! Zero agricultural emissions! Who needs food?" His serious point was that cutting emissions by cutting production would just create leakage, outsourcing pig farms and dairies along with their production and emissions abroad. He wanted Danish farmers to test other ways to go green, like using Pivot Bio's microbes as fertilizer, or Terviva's pongamia meal as feed.

His boldest idea was for Denmark to adopt a new approach to climate accounting, because the standard approach assumed land was practically free. It only counted direct emissions from farms, ignoring the opportunity costs of using farmland for growing food rather than storing carbon. So it made Denmark's pork and dairy operations look like climate menaces even though their productivity relieved pressure on nature, just as it made degraded Brazilian ranches look climate-friendlier than upgraded ranches that spared the Amazon. If Denmark embraced WRI's carbon-opportunity-cost accounting, it would reward efficient production that narrowed the global food and land gaps, encouraging Danish producers to pursue even higher yields rather than shoot their pigs.

"Bad accounting destroys the world," Searchinger said. "If Denmark can show how to fix it, you can help save the world."

He suggested a compromise where Danish farmers would get to produce even more food to keep up with rising global demand, while the government would help them increase their yields enough to do it with one-sixth less land. Then Denmark could restore one million acres of forests and wetlands while doing its part to feed the world. It could demonstrate that developed nations don't have to sacrifice their home environment to help the global environment.

The next day, Searchinger was making that pitch to a group of Danish enviros when an elderly Greenpeace activist interrupted: "Wait, you're telling farmers to make *more* pork and dairy?" His colleagues piled on, berating Searchinger for pushing factory-farmed meat instead of plant-based diets.

Searchinger tried, with characteristically limited success, to conceal his annoyance. He knew a thing or two about meat and dairy, and yes, he agreed Danes should eat less. But that didn't mean they should produce less. Even if every Dane went vegan, the rest of the world would still need more pork and milk, and efficient Danish producers could help make it as sustainably as possible. If Denmark unilaterally cut production, it would barely make a dent in global demand; it would just encourage less environmentally responsible countries to expand production—more Econ 101.

"You can't shift consumption by shifting production," he said. "That's not how it works!"

The Danish greens did not appreciate an American with a farm-lobby contract calling for even more pigs and manure in their backyards. They dreamed of restoring rural Denmark to its preagricultural glory, when it was mostly a deciduous forest teeming with elk, bears, and other wildlife.

"I want our birds back!" the Greenpeace activist vented.

Nobody really embraced Searchinger's compromise. At the event where he released his report, the head of the ag council and a leading Danish environmentalist praised its innovative thinking, but the farmer complained about its call to rewild farmland, and the enviro complained that it did not call to reduce production. Agricultural reform stalled for the next four years.

The model nation didn't even take all of his advice about biomass power, despite the lovefest at Borgen. It did phase out some subsidies, but it did not ban the burning of whole trees, and it retained the carbon-neutrality loophole. Searchinger was disappointed, but he had learned from years of policy defeats that getting a quarter of a loaf was better than nothing.

Then in June 2024, Denmark Denmarked. The government announced a consensus deal, backed by the agricultural as well as environmental lobbies, to phase in the world's first tax on farm emissions, starting around $100 per cow per year. The agreement also called for the gradual return of one million acres of farmland to nature, just as Searchinger had proposed, along with subsidies to help farmers adopt climate-friendly practices, as well as the world's most aggressive national effort to promote plant-based eating. It didn't include the accounting reforms he had advocated—he barely allows himself to imagine full-loaf victories anymore—but it didn't aim to roll back production, either, so it mostly reflected his vision.

It could set a powerful global example. Economists love the efficiency of emissions taxes, which give emitters an incentive to reduce emissions without dictating how, and impose costs on activities that impose costs on society while producing revenues society can use to make things better. They're an elegant solution to the "tragedy of the commons," the notion that collectively, everyone benefits from a stable climate, but individually, everyone has incentives to emit when emissions have no price. Countries that have enacted carbon taxes to discourage the burning of fossil fuels, including Denmark, have reduced their energy emissions.

However, the politics of emissions taxes are often hideous. France repealed a tax hike on fossil fuels after a revolt by furious drivers, and New Zealand canceled an imminent "burp tax" on methane emissions from livestock after a revolt by furious farmers. The politics of agricultural emissions taxes seems especially hideous, since policies that hike food prices can bring down governments. One American pollster told me meat taxes were the most unpopular policy he ever surveyed, "up there with veterans' benefits for ISIS." Denmark's tax won't take effect until 2030, so it could still face a New Zealand–style backlash.

Even in Denmark, change is hard. But it's a reminder that change is possible.

FIRST THINGS FIRST

I'm trying to argue against Debbie Downerism, even though I've reported a lot of downers.

To recap: Biofuels and biomass power, supposedly climate saviors, are climate disasters. Carbon farming and vertical farming are wildly overhyped. Plant-based meat has floundered in the market, while cultivated meat hasn't really made it to market. Genetically modified and edited crops and livestock still face all kinds of political and cultural obstacles. And a slew of other promising solutions—methane-suppressing feed additives, nitrogen-controlling biofertilizers, high-tech fish farms, high-yield pongamia—are struggling to scale. The eating-the-earth problem is getting worse, and not even Denmark will move the needle much before 2030.

I'm sorry about all that. I wish I had more immediate progress to report. But I warned up front that it wouldn't be easy to get the world's farmers and eaters to change their ways. Entirely new crops, foods, and fertilizers were never going to be invented, approved, and adopted worldwide overnight. There's no silver-bullet solution that can give the world 50 percent more calories in a hurry without using more land. The politics of more sustainable agriculture and more sustainable diets was always going to be brutal.

Remember, though, fossil fuels also looked unsolvable until they didn't. The news about them was almost all bad until it wasn't. In 2007, when Searchinger began questioning biofuels, not only were they considered the only plausible alternative to fossil fuels for transportation, they were just about the only commercially available alternatives to fossil fuels, period. Things can change fast. He still gets bummed out when the eating-the-earth problem isn't treated as an urgent problem, like he's Chicken Little yelling about a quantifiably falling sky, but he's more convinced than ever it's a fixable problem—if humanity gets serious about fixing it.

One reason he's so optimistic is that about half of it is a ruminant problem. We can eat much less of the earth without much social or economic pain if rich countries can keep shifting their diets from beef toward chicken and pork, and if beef substitutes can get tasty and cheap enough to compete with beef. It's not as if our civilization would crash if we started eating more Chick-fil-A and Impossible Whoppers and fewer Big Macs. On the supply side, tripling the productivity of degraded tropical cattle pastures could also free up hundreds of millions of acres—and that's already being done on ranches like Nossa Senhora on the Amazon frontier. It just needs to be done a lot more.

What makes Searchinger even more optimistic about the eating-the-earth problem is all the groundbreaking work being done on potentially transformative solutions—so far, with virtually no financial or political support.

One of his favorite examples is an obscure agronomy hack that could create a humongous source of new animal feed without a single acre of new land: chemical treatments that make crop residues like cornstalks more palatable for livestock, so that farmers can essentially double-crop with a single crop. He believes residues could one day replace one-third of all feed crops, which could free up hundreds of millions more acres of croplands. The process isn't economical yet, and he thinks it's pathetic that a scientist in Nepal who's leading a lonely effort to transform rice straw into high-quality feed has a research budget smaller than the revenues of his local coffee shop; farmers grow even more rice straw than rice. But he's hopeful our big-brained species will eventually recognize how desperately we need solutions like that on a planet with limited land, and will find ways to get them to scale.

The bad news of this book is how little has been done to fix the land sector, and how tough it will be to replicate the progress happening in the energy sector. The good news is that it's become much clearer what ought to be done.

Sorry again, but the first thing that ought to be done is rather anticlimactic for a book about the climate problem beyond fossil fuels: We need to accelerate the transition away from fossil fuels, because sorting out food and land is clearly going to take a while.

Climate hawks often quote President Kennedy's stirring pledge to go to the moon not because it was easy but because it was hard, but that never made sense. We went to the moon *even though* it was hard. We should fix the climate even though it will be even harder, but we should start with the easiest fixes. That means decarbonizing energy, especially electricity, even faster than we're doing it, without indulging bioenergy fantasies that could make food and land unfixable.

It's just math. The electric grid is the only sector of the global economy on track to meet the Paris goal of reducing emissions 43 percent by 2030, so it will need to do even better than that to offset the laggards. Fortunately, clean power keeps getting cheaper, while dirty coal-fired power plants keep getting easier to replace—and the fact that we're already doing it suggests we can do more of it, even if Trump would like us to do less of it. Beyond the grid, global emissions have kept rising, and food and farming solutions like fungi-based protein, RNA biopesticides, and Photosynthesis 2.0 won't be ready to improve the numbers much before 2030. They'll take time to get affordable, desirable, and ubiquitous. Since zero-emissions electricity is already affordable and desirable, making it more ubiquitous is the quickest current path to lower emissions.

Here agriculture can help right away, because it emits as much fossil carbon as the nation of Japan. Farmers are already helping by installing energy-efficient irrigation and refrigeration systems; rooftop solar on their barns; and "agrivoltaic" projects in fields where sheep can graze or shade-tolerant crops can grow under solar panels. Wind turbines are also sprouting throughout farm country, taking small patches of land out of production but producing so much supplemental income that some farmers who skipped gender-sensitivity training call them "the second wife." And John Deere is preparing to roll out its first electric tractors—in green, naturally.

Governments can accelerate this trend—and the Biden administration did, pouring billions into rural renewables. But the improving economics of fossil-free agricultural energy is colliding with culture-war politics, as hundreds of rural Republican counties are restricting the use of farmland for solar and wind projects, a trend likely to accelerate under a president who claims windmills cause cancer. The local restrictions are often justified with arguments about excessive land use—arguments rarely heard in farm country about corn ethanol even though it's 100 times more land-intensive than solar—but they're mostly motivated by tribal notions of green energy as a leftist plot. While they won't stop the energy transition—the Republican drill-baby-drill state of Texas now leads the nation in large-scale renewables—they can slow it down.

That's the other principle for short-term action, the Hippocratic climate oath: First, do no harm. We can't make the eating-the-earth problem much better quickly, but we should try to stop making it worse.

In the summer of 2022, the most ambitious climate legislation in American history came back from the dead. West Virginia Senator Joe Manchin, a notoriously reliable coal advocate, astonished Washington by embracing President Biden's Inflation Reduction Act and its record-breaking clean-energy investments. His flip-flop looked like a turning point for the planet. Climate hawks rejoiced.

Except for one climate hawk, who sent me an emergency email from an ancient Transylvanian town where he was visiting some unusually biodiverse grasslands. Searchinger wanted to make sure I was aware the planet was doomed. His life's work had been wasted, and the only honorable thing to do, he said in jest but not ha-ha jest, was to commit seppuku.

What terrified him was the Biden bill's new tax credit for "sustainable aviation fuels," which, nearly 15 years after he first documented the folly of using farmland to grow fuel, looked like an unprecedented bonanza for corn ethanol and soy biodiesel. According to his math, replacing all U.S. jet fuel with crop-based fuels would require all U.S. cropland and then some. The carbon losses from land-use change would be three times the carbon savings from fossil fuel replacement. The biggest environmental victory of the 21st century looked like an egregious violation of the Hippocratic oath.

The next day, he called to say he might have slightly overreacted. After studying the fine print, he now believed the planet was only possibly doomed.

The bill required the various biofuels to cut emissions 50 percent to qualify for tax credits, and while he knew emissions analysts had ways of getting to yes, he calmed down once he realized the credits were only truly generous for biofuels that cut emissions much more than 50 percent. It would require new levels of analytical acrobatics to get crop-based fuels that far past yes, so he decided the Democratic Party's historic effort to fix the climate wouldn't necessarily destroy it.

Seppuku wasn't off the table, though. Biofuels advocates launched a fierce behind-the-scenes campaign to get Bidenworld to use Michael Wang's GREET model, which now derived its land-use numbers from the acrobatic Purdue model that had been exposed for "systematic optimism bias" in California. The industry was putting its faith in GREET to jump-start a new government gravy train, and its campaign was penetrating the highest levels of the White House.

At one cabinet meeting, ethanol-loving Agriculture Secretary Vilsack handed a mystified Treasury Secretary Janet Yellen a page of ethanol-lobby talking points about GREET, telling her the issue was high-priority. Yellen, whose high-priority issues usually involved economic crises, had to ask her team: "Can someone tell me what this is about?" Meanwhile, a new "Sustainable Aviation Fuel Lifecycle Analysis Interagency Working Group" led by Biden climate czar John Podesta began holding weekly meetings to debate a model everyone knew was only being debated because farm interests liked its slant.

"Nobody talked about the elephant in the room," one administration official recalled. "We'd just argue around in circles. It was theater of the absurd."

When Searchinger looked into the latest Purdue land-use modeling behind GREET, he found its optimism bias had gotten even more flagrant. It treated land as almost free, creating a dreamworld where all of Iowa's food production could be replaced with virtually no effect on global land use. He and his Yale economist friend Steve Berry released a report shredding its Miracle Yield assumptions, invented numbers, and Reverse Murphy's Law conclusions. They showed it produced geographically impossible results—huge land areas that simply disappeared during model runs—which were conveniently tweaked through "hand of God adjustments," like revisions to the score after the game was over.

Still, the political elephant remained in the room. Biden was all in for biofuels, declaring at an Iowa ethanol plant that "we want to see facilities like this all over the Midwest!" Not only had his administration rejected environmentalist pleas to reform the Renewable Fuels Standard, it had sucked up to the corn lobby with waivers allowing extra ethanol in gasoline. Podesta made it clear the working group needed to get to yes on GREET, too.

Since even GREET didn't look lenient enough to get ethanol to yes, Vilsack also pushed for it to be "updated" to give farm-grown fuels extra credit when they were farmed with supposedly climate-smart practices like no-till and cover crops, or brewed in refineries that captured their carbon. Conveniently, the Inflation Reduction Act subsidized no-till, cover crops, and carbon capture, so the update would let the bill's tax dollars be used to help biofuels qualify for the bill's tax credits—on top of all the other federal support that made biofuels viable.

The final decision was a mixed bag. The administration announced it would use an updated GREET model, and Vilsack proudly let the political elephant out of the room, hailing the decision as "a great beginning as we develop new markets for . . . home-grown agricultural crops." But Searchinger's allies inside the administration managed to attach a few strings, and Senator Grassley actually denounced Biden for making it too hard for Iowa farmers to qualify for the law's cash—even though Grassley, like every other Republican, had voted against the law.

Searchinger doubts corn or soy will fuel many American planes soon, so American biofuels might not do any more harm than they're already doing. Globally, though, the push to decarbonize flight is injecting new hope into biofuels markets, and the land-use impact of "sustainable aviation fuels" could be ghastly. He estimates that replacing just one-fourth of all jet fuel with crops would require a tripling of global vegetable oil production. At least 40 percent of the world's cropland, and possibly as much as all of it, would have to grow fuel.

Unless we're planning to go shopping for a new planet, that won't work.

"A 12-year-old kid could do these calculations," Searchinger said. "But people don't, so we end up with this madness."

Global institutions are still calling for shockingly aggressive expansions of biofuels as well as biomass power, especially BECCS projects designed to capture carbon from wood-burning plants. The International Energy Agency's net-zero plans call for doubling bioenergy by 2030 and expanding BECCS 100-fold, an even ghastlier scenario for global forests and Searchinger's blood pressure. Drax has proposed to retrofit its Yorkshire plant into the world's largest BECCS facility, which could cost British ratepayers a breathtaking $50 billion, and it's now scouting locations for the first pellet-burning BECCS plants in the U.S.

So far, the global boom the boosters keep predicting has not materialized. Bioenergy has flatlined around 5 percent of all energy production, partly because the dumb algae that invented photosynthesis didn't make it work as well as solar or electric cars, partly because Searchinger and his allies have tarnished its reputation. Still, zero percent would be better, because most bioenergy is harmful. And there's just as robust a market for harmful food and farming ideas.

• • •

The worst idea in recent years came from Sri Lanka President Gotabaya Rajapaksa. In 2021, he abruptly banned fertilizers, pesticides, and other agrochemicals, declaring his island nation needed to get "in sync with nature." Agronomists warned the ban would cripple production of rice, Sri Lanka's staple crop, and tea, its top farm export. But Rajapaksa insisted that going all-organic would make its food system "more sustainable, resilient, and inclusive."

Nope. Rice yields plummeted, creating painful shortages and sky-high food inflation. Tea yields also plummeted, crushing exports and draining foreign currency reserves. Rajapaksa had to reverse the ban after six months. Mass protests still forced him to resign and flee. Hippocrates would've been appalled.

The Sri Lanka fiasco has not stopped the UN, green groups, Big Ag and Big Food corporations, Michael Pollan–reading and Joe Rogan–listening natural-food types on the left and right, and the regenerative movement itself from advocating a worldwide shift to lower-input, lower-output agriculture. The Global Alliance for the Future of Food—backed by the mighty Rockefeller, Walton, and Packard Foundations, and led by the *Diet for a Small Planet* visionary Frances Moore Lappé's daughter—has called for $4.3 *trillion* in investments over the next decade on a transition to regenerative agroecology, four times what the world is expected to spend on space exploration. The opportunity costs would be almost unfathomable.

Governments can do real harm when they enshrine that push for low-yield farming into policy. The E.U. has set a goal for 25 percent of its farmland to go organic by 2030, a modified Sri Lankan strategy to reduce pesticides at home by outsourcing production and deforestation abroad—less food here, fewer forests there. Biden's climate-smart agriculture initiative also focused on regenerative practices, especially no-till and cover crops that are good for soils but not yields, and grass-fed beef that's downright climate-dumb. Meanwhile, congressional Republicans want to kill the initiative—not because of reasonable questions about whether it's really helping the climate, but because they don't want farm aid used to help the climate. Robert Kennedy Jr. has vowed to work inside Trumpworld to promote much more regenerative agriculture, using Sri Lanka–style rhetoric to denounce the current industrial system for promoting herbicides, insecticides, monoculture crops, and concentrated feedlots.

The first step out of a hole is to stop digging, but policymakers keep doing things that widen the food, land, and emission gaps. A few more policies that do not pass the Do No Harm test:

- Two dozen countries, including Russia and most of the E.U., have banned GMO crops. Five dozen others, including China, have imposed major restrictions. Not only can these scientifically and morally dubious policies suppress lifesaving innovations like Golden Rice, they can deny farmers access to seeds that can increase their yields and help them adapt to a more turbulent climate.
- Governments continue to prop up their cattle industries with subsidies, exemptions from pollution rules, indemnities for losses from disease and disasters, and an array of other benefits that make beef production cheaper and the land gap larger. For example, the U.S. government practically gives away public land to ranchers, charging monthly grazing fees of just $1.35 per cow and 27 cents per sheep. That's one-twentieth the cost on private rangeland, and much less than my family spends feeding our dogs. It's a taxpayer rip-off that degrades grasslands in the West, while rewarding some of America's least efficient cattlemen.
- USDA also mandates that public schools serve cow milk with every meal, even though kids throw away 40 percent of it, wasting 40 percent of the land, water, and fertilizer used to produce it. The mandate illustrates the clout Big Dairy had at USDA even before Vilsack, its former lobbyist, became secretary again. In 2018, the department actually sent a memo, helpfully drafted by dairy executives, warning schools to stop reminding students they can drink water instead of milk when they're thirsty.
- I first heard Searchinger denounce subsidized crop insurance as a wetland killer when we first met at Union Station a quarter century ago. It's now the most expensive U.S. farm program, costing as much as the entire EPA budget, financing environmentally damaging

decisions every day. Even USDA's research arm found that farmers who know they'll get paid if their harvests get ruined tend to cash in by overcultivating, plowing up flood-prone low-yield farmland they would otherwise leave fallow to store carbon.

- Forestry advocates and some climate advocates are now pushing to use more wood instead of carbon-intense concrete in skyscrapers and other buildings, but cutting even more trees for construction could end up further depleting the world's forest carbon sinks. Searchinger has published several papers showing that scientists are dramatically underestimating the climate impacts of logging, mostly because they're continuing to assume that harvesting wood is carbon-neutral, breathing new life into the biomass loophole.

Do No Harm isn't always a simple standard, and reasonable people can often disagree about the trade-offs of land-related policies. California has a new animal welfare law guaranteeing pigs enough room in their stalls to turn around, which will make pork production a bit less efficient, but even Searchinger thinks the reduction in cruelty is worth a small reduction in yield. He's also open to replacing the scourge of single-use plastic with biodegradable straws or paper bags—even though the substitutes are made from crops or trees, and their carbon escapes to the atmosphere, while the carbon locked in plastic pollution doesn't.

These issues aren't always complicated, though. When governments do egregiously harmful things, it's usually because of politics.

"Today, Florida is fighting back," my perpetually aggrieved Republican governor, Ron DeSantis, announced in May 2024, "against the global elite's plan to force the world to eat meat grown in a petri dish." DeSantis was signing party-line legislation banning the sale of cultivated meat, even though it was not for sale in Florida, or, at the time, anywhere else on Earth. He said he was striking a blow against the "inhumane" and "authoritarian" globalists of the World Economic Forum, who, he complained, had the audacity to describe meat consumption as a "source of greenhouse gas and climate change."

"We will save our beef!" DeSantis vowed.

Meat consumption, it's hopefully clear by now, is in fact a source of greenhouse gas and climate change. It's less clear why growing beef from cells would be less humane than butchering cows, or why letting consumers choose whether to buy a food would be less authoritarian than banning it. But logic did not seem to apply to this red-meat issue. Cultivated-meat executives trekked to Tallahassee to argue that principled free-market conservatives ought to support innovation and competition, but the Republicans weren't doing principle. They were taking performative umbrage at a woke threat to a Republican interest group.

"We've put down the marker very clearly: We stand with agriculture! We stand with cattle ranchers!" DeSantis said.

The embrace of the nanny state in defense of freedom was sort of amusing, and the own-the-libs theater inspired a lot of goofy puns about high steaks and partisan beefs. I wrote a story headlined "Ron DeSantis Is Attacking the Greatest Food Innovation Since the Corndog." But cultivated meat has serious potential to save forests and cut emissions, so the way it got sucked into the vortex of politics before it even got to market—special-interest farm politics as well as shirts-and-skins partisan politics—was a seriously bad omen. Partisan trash talk has already made it harder for products like plant-based meat to compete and scale, but partisan cancellations of products like cultivated meat before they even have a chance to compete could make it impossible for them to scale. On Rogan's podcast, Vice President Vance once called cultivated meat "nasty" and "disgusting."

Agri-pandering is by no means a purely right-wing or purely American form of political correctness. France's center-left prime minister sounded just like DeSantis when he caved to the farmers blockading highways with tractors and spraying manure at government offices to protest cuts in their diesel subsidies: "We will put agriculture above everything else!" Italy banned cultivated meat, too, a move its agriculture minister described as a defense of "agricultural entrepreneurs," as if subsidized farmers, rather than futuristic biotech startups, were the real risk-takers.

But if the instinct to let farmers control farm policy spans the political spectrum, opposition to all climate action as evil globalism is unique to the increasingly powerful populist right, and mutual distaste for green ag policies is cementing its alliance with farm groups. U.S. Republicans who once denounced wars on coal, light bulbs, and gas stoves have pivoted to denouncing the war on cows. Trump has hailed the Dutch farmers who marched to The Hague to protest nitrogen regulations during the *stikstofcrisis* for "courageously opposing climate tyranny." In countries where the right opposes all climate action and the ag lobby wields veto power over ag policies, it will be extremely tough to avoid harmful ag and climate policies.

And avoiding harm is just the start of what the world needs to do.

It also needs to do stuff that helps, stuff that can narrow the food, land, and emissions gaps. As Searchinger learned as a young lawyer in Pennsylvania state government, it's tough to do stuff, at least good stuff, before you know stuff. So the next task for the world will be to learn stuff.

INVEST FOR SUCCESS

Consider gassy cattle.

Not the T-shirts with flatulent heifers bragging about global warming, or the dad jokes about cow farts coming from the dairy air, or the partisan lies about bovine bans. Consider the actual problem: Ruminants are the leading source of methane from agriculture, passing 3 percent of all greenhouse gases out their front and rear ends. They're like coal plants with tails, except the U.S. has 200,000 times as many cows as coal plants, which makes controlling their emissions tougher.

Especially when we're not yet sure how to do that.

After Biden took office, his climate team asked the Bezos Earth Fund what it could do about agricultural methane, so the Bezos team asked Searchinger to write up some ideas. For his burps and farts section on "enteric methane," he sought help from Ermias Kebreab, the UC Davis animal scientist whose big boys emitted 80 percent less with traces of seaweed in their feed. It would take more than a few low-budget studies tracking a few dozen cattle for a few months to reinvent feedlot management, so Searchinger wanted to propose longer-term global studies of enteric methane solutions. He asked Kebreab how much that would cost.

At least $4 million, Kebreab replied.

Four million bucks? To investigate a phenomenon that created more emissions than global aviation and shipping combined? It sounded like the *Austin Powers* joke where Dr. Evil threatens to destroy the world unless he's paid ONE MILLION DOLLARS. Searchinger suspected such a comically humble request would be circular-filed and forgotten. Instead, he pitched a $200 million burps-and-farts research blitz.

More good news: At the Dubai climate summit in 2023, the Global Methane Hub announced an "Enteric Fermentation Research & Development Accelerator" that was almost exactly what Searchinger proposed, funded by $200 million from the U.S. and other governments along with the Bezos fund and other charities. He often felt like he was brainstorming into a void, so it was nice to have the White House and one of the world's richest men behind one of his ideas.

He'd love to see similar research and development accelerators for chemical treatments that convert crop residues into animal feed; fertilizer substitutes like Pivot Bio; pasture improvements like the purple-flowered clover that helped cattle gain weight in Brazil; and most of his other menu items. Energy R&D fueled the renewables revolution, and a recent study pegged the cost of reducing emissions through agricultural R&D at just $12 per ton, excellent bang for the buck.

Now consider livestock poop, the second-largest source of agricultural methane after cow burps. Searchinger and two WRI colleagues had the privilege of reviewing the manure management literature—scintillating topics like "flocculation with mechanical separation"—and identified nine strategies with the potential to reduce emissions for under $50 a ton. One fix, adding acid to manure tanks, seemed to pencil out at $2 a ton. The obvious policy move would be to finance large-scale tests of the cost-effective strategies at feedlots and dairies.

However, the team also identified one extremely ineffective strategy: "anaerobic digesters" that convert manure into renewable "biogas." And that's the main strategy getting subsidized, in California and much of Europe. Denmark spends $250 million a year on digesters, even though Searchinger found its farmers could reduce more emissions from manure by simply cleaning out their barns every day. He thinks digesters often *increase* emissions, because farmers often feed them surplus grain, which in land-use terms is like craft-brewing biofuels.

Clearly, the world could use an R&D accelerator for shit, too, to help redirect limited dollars from harmful to helpful solutions. The problem is that while everyone supports agricultural R&D, it's almost no one's top priority. Over the last two decades, USDA's research budget has declined by a third, because farm lobbyists don't fight for it the way they fight for digesters, row-crop subsidies, and biofuels mandates that shovel cash directly to their clients. Trump spent 50 times more bailing out farmers hurt by his trade wars than Biden's climate bill spent on ag research. In 2023, Congress invested 22 times more in clean energy innovation than ag innovation, while the Department of Energy's blue-sky research agency received 470 times more funding than a similar USDA agency.

Globally, only a small portion of public-sector and private-sector ag R&D addresses the climate—and all of it combined is roughly equivalent to Apple's R&D, as if farming were roughly equivalent in importance to a better iPhone camera. Searchinger considers this such an egregious missed opportunity that it actually bolsters his optimism, since it suggests how much progress we could make if we stopped missing it.

Another one of his favorite ideas is "biological nitrification inhibition," a nerdy concept with amazing potential to ease the global nitrogen crisis. The CliffsNotes version is that soil bacteria break down the nitrogen in fertilizer (which comes in the relatively harmless form of ammonium) into nitrates (which leach into water bodies) and nitrous oxide (which traps heat in the atmosphere). Enter Guntur Subbarao, a self-effacing Indian agronomist working in Colombia for a Japanese research institute, on a budget so paltry Searchinger didn't think it qualified as "shoestring." Subbarao's team helped develop a high-yielding strain of wheat that prevents soil bacteria from breaking down ammonium—so the nitrogen can't escape into the environment, and instead sticks around helping crops grow. He was like a modern Norman Borlaug, without Rockefeller Foundation funding.

When Searchinger read Subbarao's work, he realized it could change the nitrogen game—helping farmers get more out of their fertilizer, limiting emissions and pollution, maybe even boosting yields. As the self-appointed marketing department for underappreciated scientists, he coauthored a paper with Subbarao on the "more ammonium solution," hoping to publicize efforts to test the new wheat and breed similar varieties of corn. It gnawed at him that the world was spending hundreds of billions of dollars on fertilizers that created hundreds of billions more dollars of damage, while stiffing research that could help clean up the mess.

More good news: In 2024, the Novo Nordisk Foundation, the world's richest charity, awarded Subbarao's nitrogen initiative a $21 million grant, by far the largest investment of its kind. Still, that's less than one day's worth of U.S. farm subsidies, or of Novo Nordisk's earnings from its Ozempic weight-loss drug.

By now, this should be a familiar theme: tiny positive steps toward distressingly distant goals. The starkest example is the recent surge in government investment in meat and dairy substitutes, which, despite their struggles in the marketplace, are still the only solution that can chip away at animal proteins the way clean energy is chipping away at fossil fuels. If public funding can help alternative proteins get good and cheap enough to outcompete high-emissions incumbents, like it did for alternative energy, it can help prevent more of the earth from becoming an animal farm. The key is helping them overcome what Bill Gates calls "the green premium," the higher initial cost of cleaner new technologies.

Before 2020, the alt-protein sector had attracted only $100 million in public funding, seed money that included a USDA research grant that led to the founding of Beyond Meat and Dutch government support for Mark Post's lab that helped spur the invention of cultivated meat. But governments pumped in another $1.5 billion over the next four years. The U.S. government spent twice as much in 2023 as it spent before 2023. It has created a precision fermentation center at the University of Illinois, a cultivated-meat institute at Tufts, and a "Cornucopia" program for microbial foods at the Pentagon research agency that helped develop GPS and the Internet.

Still, USDA's investments were only 1 percent of its spending on COVID relief for livestock producers, and nowhere near Bruce Friedrich's dream of an alt-protein Manhattan Project. The Good Food Institute's Advancing Solutions for Alternative Proteins initiative—appropriately, ASAP—has identified dozens of urgent research needs, as basic as better fat substitutes, as advanced as "mapping the secretome of animal myoblasts, adipocytes, and other cells in cultivated meat." There's so much to learn, yet for every public or private penny ever invested in alt-proteins, $10 was invested in alt-energy in 2023 alone.

"We're getting phenomenal traction, but the money is still practically nothing," Friedrich said. "If we're going to get out of the Valley of Death, we'll need funding for science and infrastructure."

Notice that he said science and infrastructure, not just science. He wants governments to subsidize alt-protein factories and bioreactors as well as research and development, just as they've subsidized solar-panel factories and solar farms.

Those investments are costlier, riskier, and far more controversial than R&D. In the U.S., more than a decade after a solar manufacturer defaulted on an Obama administration clean-energy loan, "Solyndra" is still a one-word argument against green subsidies. But because the politics of taxing and regulating high-emissions industries is so daunting, subsidizing low-emissions alternatives is often the most realistic way to address green premiums and the tragedy of the climate commons. Solar, wind, battery storage, and electric cars never would have gotten competitive enough to scale without government support for clean-energy deployment.

The Boston Consulting Group concluded that similar support for alternative proteins would get the best emissions reduction return of any climate investment—3 times better than decarbonizing cement, 11 times better than electric cars. Just as governments have installed solar on public buildings, they could serve plant-based meat and milk at schools and other public cafeterias. Just as they've compensated consumers who buy energy-efficient appliances or Teslas, they could compensate farmers who plant nitrification-inhibiting wheat or pongamia.

The world desperately needs to figure out what works, so R&D is necessary. But the world also needs to deploy what works, so R&D won't be sufficient.

• • •

Now consider rice production, the third-largest source of agricultural methane after cow burps and manure. In his sustainable food report, Searchinger discussed several water management techniques that reduce flooding on rice fields, which could reduce their methane emissions. And USDA approved an $80 million climate-smart grant to pay rice growers to try some of them.

They seem to be paying off. Jim Whitaker, a fifth-generation farmer in Arkansas, has cut his methane emissions in half with one technique, "alternate wetting and drying," without sacrificing any yield.

"It's a huge deal," Whitaker said. "It ought to be happening everywhere."

Cutting methane emissions in half from all 400 million acres of rice fields on Earth would be like erasing Mexico's greenhouse gases. Yet again, though, there's been no global deployment blitz. In fact, USDA only allowed Whitaker to enroll a couple hundred acres for the climate-smart grant, about 1 percent of his land—even though there's no acreage limit for farmers seeking funding for cover crops, no-till, and other carbon sequestration strategies with much more questionable emissions benefits. Most farmers won't splurge on green-premium solutions unless they're forced to or paid to, and it's tough to force farmers to do anything.

Searchinger would like to see funding to help scale up any solution that can reduce emissions for less than $50 a ton. Governments and foundations could help finance the pasture improvements that tripled stocking rates in Brazil; silvopasture systems where cattle graze carbon-storing vegetation like *Leucaena* and *Moringa*; and pongamia plantings in the tropics. Subsidies for factories—hard to finance with impatient venture capital—could scale up biofertilizers, biopesticides, and meat and dairy substitutes, if politicians were willing to weather the inevitable Solyndra-style failures. Insect protein has become a snarky right-wing talking point, and it's true that most of us aren't interested in eating bugs, but startups are converting insects like crickets and black soldier flies into incredibly efficient and sustainable fish and livestock feed.

There's still a lot to learn about what works, but there are already plenty of proven ways to improve the status quo. The nonprofit ReFED has identified 40 investable solutions just for food waste, as high-tech as hyperspectral imaging for spoilage detection and AI-driven pricing that discounts groceries before they go bad, as low-tech as education campaigns and composting.

The best bang for the buck in the land sector might be restoring drained peatlands, which generate 2.5 percent of all global emissions as they decompose and release their carbon. Searchinger once wrote a memo to a foundation suggesting its billionaire funder could almost singlehandedly solve a multi-gigaton piece of the climate puzzle by buying out farmed peatlands and rewetting them: "It is hard to imagine any other area in which a single philanthropist could do more to directly mitigate the climate, or such an important field in which philanthropy is absent." He cited evidence that returning farmed peatlands to Mother Nature could reduce emissions for less than $10 a ton, and in some cases just a few cents a ton.

So far, though, that billionaire has not invested in peatlands. Other than the do-gooders in the Danish government, hardly anyone is investing in peatlands.

The sad truth is that hardly anyone is making any major investments in decarbonizing the land sector. The World Bank found that only 2.4 percent of climate finance is flowing into food and farming. The UN's mechanism for funding forest conservation, REDD+, has generated just a few billion dollars since 2008, when the UN says it needs $200 billion *every year* to meet its climate goals.

The food, land, and emissions gap won't be closed unless we can close this money gap, too—and it turns out that Searchinger has a couple of big ideas about how to do that. But money, while also necessary, will also be insufficient. Before we get to his fundraising ideas, there are a couple of twists that will make defending nature from agriculture even harder than the last 300 pages have suggested

WICKED PROBLEMS

Let's step back for a broader recap.

Agriculture is eating the earth, overrunning the natural landscapes that sustain biodiversity and the climate to grow the food that sustains us. It's on track to eat even more of the earth as our population increases and more of us can afford meat. The math is remorseless: We can't stop deforestation until we stop expanding our agricultural footprint, so we'll need to reduce our demand for new farmland—by eating less beef, wasting less food, using less bioenergy, and making more food on existing farmland. Since we're not sure how to do all that, we'll need an all-out effort to figure out what works, then deploy what works. And that will require an extraordinary global financial commitment—not the billion here and billion there that famously starts to add up to real money, but orders of magnitude more.

The implicit argument is that no matter how hard we try to protect forests and wetlands, we'll fail if we keep eating more of the earth. It's Econ 101 again: When demand for farmland exceeds supply, nature takes the hit. It's admirable that most countries have embraced the UN "30x30" goal of conserving 30 percent of the planet's land by 2030, but if consumers keep eating land-intensive diets and producers don't jack up yields to meet global food demand, forests and wetlands will be screwed no matter who embraces what. Brazilian farmers and ranchers will tear down the Amazon and Cerrado to feed the world. African smallholders will do what they have to do to feed their families. Our demand for calories and protein will find a way around our desire to protect nature and the climate, just as a river finds a way around rocks.

That brings us to the first twist, a bummer Searchinger has tried to bring to the world's attention: Even if we do manage to stabilize our agricultural footprint, we still won't end deforestation, because zero *net* deforestation isn't zero deforestation. Even if we stop shrinking the earth's overall tree cover, trees will still fall and ecosystems can still get pillaged. He calls this "the land-shifting problem." In the long run, reducing global demand for new farmland will ease global pressure on nature, but in the short run, 600 million farmers will continue making local land-use decisions every day, bringing new land into production and taking depleted land out of production.

One reason this is such bad news is that global agriculture is shifting south, toward tropical forests and wetlands that are the world's most valuable carbon sinks. From a satellite perspective, converting an acre of North Carolina soybeans to a pine plantation and an acre of Amazon rainforest to soybeans might look like a wash, but it isn't from a habitat or climate perspective. Since some land clearing is inevitable, it really matters which land gets cleared.

Ideally, agricultural expansion would be limited to natural lands with low carbon opportunity costs, and nature would reclaim unproductive farmland. Forests have already reclaimed many abandoned fields in New England and western Europe, while China is actively reforesting low-yield cropland. The problem is, trees aren't regrowing nearly as fast in temperate zones as they're being pulverized in the carbon-rich tropics—and much of the regrowth is tree plantations with much less biodiversity than the native forests being lost. Searchinger finds the literature absurdly bullish about our potential to reforest marginal farmland and farm marginal natural land, because land is never free, and "marginal land" is rarely marginal. Studies identifying vast swaths of surplus land for "nature-based solutions" often assume that pastures can be rewilded without being replaced, or that land ill-suited for growing crops will somehow be well-suited for growing trees.

The second twist is a related problem that makes land shifting even worse, a catch-22 he calls "the wicked problem." On a global basis, agricultural expansion is the most dangerous threat to forests, which is why higher yields on existing farmland is so vital. But on a local basis, when farmers can grow more food per acre, they can usually make more money by clearing more acres. This helps explain the rapid expansion of high-yield soy and oil palm in Brazil and Indonesia. Tropical ecosystems face long-term doom if global farm yields don't increase, but they can face short-term threats when local farm yields do increase.

Searchinger's slogan for solving the land-shifting problem, the wicked problem, and just about the entire food and land problem is Produce and Protect—or in longer form, Produce, Protect, Reduce, and Restore. It won't be enough to produce more food per acre and reduce our demand for meat, biofuels, and other land-hogging products; we'll also need to protect key habitats and restore some unproductive farmland to nature. Not only will we have to eat less of the earth, we'll have to make sure the yield improvements necessary to achieve that long-term goal don't induce the short-term destruction of key ecosystems.

Right now, we're not even trying.

The main global initiative for financing climate progress in the land sector, REDD+, is widely viewed as a New Coke–level debacle. It's supposed to raise money from rich countries to stop deforestation in poor countries, but it's failed at both the raising and the stopping. It's also gotten mired in scandals over phony forest carbon offsets in the private carbon markets it helps support. Its basic strategy is to pay farmers and nations not to cut forests, but since forests cover one-third of the land on Earth, that strategy can't work without crazy money. And since it hasn't attracted crazy money, its defenders, like communism's defenders, often argue that it hasn't really failed, it just hasn't been tried.

Searchinger is not as harsh as many REDD+ critics who want to scrap it entirely. He describes it as Churchill described democracy, the worst system except for all the others. He agrees with its basic premise that poor nations will only protect forests if they're paid, and that rich nations that already cleared most of their forests should do the paying. He doesn't have a moral problem with letting big emitters buy offsets for their emissions.

He just doesn't think REDD+ in its current form can produce good enough results to reduce our eating of the earth. He sees three crippling problems: It's preposterously underfunded. Its strategy isn't Produce and Protect, just an ineffectual Protect. And it's a purely voluntary system. Even if REDD+ somehow attracts enough funding to start offering generous carrots, he's convinced that unless they're accompanied by credible sticks, forests will keep falling.

There are just too many misaligned incentives. Paying a farmer or rancher not to cut a forest runs into the problems of permanence (he might cut it later, or it could burn), additionality (he might never have intended to cut that forest), and leakage (a different forest might get cut instead). Landowners with the weakest desire to cut forests have the strongest incentives to sell sketchy credits for not cutting them; buyers want the cheapest credits, not the highest-quality ones; and the third-party verification outfits that are supposed to ensure the quality of credits can lose business if they refuse to certify sketchy ones. Searchinger sees leakage and additionality as Scylla and Charybdis: If you pay someone not to cut over here, someone will probably cut over there—and if not, the first guy probably wouldn't have cut in the first place. Meanwhile, the more forests get protected, the greater the economic payoff for cutting unprotected ones, and the more expensive it becomes to pay remaining forest owners not to cut.

There just isn't enough money on Earth to keep giving it to everyone who agrees not to harm the environment. And it's futile to keep paying people to do helpful things if you can't stop harmful things. The trendiest new carrots-only approach is the Trillion Trees initiative, a global afforestation effort launched at Davos that even Trump has endorsed. But while planting a trillion trees could help the climate, as long as they're not planted on productive agricultural land, they won't help much if they can't be protected, or if other trees keep getting cleared. In exile from Impossible Foods, Pat Brown has pursued a new dream of getting carbon markets to reward the rewilding of cattle pastures, but again, how do you pay landowners for getting rid of cows over here without inducing more cows to leak over there?

Searchinger's solution is a global quid pro quo where crazy money to help Produce is tied to binding commitments to Protect, a mandatory REDD+ where the international community gives lucrative carrots to forest-owning countries that use credible sticks to crack down on deforestation within their borders. Much of the money would go to help their agriculture sectors make more food on existing farmland, by financing better seeds, fertilizers, tractors, and other yield innovations, but on the condition that their farmers can't clear new land. The funding could also help fund satellite monitoring, wildfire management, and environmental enforcement, but the governments would stop getting paid if they failed to protect their forests, or if they built roads and other infrastructure promoting agricultural expansion in sensitive areas.

Brazil has shown the potential and the peril of that kind of tough love. During Lula's first reign, it shut down farm credit in states with excessive deforestation, creating incentives for farmers to stop clearing forests. (It also created incentives for farmers to monitor each other, because bad actors could jeopardize all their loans.) But the international community failed to supply the financial love it had promised for Brazil's toughness. Anger in the countryside over those broken promises undermined political support for conservation and helped elect Bolsonaro, whose enthusiasm for developing the Amazon led the international community to cut off aid entirely. Lula's return to power has reduced deforestation in the Amazon, but not in the Cerrado.

Even under Bolsonaro, technocrats in Brazil's agriculture ministry pursued a sensible sustainable intensification strategy that rewarded farmers and ranchers for making pastures more productive, using biological alternatives to chemical fertilizers and pesticides, and taking other actions to increase yields and reduce impacts. But the ministry's head of innovation, Renata Bueno Miranda, said that without global support, the government couldn't pay farmers nearly as much to pursue sustainable intensification as they could make through deforestation.

"Everybody tells Brazil what to do, and nobody gives us the money to do it," said Miranda, a veteran of both the Lula and Bolsonaro administrations. "In the '70s, we told our people agriculture was a national priority, go cut the trees. Now you want us to tell them to hug and kiss the trees. Fine. But they've got to get paid!"

Searchinger hopes to achieve that with his first big money idea, which would not only fund forest protection but defund forest destruction: Instead of requiring airlines to pursue sustainable aviation fuels that aren't sustainable, just tax airline emissions. He says a $100-a-ton global carbon fee on air travel, which would increase the cost of a transatlantic flight by $100, could raise $100 billion a year—and unlike farm-grown jet fuels, it wouldn't trigger devastating land-use changes or compete with the food supply. Airlines would still have incentives to pursue truly sustainable alternatives that could cut their emissions in the long term, like electric planes for short flights, but they wouldn't have to pursue fake energy solutions that make the food and land problem worse. He says a similar emissions tax on the shipping industry, which is also considering a potentially disastrous shift to biofuels, could raise the other $100 billion a year the UN says it needs to curb deforestation.

Searchinger's other climate finance idea would focus on wealthier countries: Redirect some of the cash already avalanching into the agricultural sector to R&D and deployment of eating-the-earth solutions. Governments provide $600 billion in agricultural support every year—half through market barriers that raise prices for consumers, half through subsidies—and most of the subsidies are handed to farmers with few or no strings attached. Rewarding farmers for field-testing innovative ways to cut emissions and intensify production, growing eco-friendly crops, and rewilding unproductive fields and pastures would make much more sense than rewarding them for merely farming or owning farmland.

In the past, the rich countries with the most farm subsidies have not been eager to link them to climate mitigation. But Denmark is a reminder that past is not always prologue. And while the world does not seem eager to impose a $200 billion-a-year tax on air travel and shipping to protect forests, either, that would amount to just 0.2 percent of global economic output, the size of the global video game market or the U.S. Air Force budget. Governments are already spending twice that much on clean energy, and as energy emissions start to dip, food and land emissions will get harder to ignore. Denmark went after its farm sector because it was already cleaning up its other sectors—and its farmers, recognizing they were becoming outliers, cut a deal. The E.U. has already banned imports of agricultural commodities from recently deforested land, which mostly just encourages countries to send their commodities from recently deforested land elsewhere; maybe it could be pressured to adopt more effective reforms, like tariffs on agricultural commodities from countries with any ongoing deforestation.

The world has been slow to focus on the land sector's climate problems, and with climate-skeptical populists gaining power in the United States and elsewhere, those problems seem likely to continue to fester. It certainly doesn't help that when non-skeptical global policymakers have focused on farms and forests, their preferred solutions—biofuels, biomass power, regenerative carbon farming, REDD+—have often been ineffectual or counterproductive. At some point, though, if they're serious about fixing the general climate problem, they're going to have to get more serious about the one-third of it that's still getting worse. There's no doubt that it would be expensive to launch an all-out global effort to research food and land solutions, deploy the best ones, and finance Produce and Protect deals generous enough to save forests. But letting climate change spiral out of control would be expensive, too.

The key, as always, will be to get the incentives right—so farmers can make more money by making more food with less land; forests are worth more standing and storing carbon than logged and burned; and nations and corporations that want to shrink their carbon footprints get rewarded for shrinking the eating-the-earth problem. None of that will happen without better funding and better policy. But it also won't happen as long as land is considered free.

That's the final twist. If climate policies and the emissions analyses behind those policies keep ignoring the carbon opportunity costs of using land, they'll keep promoting bioenergy and low-yield agriculture rather than sustainable intensification and ecosystem restoration. When steps forward are counted as steps backward and vice versa, you can't expect forward progress.

Sorry yet again, because this is anticlimactic, too: The last piece of the policy puzzle will be better accounting rules. They're not sexy, but they can determine whether problems get solved.

The accounting rule book that most Fortune 500 companies use to calculate their emissions is called the Greenhouse Gas Protocol, and in 2019, the organizations that oversee it decided to fix its approach to the land sector. That made sense, because the Protocol treated land as mostly free. It only assigned climate costs to clearing new farmland, not to using existing farmland, even though the inefficient use of existing farmland leads to the clearing of new farmland. It did not reward higher yields or penalize indirect land-use change, so it created perverse incentives for food and ag businesses that wanted to reduce their climate impact.

Rewriting the Protocol was a great opportunity to steer the Cargills, Tysons, Nestlés, and Walmarts of the world toward practices and products that could ease pressure on nature, while sending a message to policymakers that land use mattered. It was a particularly great opportunity for Searchinger, because it just so happened that one of the two organizations that oversaw the Protocol was the World Resources Institute, where he was technical director for land issues.

And he blew it.

He met with the Protocol's staff early on to explain why land-use accounting needed reform, and how entire industries like biofuels and biomass depended on basic accounting errors. But he didn't get involved in the reform process. He was busy. It looked painfully bureaucratic, with elaborate clusters of working groups and technical committees. His protégé at WRI, Richard Waite, an upbeat wonk who coauthored his food report and embraced his land-use ethic, was assigned to coordinate the process, so he figured it would turn out fine.

Over the next couple years, he got sporadic updates from Waite—who, as coordinator, had limited ability to weigh in on the substance—and he got a sense that industry executives and consultants were dominating the process. Still, he didn't focus on the details. When the process produced a 429-page draft guidance in the fall of 2022, he only skimmed it, and found it mostly impenetrable. He rarely prints out documents, because he hates to kill trees, but this one had so many bewildering cross-references that it didn't make much sense on a screen.

In the spring of 2023, he finally printed it out, stuffed it into a three-ring binder, and spent an afternoon paging through it in his Princeton office, flipping back and forth through the references as if he were reading a Choose Your Own Adventure story. It was mostly opaque legalese interspersed with jargon-filled boilerplate, but he had the legal experience to catch how ambiguities in a quasi-legal rule book could be exploited. And he caught a bunch. Every hundred pages or so, he'd spot a seemingly innocuous sentence or two written in clunky coded language that subtly threatened to destroy the world's forests. It was as if Cher's *Moonstruck* character not only buried her affair with her fiancé's brother in an unreadable 429-page confession, she called it a "prenuptial indiscretion involving the future non–blood relative referenced in section 9(d)."

He quickly realized the new draft had failed to fix the Protocol's old problems. It still treated existing agricultural land as free, so it still blessed most bioenergy as carbon-neutral. It didn't distinguish between an efficient farm that made lots of food with a little land and an inefficient farm that made a little food with lots of land. It only counted emissions from agricultural production, so it would still encourage efficient producers to shoot (or at least sell) their pigs, while pushing food companies and agribusinesses toward low-yield regenerative products that would expand the food and land gaps.

When he got to the draft's treatment of forests, he realized it would also create a slew of new problems. It would let forest owners claim emissions reductions for trees growing on their land whether or not they did anything to promote that growth. So timber companies could cancel out their emissions from cutting trees just by owning trees, while meat conglomerates and fast-food chains could claim offsets just by buying shares in timber companies. The draft would classify most wood as not only carbon-neutral but carbon-negative, which meant that the more lumber and IKEA furniture a company purchased, the more it would reduce its emissions, even if it decided to burn everything it purchased in a bonfire.

He felt sick. If WRI gave its seal of approval to accounting rules that treated land as free, companies would have no incentive to invest in real solutions like pongamia or pasture intensification. It would contradict WRI's public positions and his body of work. And it would give policymakers a powerful new justification for promoting fake solutions like grass-fed beef, corn ethanol, and biomass plants; when enviros argued against them, defenders could just point to WRI's own guide.

He knew he would get accused of a process foul if he tried to undo a consensus built over three years of meetings without him, just as he had been yelled at after challenging consensus approaches to Everglades restoration and biofuels analysis. But the process wasn't quite done yet. And he believed it would be a moral foul for WRI to sell out the climate and undermine everything it had published in *Creating a Sustainable Food Future*.

So Searchinger belatedly raised alarms with his bosses, churning out memos backed by excerpts and citations from the Paris accords and the scientific literature. He also jumped into the final stage of the process, a four-day workshop at WRI's offices led by facilitators from the Consensus Building Institute. The goal was supposed to be to hash out language everyone in the room could accept, but since most of the room had ties to industries hoping to greenwash their products, his goal was to prevent decisions until the scientific world knew what was happening.

First, he had to endure four days of hell. He and William Moomaw, the Tufts ecologist who had fought bioenergy in Massachusetts and on the IPCC, were the main voices arguing that land was not free, and Moomaw left early. Industry types kept accusing Searchinger of trying to ruin the consensus to push a personal agenda, as if he derived a financial benefit from carbon-opportunity-cost accounting that tied climate rewards to actual climate action.

"It was so discouraging, all those corporate people trying to get credit for fixing the climate without doing any work," Moomaw recalled.

Searchinger felt like he was arguing in the Twilight Zone, making indisputable points that were not only disputed but rejected. He repeatedly tried to explain that a foundational principle of carbon accounting, emphasized throughout the Paris accord, was that only forest growth caused by human activity counts as a legitimate carbon removal. Forest owners shouldn't get rewarded for carbon removal that would have happened without them, especially now that the excess carbon in the atmosphere was fertilizing faster forest growth worldwide. Nevertheless, the question was put to a vote, and a breakout group dominated by forestry interests decided the guidance shouldn't even try to determine whether human activity had anything to do with forest growth. It felt like Lincoln's riddle about how a dog still has four legs even if you call a tail a leg, except that if the Protocol incorrectly called wood harvests carbon-negative, more wood would get harvested and the planet would get hotter.

The whole workshop felt like an exercise in calling tails legs. At one point, Waite had to ask if the group really wanted meat to be considered carbon-negative—so that buying more of it would be deemed climate-friendly—if the livestock's owner happened to own a nearby forest. Most of the room was fine with that. But Searchinger wasn't, and the nice thing about consensus building was that his dissent could prevent closure. He managed to keep the key issues unresolved.

After the workshop, the cavalry arrived. More than 200 scientists signed a letter warning that the Protocol would be fundamentally flawed if it made biofuels carbon-neutral and failed to fix its approach to wood. The

European Academy of Sciences also sent a harsh critique, as did dozens of green groups.

Over the next few months, WRI's leaders and their partners on the Protocol, the industry-led World Business Council for Sustainable Development, agreed to punt the forestry issues to a new process with more independence and more scientific input. The agricultural issues were also unresolved as this book went to press; it was not yet clear whether the revised Protocol will recognize the carbon benefits of more efficient land use, or whether companies that use more land will be required to count the carbon opportunity costs.

The big question is whether land will still be considered free. That's the key. As more governments start limiting emissions, and more corporations start paying attention to their carbon footprints, rules that don't penalize additional land use or reward efficient land use will just encourage more land clearing. Establishing the principle that land isn't free won't solve the eating-the-earth problem, but it's yet another prerequisite. Every piece of ground needs to be valued as a potential food source or a potential carbon sink.

Searchinger used to shorthand his land-use philosophy with that twist on Bush's education slogan: No acre left behind. His new slogan is a Monty Python reference: Every acre is sacred. It's a play on "Every Sperm Is Sacred," a satire of prudish Catholicism sung by an angelic children's choir in *The Meaning of Life*. He likes the religious overtones, the implication that wasting land is sinful.

The challenge will be to treat every acre of farmland and every acre of nature as sacred, because God isn't making more of them. That's the gospel he believes can save the world. It's about using land efficiently—by maximizing yields, growing less energy, eating less beef, wasting less food, and resisting the delusion that soil carbon will save us. It's about research and deployment, Produce and Protect. We now know what we need to do to stop eating the earth.

But like that Danish politician said: Now we've got to do it!

EPILOGUE
CAN WE DO THIS?

IT'S HARD TO SOLVE A CRISIS

When America shut down in March 2020, Searchinger temporarily set aside his work trying to help solve the climate crisis. He decided, honorably if not exactly humbly, that he needed to help solve the coronavirus crisis.

He wasn't a public health expert or an economist, but a lack of credentials had never stopped him from diving into bioenergy or food. And he was once again convinced that everyone was missing something hidden in plain sight.

Congress was finalizing a stimulus bill to shower money on furloughed workers and shuttered businesses. That seemed like a no-brainer. But even in the pandemic's early days, he didn't understand why nobody was talking about how to reopen the economy after that emergency blast of cash. The daily cost of keeping the workforce at home was so incomprehensibly vast that anything Washington could do to speed up its return to work seemed like another no-brainer. Sitting in his own Takoma Park home, trying to imagine the entire economy in lockdown until vaccines were ready, he kept thinking: *This is not sustainable.*

It also seemed fixable. So he thought he should try to fix it.

"Something obvious was missing," he said. "It seemed like I could help."

The missing link was ubiquitous COVID testing—not just for the sick, but for all workers, all the time, so they could go back to work safely. When he ran his theory past actual experts, they agreed. Mass testing was common sense.

But Trump didn't want mass testing, for the perverse public-relations reason that it would increase the number of confirmed infections. "So I said to my people, 'Slow down the testing, please,'" he blurted out publicly. Congress was less flagrantly indifferent to the spread of the virus, but almost all the spending in the $2 trillion CARES Act it was rushing to pass was for immediate relief.

"They just wanted to give people money," Searchinger recalled. "It was amazing how little they were talking about what happens next."

So less than two weeks into the lockdown, he pitched *The Washington Post* an op-ed calling for a World War II–style mobilization to manufacture and administer tens of millions of tests every week to get workers back to work. The *Post* published it the next day. It was a bit odd for a climate guy to propose such a radical public health policy in such a prominent media outlet.

But, it should be clear by now, not too odd.

The next morning, when Searchinger walked outside in his bathrobe to get his paper, his next-door neighbor congratulated him on his column. That neighbor happened to be Jamie Raskin, the liberal congressman and constitutional lawyer who would become famous the next year for leading Trump's impeachment for insurrection. The two like-minded crusaders had a socially distanced, extremely Beltway chat about how vital it would be to restart the economy quickly and responsibly, not only for the country but for the Democratic Party, which they feared would get blamed for long lockdowns. Raskin agreed to talk to Speaker Pelosi about testing.

But the bipartisan CARES Act passed three days later without new money for tests, much less a new Manhattan Project for tests. Raskin thought ignoring the testing problem would be a substantive and political disaster, so during their next chat across their yards, he asked Searchinger to draft new legislation addressing it. Searchinger spent the next month basically working as an unpaid Raskin aide.

"I've got NIH and FDA in my district, so I represent some of the greatest scientists in the world," Raskin said. "But I was lucky to have Tim next door."

This isn't a pandemic book, so I won't recount the whole saga of the Reopen America Act that Searchinger wrote and Raskin introduced. It laid out a plan that would have empowered states to launch an all-out testing push at federal expense. Unfortunately, Republicans thought enacting a plan would jam Trump. And while Raskin helped persuade Democratic leaders to take testing seriously, they preferred throwing cash at the problem to taking ownership of a strategy to fix it. Pelosi bluntly told one of his allies: "Congress doesn't do plans. We do money."

The next COVID package in April did include $25 billion for testing, but no plan—and that was the last package of 2020. Election-year bipartisanship had passed its expiration date, so America muddled through the crisis with a spotty testing regime focused on the sick. Much of the economy reopened before vaccines were available anyway. More than a million Americans ultimately died.

I know, I know, that's another bummer of a story. And it raises a bummer of a question: If we couldn't come together to fight a virus that was trapping us at home and killing thousands of us every day, how the hell are we going to solve a slower-burning collective action problem that won't visibly improve even if we collectively act? How do we win a global war against an invisible enemy when the president of the United States won't fight? There's no vaccine for climate change, and even if there were, how many of us would take it?

Searchinger sometimes wonders whether his professional eating-the-earth advice will end up driving any more change than his amateur pandemic advice. Most of the world isn't Denmark, and the global trajectory has not improved since he published his sustainable food report in 2019. Population and beef consumption are growing a bit slower than expected, but cropland and rangeland are still expanding as quickly as he feared.

There's still a lot of groupthink, status quo bias, and flat-out wrongness in the land sector, and the IPCC still often assumes land is free. The world certainly isn't treating food and land as a five-alarm emergency; one recent study found that only 7 percent of news stories about climate change even mention meat or livestock. And Trump's return to power means the end of U.S. government efforts to reduce emissions for at least four years; it could also mean federal obstruction of cultivated meat and other climate-friendly technologies. It's fortunately unlikely that Robert F. Kennedy Jr. will persuade Trump to launch a Sri Lanka–style campaign against agrochemicals, but it's quite likely that his administration will allow Big Ag to pollute and emit as much as it wants.

In general, Searchinger has to admit there's some evidence for the Terminator's view that it's in our nature to destroy ourselves. Atmospheric carbon levels keep soaring—from 317 parts per million at his birth to 383 when he started working on climate to a recent high of 424, far past the 350 threshold needed for climate stability. It's hard to deny that we often suck.

But we don't always suck. The willingness of former bioenergy advocates like the scientist Alex Farrell, environmentalist Nathanael Greene, and public servant Ian Bowles to draw new conclusions when confronted with new evidence reinforced his faith in the power of facts and logic. He's also optimistic about humanity's talent for solving problems through innovation; after all, COVID would've been even deadlier if scientists hadn't developed vaccines remarkably quickly. Maybe some of the geeks working on alternative proteins, gene-edited crops, and bioengineered fertilizers will have similar success. Searchinger once got unexpectedly emotional at a nitrification inhibition workshop in Japan, as he watched a few dozen underfunded agronomy obsessives from all over the world working together on obscure climate solutions that would never make any of them rich or famous. They made him genuinely proud of our species.

COVID also inspired, at least early on, genuinely prosocial behavior, as most people outside the Oval Office tried to do their part to stop the spread. Like World War II or the stories you always see about neighbors helping neighbors after natural disasters, the pandemic was a reminder that we're at least capable of sacrificing for the common good during a crisis. And even though our drastic emissions reductions during the lockdowns were an involuntary consequence of our inability to drive or fly, nobody can ever again claim we're incapable of drastically reducing emissions by changing our behaviors.

I'm not supposed to say that. It's no longer considered cool in environmental circles to suggest that climate progress depends on our personal choices. We're supposed to focus on collective action, not individual behavior.

A typical *Times* essay warned: "Worrying About Your Carbon Footprint Is Exactly What Big Oil Wants You to Do." The climatologist Michael Mann made a similar case in eating-the-earth terms in *USA Today*: "You Can't Save the Planet by Going Vegan." A *Vox* headline actually proclaimed: "I Work in the Environmental Movement. I Don't Care If You Recycle."

This green consensus against individual responsibility is quite new. The U.S. environmental movement used to echo Pogo's lament from an antipollution poster for the first Earth Day in 1970: "We have met the enemy and he is us." The iconic Crying Indian ad the next year also emphasized personal virtue, featuring a tear trickling down a Native American's cheek as a bag of garbage chucked out of a passing car landed at his feet. "People start pollution," the narrator intoned, "and people can stop it." This antipollution zeitgeist helped build popular support for all those Earth Day–era environmental laws that persuaded Searchinger he could make a difference as an environmental lawyer.

But the new environmentally correct message is that only systemic change can stop climate change. Guilt-tripping ordinary people about leaving lights on, eating meat, and other individual actions with modest impacts is now considered counterproductive, a distraction from corporate and government actions with disastrous impacts. Mann's op-ed pointed out that the ad with the Crying Indian (actually an Italian American actor with a feather in his hair) was secretly financed by business groups eager to shift responsibility for pollution to the public. He spoke for many activists when he dismissed the modern focus on individual carbon footprints as a new Crying Indian campaign designed to shift responsibility for carbon pollution, "playing into the hands of the polluting interests."

It's understandable that enviros, after being caricatured for years as annoying scolds pestering us to stop using plastic straws, would prefer to make the enemy Big Oil and Big Meat rather than us. And it's true that individual emissions are tiny drops in a colossal atmospheric bucket. The solar panels on my roof have prevented 200 tons of emissions over the last seven years, a period when America's energy emissions were about 35 billion tons. My electric car saves 500 annual gallons of gasoline, when U.S. drivers burn 130 billion annual gallons. There are a billion cattle on Earth, and my decision to cut out beef because I felt like a journalistic hypocrite won't make a noticeable dent in global emissions if I live another million years.

But drops fill buckets. Neither ExxonMobil nor the Republican Party is driving America's gas-guzzling SUVs to the mall. Americans are. And neither JBS nor McDonald's is shoving three burgers a week down the average American's throat. The plunge in emissions during the lockdowns was an unwanted reminder that emissions come from us—the energy we use, the food we eat, the products we buy. You can't save the planet by going vegan, but it helps—and even though behavioral change alone can't fix the climate, the climate can't be fixed without it.

Especially the food part of the problem. When the scientific nonprofit Project Drawdown ranked the top 100 solutions for cutting emissions, reducing food waste and shifting toward plant-based diets came in third and fourth. I try not to be an annoying scold—I fly too much to criticize anyone else's carbon footprint—but I can confirm that it's possible to waste less food and eat less beef, the two most powerful daily changes we can make for the climate. My family hasn't wasted any food at home since the Bay Area startup Mill Industries gifted me a "food dehydrator," a sleek-looking garbage bin (created by the inventors of the Nest thermostat) that dries and shrinks our kitchen scraps into chicken feed without stink or schlep. By keeping food waste inside the food system, it saves soybeans as well as the land needed to grow them. And while I don't delude myself that I've saved the planet by quitting steak, an average American switching just half his beef intake to chicken would save four transatlantic flights' worth of emissions a year. We also stopped feeding our dogs beef and lamb; one study of pet food's "environmental paw print" found it generates nearly 3 percent of agricultural emissions.

It's true that the large-scale policy changes featured in the last chapter—massive investments in researching solutions and deploying the best ones, Produce and Protect deals to prevent deforestation, accounting rules that recognize land is not free—will be more important than the individual choices we make every day. But policy changes can also encourage individual changes. Mill's bins still cost too much to go mainstream, but subsidies for early adopters could help them scale, while "pay-as-you-throw" trash collection could reward Mill owners for reducing their rubbish. Even if meat taxes remain as unpopular as veterans' benefits for ISIS, governments could still promote climate-friendlier eating through carbon pricing that accounts for the opportunity cost of land and stricter regulation of animal agriculture—and if that's still politically toxic, more aggressive support for alternative proteins and food-labeling laws requiring carbon disclosures alongside nutritional disclosures could also nudge us toward eating less of the earth.

In any case, it's strange for climate hawks to argue that greenhouse gases are such an urgent threat to humanity that we desperately need systemic change to cut emissions—and also argue that we shouldn't worry about our personal emissions. Every ounce of carbon dioxide traps heat in the atmosphere. Every acre is sacred. When you tell people their personal emissions don't matter, they might believe it, and might assume the climate crisis isn't that urgent a crisis.

The Earth Day movement understood the connection between the personal and political, the idea that consistent progress required a broadly shared ethic that it's wrong to desecrate our planetary home. Even the manipulative Crying Indian ad helped build a consensus against pollution; you don't see much garbage chucked out of cars anymore, and all politicians at least pretend to support clean air and water. It will be hard to build that kind of consensus against climate pollution while dismissing the importance of individual climate action.

It will be even harder to build the kind of land ethic Searchinger envisions, a consensus that we need to maximize the efficiency of every acre. Most people don't think about the climate or the global environment when they order a sandwich, or look out a plane window at all those agricultural rectangles and circles. The new ethic would have to be grounded in science, economics, and logic rather than our intuitions about what kind of food and farming feels wholesome. Most of us would have to shed assumptions that pasture-raised beef must be more sustainable than industrially farmed fish, and small organic farms must be greener than intensive agriculture with higher yields. When large swaths of the international community, corporate community, and scientific community are pushing an opposing ethic that values how naturally farms are managed rather than how much nature they spare, reform can seem unimaginable.

Again, though, radical change always seems unimaginable until it isn't, and reform of an unsustainable status quo is worth pursuing even when it's hard. Searchinger had no business challenging the scientific and political consensus over biofuels and biomass, and it took just as much chutzpah to chart a new course for food and farming without a scientific or agronomic background. But he figured he ought to do what he could to try to help. We ignored the eating-the-earth problem for too long, but the Chinese proverb that the best time to plant a tree was 20 years ago, and the second-best time is now, is literally true in the land and climate arena.

THE TEST

One last story, a happier one.

In December 2023, Searchinger returned to the Everglades for the first time in years, revisiting the beleaguered ecosystem where he started his career as an environmental lawyer. He stood on a boardwalk jutting into a marsh, gazing out at a vast panorama of cattails, bulrushes, and other wetland grasses that arched in the breeze like a congregation at prayer. This was not one of the natural Everglades marshes he once fought to save—although it was full of alligators, and an endangered snail kite circled overhead searching for snails. This was one of the manmade filter marshes he once fought to build, in order to clean up polluted runoff from sugar farms upstream before it reached the natural Everglades.

What had once been an impossible dream was now a highly engineered expanse of muck and mire nearly the size of Manhattan. It was called "Stormwater Treatment Area 1-West," and it was working exactly as designed, absorbing nutrients from the sugar industry's fertilizers in its vegetation and soils. The water flowing into the River of Grass wasn't perfectly clean yet, but it was getting close.

"So awesome," he said with a broad smile. "And so ironic."

What was awesome was the cleaner water, a rare three-quarters-of-a-loaf victory for the patron saint of almost-lost causes. It was a vindication of the faith he placed in the law more than three decades earlier, when Big Sugar as well as the activist Charles Lee and Governor Chiles wanted to shift the Everglades water-quality battle into the political arena. A lot had gone wrong since then; the larger Everglades ecosystem restoration project he had attacked as an $8 billion Rube Goldberg mess was now a $25 billion mess, and a right-wing Supreme Court had wiped out many of the wetlands protections that had been the focus of his legal career. But there had been steady progress on Everglades water quality, because the consent decree imposed after Chiles surrendered in court had required the construction of 100 square miles of filter marshes like this one.

What was ironic was that those constructed marshes also happened to be excellent carbon sinks, so the Everglades nutrient-removal project was turning out to be an inadvertently excellent climate mitigation project. Since 1880, the draining and farming of Everglades peatlands had released over a gigaton of carbon, the equivalent of a year of emissions from all U.S. factories. This part of the northern Everglades had lost six feet of spongy black peat soils in a century. Now Stormwater Treatment Area 1-West was very slowly starting to reverse that process, rebuilding a few millimeters of peat every year. Peatland restoration was another one of Project Drawdown's top climate solutions, and Florida's court-ordered filter marshes accidentally amounted to one of the largest peatland restoration projects in world history. It was especially ironic considering the low-lying Everglades is one of the ecosystems most imperiled by climate-driven sea-level rise.

"We never even thought about climate change when we were pushing for this," Searchinger said. "It's pure coincidence. It's luck! But it's huge."

We'll need policy changes and personal changes to solve the eating-the-earth problem, but we'll also need some luck. We're already lucky that no one has figured out how to make cost-effective cellulosic ethanol; otherwise, hundreds of millions more acres might be growing fuel today, and deforestation might be far worse. In the future, appetite-suppressing drugs like Ozempic could unintentionally reduce global food demand, while a climate-driven expansion of the lone star tick, whose bite transmits an allergy to red meat, could coincidentally reduce beef demand. All kinds of technologies I haven't even mentioned could transform the landscape if someone manages to make them work at scale—alternative proteins fermented from air, a slew of AI applications designed to turbocharge genetic engineering, "boosted breeding" that aims to increase crop yields by accelerating evolution, even innovations unrelated to food and land that aim to capture carbon without trees. There's no way to know what we don't yet know.

That's a long-winded way of saying that shit happens, but shit does happen. To Searchinger, the hidden-in-plain-sight lesson of the filter marshes is that it pays to keep working and fighting the good fight, because maybe something good will happen. Maybe it won't, but if you don't keep working and fighting, it definitely won't.

Searchinger has now been a land and climate researcher for as long as he was an environmental lawyer, and some good things are starting to happen. In 2024, the European Union, which had already limited crop-based biofuels for vehicles, ruled them ineligible for its new subsidies for sustainable aviation, a near-admission that they were never sustainable in the first place. The green-ribboned wood wranglers of Enviva declared bankruptcy, foiled by biomass power's failure to outcompete solar and wind. New York City began making plant-based meals the default option for patients in its public hospitals unless they specifically request meat, an actual needle-moving policy shift that has reduced the carbon footprint of hospital meals by more than a third, and is already inspiring efforts to expand the program elsewhere. Starbucks also stopped charging extra for plant-based milk, a major step toward the mainstream. And Brazil announced a goal of intensifying grazing yields on 100 million acres of degraded pastures, an area larger than Montana. It's progress. Rodney Dangerfield would be impressed.

The climate, as the legendary Florida activist Marjory Stoneman Douglas used to say about the Everglades, is a test; if we pass, we may get to keep the planet. It's a test of our capacity for self-restraint, our ability to prioritize the future over the now, our willingness to course-correct. But it's not a pass-fail test. The earth won't implode if we only cut emissions 89 percent by 2040, and everything won't be hunky-dory if we meet the Paris goal of 90 percent. We should try to limit warming to 1.5 degrees, but since we'll probably fail, 1.6 degrees would be less damaging than 1.7, which would be less damaging than 1.75. As Josh Tetrick likes to say, perfect isn't on the menu, but better is good. It's certainly better than worse.

The binary idea that we're destined for extinction if we don't meet this target or stop that pipeline drives a lot of climate fatalism as well as climate anxiety. *The Onion* nailed the typical climate-civilian response: "Yeah, Yeah, Nation Gets It, We're Rapidly Approaching End of Critical Window to Avert Climate Collapse or Whatever." But that mentality is simultaneously too alarmist about theoretical calamities—we're no more destined for extinction from climate change than we were from COVID—and too dismissive about the need for hard work to avoid real harm. We've already passed the window to avert the collapse of the California town of Paradise, which was wiped out by a climate-accelerated wildfire, and more collapses are coming, especially for the flood-prone and drought-prone global poor. The case for action is that the less carbon we emit, the less global temperatures will rise, which will mean fewer Paradises lost and fewer victims.

Maybe that's not as inspiring as the apocalyptic warnings that it's game over for the climate if we don't transform our economy and our food system this instant, but there's no such thing as game over for the climate. The game will go on. We should try harder to keep the only planet with tacos and tennis and our children as hospitable as possible for our species and other species. Even if perfect isn't on the menu, there's a lot we can do to make things better, and better is better than worse.

Yes, change is hard—political change, structural change, behavioral change, any change. Inertia is a powerful force. There's a lot of money and influence backing the status quo. Nothing we need to do to fix our food and land problems will be easy—not making meat and dairy substitutes as delicious and affordable as the animal versions, not supersizing yields by reengineering crops or intensifying pastures or reinventing photosynthesis, not mobilizing rich-nation money to rein in poor-nation deforestation. But in less than two decades, Searchinger has driven real intellectual change, exposing the dangers of biofuels and biomass power, ferreting out the flaws of models and accounting systems that treat land as free, drawing up a plan for our voracious species to stop eating so much of the earth. He's provided a guide for how we can participate in our own rescue.

He's a smart guy, but as he's happy to admit, his change-the-world insights about opportunity costs and leakage weren't all that brilliant. It didn't take a genius to see that burning trees is worse for the climate than not burning them, or that storing a little more carbon in soils won't help the climate if it requires dumping a lot more nitrogen in soils. What really set Searchinger apart was his pain-in-the-ass tenacity, his refusal to read the room and stop saying things nobody wanted to hear, his insistence on fighting to correct wrong answers that would be inconvenient to correct. He's always believed that the arc of the scientific and political universe bends toward truth, even if it doesn't bend as quickly as it should. But it doesn't bend on its own. It needs to be bent, and that often requires unpleasant fights.

The climate fight is not the kind of war that will end with carbon dioxide's surrender on a battleship. There will be gratifying victories and frustrating defeats, but no final victory or defeat, just better or worse. Forests will fall and new forests will be planted. Wetlands will be drained and drained wetlands will be rewetted. But there will be a cost to every acre we lose, because land is not free, and every acre is sacred.

ACKNOWLEDGMENTS

I'm supposed to start my acknowledgments by making it clear that I take full responsibility for this book, and that any errors or misjudgments are mine alone. But honestly, if you hated it, you can assign at least a bit of the blame to Tim Searchinger. Not only did he inadvertently give me the idea for it, he's taught me more than anyone else about its substance, and he's been an enthusiastic adviser as well as an infinitely patient source. He always knew that he'd have no control over it, and that not everything in it would be entirely flattering, but he seemed to care as much as I did that every word of it would be accurate and compelling. Since that first fateful call when I asked him if meat was bad for the climate, we've probably spoken a thousand more times, not including the time we spent together in Denmark, Brazil, Washington, and Florida. I sometimes struggled to follow his soliloquies about technicalities, and he was occasionally a pain in my ass, but he never told me anything that wasn't true, and he never asked me to do anything but tell the truth. I owe him a debt I can never repay, and I hope he's pleased with the result. (For anyone who isn't, his Princeton email is easy to find.)

Hundreds of other generous sources have also helped reduce my ignorance about these issues; I'm especially grateful to Bruce Friedrich, Rini Greenfield, Paul Shapiro, and Naveen Sikka. I've also learned from all the food and climate reporters and authors who beat me to this story; they're listed in my introduction to sources, but since nobody reads introductions to sources, I also want to acknowledge my debt to them in the acknowledgments.

I received generous support for my research from the University of Missouri's Watchdog Writers Group program; I'm so thankful to Christopher Leonard, who taught me so much about the weirdo craft of book-writing, and my fellow fellows, my companions in book-writing hell. Thanks also to Kenny Broad and the Abess Center at the University of Miami, as well as Danny Amron, Gabriela Karlan, Zach Marcus, Mark Ossolini, Kathryn Poppiti, and Jana Schleis.

This is my third book with Jonathan Karp's great team at Simon & Schuster, but my first without the irrepressible Alice Mayhew, who was so excited about this project before she died, although she did warn me "it better not be *Malthusian*." (It's not, Alice!) I loved working with Emily Simonson, whose sharp editing eye and good energy made the hell less hellish. I'm also grateful to my guardian angel Priscilla Painton, production editor extraordinaire Rachael DeShano, publicist extraordinaire Rebecca Rozenberg, and all the other extraordinaires at S&S. And as always thanks to my mega-agent, Andrew Wylie, as well as Jacqueline Ko and the rest of his team. I've also been lucky to work with great editors who let me workshop my ideas into print, including my former boss Steve Heuser at *Politico*, Lisa Hymas and Eric Wesoff at *Canary Media*, Eliza Barclay at *The New York Times*, Yoni Appelbaum at *The Atlantic*, and the team at *Heatmap News*.

Books are best in the pluperfect; they're nice to have written, but they suck to write. I couldn't have finished this one without the support of my super-friends: Fernand Amandi, Peter Baker, Peter Canellos, Marc Caputo, Susan Glasser, Jon Gross, Jed Kolko, Indira Lakshmanan, Richard Primus, Manuel Roig-Franzia, and the GPT305 crowd. Extra credit for three of my nearest and dearest: Gary Bass, a compassionate fellow-sufferer and guide; Tamar Haspel, my *Climavores* cohost and beloved food guru; and my bestie Mark Wiedman, whose come-to-Jesus anti-pep talk yanked me out of my Preposterocene. Gary and Tamar were also kind enough to read chapters, as were the two Peters, my friends Walter Alarkon, Dean Karlan, my awesome brother-in-law Philip Arlen, Dan Blaustein-Rejto, Frank Mitloehner, and my best and gentlest reader, the brilliant Buddha himself, the one-of-a-kind Max Dominguez Grunwald.

This book is dedicated to Max and his hilarious, badass, destined-for-greatness sister Lina Dominguez Grunwald, the brightest lights of my life. They grew from little kids to awesome mini-adults while I was working on this book, and my favorite part of the work was my daily escape from my cave when they came home. It's my greatest honor to be their dad. I hope we can leave the earth in half-decent shape for these two beautiful souls.

I should really dedicate everything I do to my amazing parents, Doris and Hans Grunwald, whose love and support have been the constant in my life. I owe them everything, and I should tell them more often how much I love and appreciate them. Love also to the rest of my wonderful family: Dave, Ruchi, Judy, Steve, Maylen, Phil again, Peggy, Jared, Carmen, and all my cousins, nieces, and nephews.

My partner on this journey, and all my journeys over the last 20 years, has been Cristina Dominguez. My love for her is still as deep as ever.

NOTES

A NOTE ON SOURCES

I interviewed nearly 2,000 people while I was reporting this book, visited 10 U.S. states and four foreign countries, and basically carried on a five-year conversation with Tim Searchinger. I don't cite those interviews in these endnotes, but they inform every page. Searchinger also allowed me to review all his incoming and outgoing emails about bioenergy from 2007 through 2009, which helped me put together the story of chapters 1, 2, and 3. I am especially indebted to the terrific data hounds at *Our World in Data*, which I relied on for most of the statistics in this book. I cite specific articles for specific facts, but I owe much of my background knowledge to the great work of many food and agriculture writers, including Daniel Blaustein-Rejto, Dan Charles, Art Cullen, Helena Bottemiller Evich, Joe Fassler, Jonathan Safran Foer, Jonathan Foley, Georgina Gustin, Frances Moore Lappé, Amanda Little, Charles Mann, George Monbiot, Oliver Morton, Ted Nordhaus, Adele Peters, Tom Philpott, Michael Pollan, Gabriel Popkin, Chase Purdy, Matt Reynolds, Elisabeth Rosenthal, Deena Shanker, Paul Shapiro, Jenny Splitter, Kenny Torrella, Richard Waite, Elaine Watson, Timothy Wise, Larissa Zimberoff, and my friend and *Climavores* cohost Tamar Haspel. I've also learned so much from so many brilliant climate journalists, including Zack Colman, Juliet Eilperin, Jeff Goodell, Elizabeth Kolbert, Bill McKibben, Robinson Meyer, Andrew Revkin, David Roberts, David Wallace-Wells, Bill Weir, and the entire team at *Canary Media*. As I mention in the notes, some of my reporting originally appeared in my articles in *Time*, *Politico Magazine*, *The New York Times*, *Canary*, and *Heatmap News*.

INTRODUCTION: THE LAND PROBLEM

2 *There hasn't been this much carbon*: Hannah Ritchie and Max Roser, "CO2 and Greenhouse Gas Emissions," *Our World in Data*. When I don't specify otherwise, I use *Our World in Data* for most this book's statistics on climate, land use, agriculture, population, and just about everything else. It's a terrific resource.

2 *The earth hasn't heated up this quickly*: Emily J. Judd et al., "A 485-Million-Year History of Earth's Surface Temperature," *Science* 385, no. 6715 (September 20, 2024).

2 *apocalypse on loop*: Al Gore collects this kind of disaster porn for his famous slideshow, which he compares to "a nature hike through the Book of Revelations." Michael Grunwald, "Al Gore: Optimist?" *Politico Magazine*, October 27, 2015.

2 *Scientists give us almost no chance*: Damian Carrington, "World's Top Climate Scientists Expect Global Heating to Blast Past 1.5C Target," *Guardian*, May 8, 2024.

3 *three-fourths of young people*: Elizabeth Marks et al., "Young People's Voices on Climate Anxiety, Government Betrayal and Moral Injury: A Global Phenomenon," *Lancet Planetary Health* 5, no. 12 (December 2021).

NOTES

5 *a soccer field's worth of tropical forest every six seconds*: Mikaela Weisse and Elizabeth Goldman, "We Lost a Football Pitch Worth of Primary Rain Forest Every Six Seconds in 2019," World Resources Institute, June 2, 2020. And primary forest loss was only one-third of the overall global loss of tree cover.

5 *If current trends hold*: Tim Searchinger et al., *Creating a Sustainable Food Future: A Menu of Solutions to Feed Nearly 10 Billion People by 2050*, World Resources Institute, July 2019. This WRI report forms the spine of chapter 6. It's not exactly a thriller—I'm probably the only layman who's read every word—but it provides a lot of the intellectual foundation for this book. The report projected nearly 15 Californias' worth of deforestation (about 1.5 million acres) from a 2010 baseline, but in this case I've updated the data.

5 *we must end all deforestation by 2030*: Intergovernmental Panel on Climate Change, *Climate Change and Land: An IPCC Special Report on Climate Change*, 2019; IPCC, *Climate Change 2014: Synthesis Report*, 2014.

7 *the Anthropocene's official start date*: Seth Borenstein, "Scientists Say a New Epoch of Human Impact - the Anthropocene - Began in the 1950s," *Los Angeles Times*, July 11, 2023.

7 *They've used ice-core data*: Kelly Tyrell, "Ancient Farmers Spared Us from Glaciers but Profoundly Changed Earth's Climate," *ScienceDaily*, September 7, 2018; Stephen J. Vavrus, Glacial Inception in Marine Isotope Stage 19: An Orbital Analog for a Natural Holocene Climate, *Scientific Reports* 8 (2018): 10213.

7 *one study of their tragic depopulation*: Alexander Koch et al., "Earth System Impacts of the European Arrival and Great Dying in the Americas after 1492," *Quaternary Science Reviews* 207 (March 1, 2019).

9 *fields and pastures have replaced*: Luciana Gatti et al., "Amazonia as a Carbon Source Linked to Deforestation and Climate Change," *Nature* 595 (July 14, 2021). About 17 percent of the Amazon is now agricultural.

10 *less than 3 percent of the world's climate finance*: William R. Sutton et al., *Recipe for a Livable Planet: Achieving Net Zero Emissions in the Agrifood System*, World Bank Group, 2024.

10 *an agribusiness sector that spends even more*: Omanjana Goswami and Karen Perry Stillerman, "Cultivating Control: Corporate Lobbying on the Food and Farm Bill," Union of Concerned Scientists report, May 13, 2024.

12 *during a 2024 podcast*: David Roberts, "The Energy Transition's 5 Supervillains and 5 Superheroes," *Volts*, May 1, 2024.

13 *a magazine story about my new green life*: Michael Grunwald, "My Life in the Elusive Green Economy," *Politico Magazine*, March/April 2018.

14 *his first scientific paper*: Timothy Searchinger et al., "Use of U.S. Croplands for Biofuels Increases Greenhouse Gases Through Emissions from Land-Use Change," *Science* 319, no. 5867 (February 29, 2008).

14 *a man-bites-dog* Time *cover story*: Michael Grunwald, "The Clean Energy Myth," *Time*, March 27, 2008.

15 *beef generated 50 times more emissions than coal*: J. Poore and T. Nemecek, "Reducing Food's Environmental Impacts Through Producers and Consumers," *Science* 360, no. 6392 (June 1, 2018), cited in Jonathan Foley, "Greenwashing and Denial Won't Solve Beef's Enormous Climate Problems," *Drawdown Insights*, October 30, 2024.

16 *it's coming for our agriculture*: "2024 Global Report on Food Crises," Global Network Against Food Crises, Food Security Information Network, http://www.fsinplatform.org/grfc2024.

17 *Agriculture has annihilated so much of the Amazon*: Scott Denning, "Southeast Amazonia Is No Longer a Carbon Sink," *Nature*, July 14, 2021; Chris A. Boulton et al., "Pronounced Loss of Amazon Rain Forest Resilience Since the Early 2000s," *Nature Climate Change* 12 (March 7, 2022).

ONE: THE GUY WHO FIGURES STUFF OUT

21 *He was reading a complex technical paper*: Michael Wang et al., "Effects of Fuel Ethanol Use on Fuel-Cycle Energy and Greenhouse Gas Emissions," Argonne National Laboratory, 1999.

22 *Henry Ford once called it*: Bill Kovarik, "Henry Ford, Charles Kettering and the Fuel of the Future," *Automotive History Review* no. 32 (Spring 1998).
23 *The U.S. industry owed its existence*: Tom Philpott, "How Cash and Corporate Pressure Pushed Ethanol to the Fore," *Grist*, December 7, 2006.
23 *barely 1 percent of America's fuel*: Caley Johnson et al., "History of Ethanol Fuel Adoption in the United States: Policy, Economics, and Logistics," National Renewable Energy Laboratory, November 2021.
24 *the emissions model*: GREET is an acronym for "Greenhouse gases, Regulated Emissions, and Energy use in Technologies."
30 *a dissident group of Dalton parents*: Margot Hentoff, "On Donald Barr, the Captain Queeg of Dalton School," *Village Voice*, May 9, 1974; Marie Brenner, "'I Had No Problem Being Politically Different': William Barr Among the Manhattan Liberals," *Vanity Fair*, December 2019.
35 *clerked for a moderate Republican federal appeals judge*: Searchinger passed up a chance to clerk for future Supreme Court Justice Stephen Breyer, a Democratic insider on the powerful D.C. Court of Appeals who was known for getting clerks plum D.C. jobs, but also for writing opinions himself. Searchinger wanted to foist his own wisdom on the world, so he clerked on a less prestigious appeals court for Judge Edward Becker.
37 *The proof arrived at a 1991 hearing*: Michael Grunwald, *The Swamp: The Everglades, Florida and the Politics of Paradise* (Simon & Schuster, 2006).
38 *he noticed a Times story*: Keith Schneider, "Bush Announces Proposal for Wetlands," *New York Times*, August 10, 1991, p. A7.
38 *He wrote his first-ever Times op-ed*: Tim Searchinger and Douglas Rader, "Bush's Cynical Attack on Wetlands," *New York Times*, August 19, 1991, p. A15.
39 *writing a 175-page report*: Tim Searchinger et al., *How Wet Is a Wetland? The Impacts of the Proposed Revisions to the Federal Wetlands Delineation Manual*, Environmental Defense Fund and World Wildlife Fund, 1992.
42 *a public-interest lawyer working the levers of power*: David Folkenflik, "Behind Scenes of Deal to Curb Bay Pollution," *Baltimore Sun*, November 7, 1997, p. A1.
44 *I spent the next year investigating*: I wrote dozens of stories about the Army Corps in *The Washington Post* in 2000. The ones referenced here include: "A River in the Red," January 9, 2000; "How Corps Turned Doubt into a Lock," February 13, 2000; "Generals Push Huge Growth for Engineers," February 24, 2000; "An Agency of Unchecked Clout," September 10, 2000; "In Everglades, a Chance for Redemption," September 14, 2000.

TWO: LAND IS NOT FREE

49 *The most authoritative study*: Alexander Farrell, et al., "Ethanol Can Contribute to Energy and Environmental Goals," *Science* 311, no. 5760 (January 2006): 506–508.
51 *a concept he introduced to the scientific community:* Timothy D. Searchinger, Stefan Wirsenius, Tim Beringer, and Patrice Dumas, "Assessing the Efficiency of Changes in Land Use for Mitigating Climate Change," *Nature*, December 12, 2018.
51 *or merely suggested the issue should be studied*: Edward Smeets, et al., "Sustainability of Brazilian Bio-Ethanol," Copernicus Institute, August 2006, http://www.neema.ufc.br/Etanol25.pdf. This paper noted in passing that land use "exceeds the scope of this research, but is deemed very important and strongly recommended for further research."
52 *This rush to convert nature*: Elisabeth Rosenthal, "Once a Dream Fuel, Palm Oil May Be an Econightmare," *New York Times*, January 31, 2007.
52 *Even an early IPCC report*: IPCC, Working Group III: Mitigation of Climate Change, 2001, http://www.ipcc.ch/ipccreports/tar/wg3/index.php?idp=0, p. 244.
54 *A literature review a decade after he coined the phrase*: Geert Walter et al., "European Commission Study Report on Reporting Requirements on Biofuels and Bioliquids," European Commission, August 2017.
54 *had argued in an unpublished manuscript*: Mark Delucchi, "Lifecycle Analyses of Biofuels," unpublished manuscript, 2006. The caveat in the Berkeley paper about how land use might be an

NOTES

issue—the caveat all six authors wished they could claim credit for—was actually inserted in response to private comments from Delucchi.

55 *A digital news startup that went live*: Jim Vandehei, "Welcome to Politico," *Politico*, January 23, 2007.

56 *ethanol plants were sprouting throughout*: Alexei Barrionuevo, "Boom in Ethanol Reshapes Economy of Heartland," *New York Times*, June 25, 2006.

57 *"The spigot of public money"*: Michael Pollan, "The Great Yellow Hope," *New York Times*, May 24, 2006.

57 *McCain, who used to call*: Jon Birger, "McCain's Farm Flip," *Fortune*, October 31, 2007; Barack Obama, *The Audacity of Hope: Thoughts on Reclaiming the American Dream* (Three Rivers Press, 2006), p. 170; Gerard Wynn, "U.S. Corn Ethanol 'Was Not a Good Policy': Gore," Reuters, October 22, 2010.

58 *NRDC even trial-ballooned the idea*: "Growing Energy: How Biofuels Can Help End America's Oil Dependence," Natural Resources Defense Council, December 2004.

58 *Silicon Valley icon Vinod Khosla*: Hanna L. Breetz, "Fueled by Crisis: U.S. Alternative Fuel Policy, 1975-2007," Massachusetts Institute of Technology dissertation, February 2013.

59 *Wind generated 0.3 percent of U.S. energy*: Energy Information Administration, *Renewable Energy Consumption and Electricity Preliminary Statistics 2010*, https://www.eia.gov/renewable/annual/preliminary/pdf/table1.pdf.

59 *In another rah-rah report*: "Ethanol: Energy Well Spent," Natural Resources Defense Council, February 2006.

63 *He found it 93 percent worse*: Timothy Searchinger et al., "Use of U.S. Croplands for Biofuels Increases Greenhouse Gases Through Emissions from Land-Use Change," *Science* 319, no. 5867 (February 29, 2008).

64 *When he ran the numbers*: Timothy D. Searchinger and Ralph E. Heimlich, "Estimating Greenhouse Gas Emissions from Soy-Based US Biodiesel When Factoring in Emissions from Land Use Change," conference paper, Lifecycle Carbon Footprint of Biofuels Workshop, January 29, 2008.

64 *cellulosic ethanol from switchgrass*: Growing switchgrass doesn't require nearly as much fertilizer, pesticide, or tractor runs as growing corn. However, cellulosic ethanol's land-use emissions are slightly worse than corn ethanol's. That's because corn ethanol produces a byproduct called "distillers grains," an animal feed that reduces the amount of food that needs to be replaced when corn is used for fuel, and therefore reduces land conversion.

64 *any biofuels that used arable land*: Searchinger did find the most promising farm-grown fuel, ethanol made from fast-growing sugarcane grown on former grazing land in the Brazilian tropics, could have a payback period of only four years. But that rose to 45 years when the displaced cattlemen carved new grazing land out of the rainforest.

68 *Farrell warned his California bosses*: Alexander E. Farrell and Dan Sperling, "A Low-Carbon Fuel Standard for California, Part 2: Policy Analysis," Institute of Transportation Studies, University of California, Davis, August 2007.

69 *A* Salon *column noted*: Andrew Leonard, "Who Do You Trust on Ethanol?" *Salon*, January 25, 2008.

69 *Chief Kotok wore a beaded belt*: Much of the reporting from my 2008 trip to Brazil first appeared in my ethanol story, "The Clean Energy Myth," *Time*, April 7, 2008.

71 *That record has since been broken*: Matt Rogers, "February Wasn't Just Warmest on Record in D.C., It Was Also Warmer Than a Normal March," *Washington Post*, March 1, 2017.

72 *The article that awarded Searchinger*: Elisabeth Rosenthal, "Biofuels Deemed a Greenhouse Threat," *New York Times, February* 8, 2008; Richard Harris, "Study: Ethanol Worse for Climate Than Gasoline," *All Things Considered*, NPR, February 7, 2008; Juliet Eilperin, "Studies Say Clearing Land for Biofuels Will Aid Warming," *Washington Post*, February 7, 2008.

72 *A trade journal ranked Searchinger*: Tom Waterman, *The Ethanol Monitor*, cited in Robert Bryce, "Ethanol Industry Produces a Top Ten Enemies List," OilPrice.com, June 17, 2010.

THREE: WEIRD SCIENCE

76 *the first scientific critique of his paper*: Michael Wang and Zia Haq, "Response to February 7, 2008, Sciencexpress Article," February 14, 2008.

NOTES   343

77   *it assumed steady increases*: Timothy Searchinger, "Response to M. Wang and Z. Haq's E-Letter," *Sciencexpress*, August 12, 2008; Timothy Searchinger, "Response to New Fuels Alliance and DOE Analysts Criticisms of Science Studies of Greenhouse Gases and Biofuels," *Science*, February 26, 2008.

77   *Some argued biofuels shouldn't be blamed*: Tom Wassenaar and Simon Kay, "Biofuels: One of Many Claims to Resources," *Science* 321, no. 5886 (July 11, 2008); Vinod Khosla, "Biofuels: Clarifying Assumptions," *Science* 322, no. 5900 (October 17, 2008); Keith L. Kline and Virginia H. Dale, "Biofuels: Effects on Land and Fire," *Science* 321, no. 5886 (July 11, 2008).

78   *a global shift away from meat*: Steven A. Kolmes, "Food, Land Use Changes, and Biofuels," *Science*, March 25, 2008.

78   *That's why a biofuels trade group*: Renewable Fuels Association, "Scientists Raise Doubts About Recent Studies on Biofuels and Global Warming," February 15, 2008.

79   *Grist declared that his response*: "Khosla's Letter to Science Backfires," *Grist*, October 24, 2008; Timothy Searchinger and R. A. Houghton, "Biofuels: Clarifying assumptions - Response," *Science*, October 17, 2008.

79   *he compiled a long list*: Tim Searchinger, "Summaries of Analyses in 2008 of Biofuels Policies by International and European Technical Agencies," German Marshall Fund, 2008.

82   *photosynthesis itself was wildly inefficient*: An awesome book about photosynthesis: Oliver Morton, *Eating the Sun: How Plants Power the Planet* (HarperCollins, 2008).

82   *Europe's biofuels industry launched*: Alex Evans, "Great Public Relations Disasters of Our Time," Global Dashboard, June 26, 2008; Ian Traynor, "EU Set to Scrap Biofuels Target Amid Fears of Food Crisis," *Guardian*, April 18, 2008.

83   *An Oxford dissertation on the mandate*: Amelia Sharman, "Evidence-Based Policy or Policy-Based Evidence Gathering? The Case of the 10% Target" (dissertation, University of Oxford), September 1, 2009.

86   *When Yale's Berry did a comprehensive literature review*: Steven T. Berry, "Biofuels Policy and the Empirical Inputs to GTAP Models," prepared for the California Air Resources Board, January 4, 2011.

87   Inside EPA *reported amusingly frantic concerns*: "Ethanol Sector Lobbies to Halt EPA's RFS Focus on Overseas GHG Emissions," *Inside EPA*, September 22, 2008.

88   *"I don't trust anybody anymore!"*: Tom Philpott, "Why Farm-State Pols Rage Against the EPA's Biofuel Stance," *Grist*, May 15, 2009.

89   *she told the man in the next seat*: Margo Oge, *Driving the Future: Combating Climate Change with Cleaner, Smarter Cars* (Arcade Press, 2016).

89   *a new "uncertainty analysis"*: Ben Geman, "EPA Rule Will Reflect 'Uncertainty on Indirect Biofuels Emissions, Fending Off Amendment," *New York Times*, September 24, 2009.

89   *In the fine print*: Richard Plevin, one of the coauthors of the original Berkeley paper, first spotted the miracle in the appendix. It's in the supporting materials of: Richard J. Plevin et al., "Greenhouse Gas Emissions from Biofuels' Indirect Land Use Change Are Uncertain but May Be Much Greater Than Previously Estimated," *Environmental Science & Technology Journal* 44, no. 21 (October 13, 2010).

90   *A paper in the* Journal of Cleaner Production: C. Malins et al., "How Robust Are Reductions in Modeled Estimates from GTAP-BIO of the Indirect Land Use Change Induced by Conventional Biofuels?" *Journal of Cleaner Production* 258 (June 10, 2020).

FOUR: DESTROYING THE CLIMATE IN ORDER TO SAVE IT

95   *If a tree falls in a forest*: Some of the reporting in this chapter first appeared in *Politico Magazine*. Michael Grunwald, "The 'Green Energy' That Might Be Ruining the Climate," *Politico Magazine*, March 26, 2021.

98   *But the studies it cited*: Marshall Wise et al., "Implications of Limiting CO2 Concentrations for Land Use and Energy," *Science* 324, no. 5931 (May 29, 2009); J. M. Melillo et al., "Unintended Environmental Consequences of a Global Biofuel Program," MIT Joint Program Report Series, Report 168 (2009).

99   Science accepted his contrarian take: Timothy D. Searchinger et al., "Fixing a Critical Climate Accounting Error," Science 326, no. 5952 (October 23, 2009).

99   a rush to burn trees for heat and power: "Renewables Roundup: Analysts Look to Eastern Europe for Biomass Generation Potential," European Daily Electricity Markets, March 12, 2010.

100   A new clean-tech industry was springing to life: Jake Jason et al., "US Pellet Industry Analysis," Idaho National Laboratory, June 2010; Ron Kortba, "Closing the Wood Pellet Gap," Biomass Magazine, December 2008; Henry Specter and Daniel Toth, "North America's Wood Pellet Sector," USDA Forest Service Research Paper, August 2009; Ronald Gonzalez et al., "Filling a Need: Forest Plantations for Bioenergy in the Southern US," Biomass Magazine, January 2009; International Energy Agency, Energy Technology Perspectives: Scenarios and Strategies to 2050, 2008. UNFCC Secretariat, Challenges and Opportunities for Mitigation in the Agricultural Sector, 2008.

101   "First in the nation!": Erin Ailworth, "Green, with Envy," Boston Globe, August 8, 2008.

102   "it takes a minute to burn a tree": Mary Stuart Booth, "A Red Flag on Green Energy Plan," Boston Globe, May 25, 2009.

103   "Although the very term 'accounting rules'": Vinod Khosla and Tim Searchinger, "Crunching the Numbers on Bioenergy Rules," Boston Globe, November 23, 2009.

103   the independent study concluded that burning wood: "Biomass Sustainability and Carbon Policy Study," Manomet Center for Conservation Sciences, June 2010.

108   "enormous pressures on the Earth's land-based ecosystems": "Opinion of the European Environmental Agency Scientific Committee on Greenhouse Gas Accounting in Relation to Bioenergy," European Environment Agency Scientific Committee, September 15, 2011.

108   He sounded similar alarms: Helmut Haberl et al., "Correcting a Fundamental Error in Greenhouse Gas Accounting Related to Bioenergy," Energy Policy 45 (June 2012).

109   critique of its support for wood-burning plants: Tim Searchinger, "Sound Principles and an Important Inconsistency in the 2012 UK Bioenergy Strategy," letter to U.K. energy and climate ministry, September 20, 2012, published here: https://www.biomassmurder.org/docs/2012-09-20-rspb-sound-principles-and-and-important-inconsistency-in-the-2012-uk-bioenergy-strategy-english.pdf; "Dirtier Than Coal? Why Government Plans to Subsidize Burning Trees Are Bad News for the Planet," Greenpeace U.K., November 2012.

109   The queen had named: Bernie Bulkin, Crash Course: One Year to Become a Great Leader of a Great Company (Whitefox Publishing, 2015).

110   the E.U. was on track: "State of Play on the Sustainability of Solid and Gaseous Biomass Used for Electricity, Heating and Cooling in the EU," European Commission Staff Working Document, Brussels, July 28, 2014; U. Mantau et al., "EU Wood - Real Potential for Changes in Growth of EU Forests," Hamburg, June 2010.

111   its 2011 report on renewables: IPCC, Renewable Energy Sources and Climate Change Mitigation, 2011.

111   the IPCC reviewed bioenergy again: IPCC, Climate Change 2014: Mitigation of Climate Change, 2014.

111   critique of rosy bioenergy analyses: Richard J. Plevin et al., "Using Attributional Life Cycle Assessment to Estimate Climate-Change Mitigation Benefits Misleads Policy Makers," Journal of Industrial Ecology 18, no. 1 (February 14, 2014): 73–83.

112   "The argument for caution": Eduardo Porter, "A Biofuel Debate: Will Cutting Trees Cut Carbon?" New York Times, February 10, 2015.

113   Congress also declared biomass: Chris Mooney, "These Experts Say Congress Is 'Legislating Scientific Facts'—and Wrong Ones, Too," Washington Post, February 16, 2016.

113   He got 800 researchers: "Letter from Scientists to the E.U. Parliament Regarding Forest Biomass," January 11, 2018, https://www.euractiv.com/wp-content/uploads/sites/2/2018/01/Letter-of-Scientists-on-Use-of-Forest-Biomass-for-Bioenergy-January-12-2018.pdf.

114   A North Carolina forest advocacy group: "The Use of Whole Trees in Wood Pellet Manufacturing," Dogwood Alliance, November 13, 2012.

114   Even an industry analysis commissioned: An Analysis of UK Biomass Power Policy, US South Pellet Production and Impacts on Wood Fiber Markets, RISI, January 12, 2010.

116 *Biomass power was still growing*: Saul Elbein, "Europe's Renewable Energy Policy Is Built on Burning American Trees," *Vox*, March 4, 2019.

FIVE: THE MENU

This entire chapter focuses on Searchinger et al., *Creating a Sustainable Food Future*, World Resources Institute, 2019. Most of the facts and statistics in the chapter are drawn from the report.

123 *Lappé's tract was often dismissed*: Frances Moore Lappé, *Diet for a Small Planet* (Ballantine Books, 1971).

124 *His wake-up call came from*: Henning Steinfeld et al., *Livestock's Long Shadow: Environmental Issues and Options*, U.N. Food and Agriculture Organization, 2006.

124 *the UN retracted its claim*: Lisa Abend, "Meat-Eating vs. Driving: Another Climate Error?" *Time*, March 27, 2010.

124 *One* Livestock's Long Shadow *coauthor*: Tom Wassenaar and Simon Kay, "Biofuels: One of Many Claims to Resources," *Science* 321, no. 5886 (July 11, 2008).

132 *eating five pounds of beef*: Timothy D. Searchinger et al., "Assessing the Efficiency of Changes in Land Use for Mitigating Climate Change," *Nature* 564 (December 12, 2018).

133 *We shifted to chicken*: Michael Grunwald, "What's the Most Climate Friendly Way to Eat? It's Tricky," *Canary Media*, June 6, 2022.

134 *"We'll take care of getting rid of beef for you"*: Tad Friend, "Can a Burger Help Solve Climate Change?" *New Yorker*, September 23, 2019.

139 *Researchers couldn't even agree*: FAO data cited in Timothy Searchinger et al., "The Global Land Squeeze: Managing the Growing Competition for Land," World Resources Institute, July 2023; Leandro Parente et al., "Shaping the Brazilian Landscape: A Process Driven by Land Occupation, Large-Scale Deforestation, and Rapid Agricultural Expansion," preprint, August 2021, https://www.researchgate.net/publication/354016799_Shaping_the_Brazilian_landscape_a_process_driven_1_by_land_occupation_large-scale_deforestation_and_2_rapid_agricultural_expansion.

143 *But the OpenAg Initiative turned out*: Tom McKay, "MIT Built a Theranos for Plants," *Gizmodo*, September 8, 2019.

SIX: IT'S THE FOOD THAT NEEDS TO CHANGE

I've relied on the Good Food Institute's annual reports on alternative proteins for industry data throughout this chapter and the next one.

149 *"Good job getting attention to your cause"*: Rex W. Huppke, "McVeigh Responds to PETA Request That He Have Vegan Last Meal," *Oklahoman*, April 16, 2001.

149 *He fit the stereotype of a vegan extremist*: Gene Weingarten, "Gene Meets His Conscience," *Washington Post Magazine*, December 30, 2011.

149 *he argued that a hen trapped in a cage*: Bruce Friedrich, "Effective Advocacy: Stealing from the Corporate Playbook," Mercy for Animals, https://mercyforanimals.org/essays/effective-advocacy/.

150 *they displayed gruesome images of factory-farm cruelty*: David Montgomery, "Animal Pragmatism," *Washington Post*, September 6, 2003.

152 *The dishes challenged his descriptive talents*: Gene Weingarten, "Below the Beltway," *Washington Post Magazine*, June 24, 2001.

152 *The first commercial meat substitute*: William Shurtleff and Akiko Aoyagi, "History of Meat Alternatives," Soyinfo Center, 2014.

152 *"not like your backyard burger"*: Molly Martin, "Cutting the Federal Fat—Meatless Burgers Invade the Clintons' White House Kitchen," *Seattle Times*, May 7, 1995.

154 *He had never held down any of those prestigious jobs*: Bianca Bosker, "Mayonnaise, Disrupted," *Atlantic*, November 2017.

154 *a newspaper column about social entrepreneurship*: Josh Tetrick, "You Can Save the Planet," *Richmond Times-Dispatch*, March 15, 2009.

156 *when Beyond debuted its first plant-based chicken strips*: Mark Bittman, "A Chicken Without

NOTES

Guilt," *New York Times*, March 9, 2012; Farad Manjoo, "Fake Meat So Good It Will Freak You Out," *Slate*, July 26, 2012; Bill Gates, "Future of Food," *GatesNotes*, March 18, 2013,

164 *It later came out that one American Egg Board executive*: Sam Thielman and Dominic Rushe, "Government-Backed Egg Lobby Tried to Crack Food Startup, Emails Show," *Guardian*, September 2, 2015.

164 *"Mark my words"*: Andrew Zimmern, "Welcome to the Future," andrewzimmern.com, June 25, 2014.

164 *But the drama only deepened*: Olivia Zaleski, Peter Waldman, and Ellen Huet, "How Hampton Creek Sold Silicon Valley on a Fake-Mayo Miracle," *Bloomberg Businessweek*, September 22, 2016; Biz Carson, "Sex, Lies and Eggless Mayonnaise: Something Is Rotten at Food Startup Hampton Creek," *Business Insider*, August 5, 2015.

171 *Soon Friedrich was declaring publicly*: Barbara J. King, "Humans Are 'Meathooked' but Not Designed for Meat-Eating," NPR, May 19, 2016.

171 *Tyson also bought 5 percent of Beyond*: Adele Peters, "Get Ready for a Meatless Meat Explosion, as Big Food Gets on Board," *Fast Company*, December 18, 2017; "The Next Agricultural Revolution," New Crop Capital, March 22, 2018.

173 *When the animal pragmatist*: Paul Shapiro, *Clean Meat: How Growing Meat Without Animals Will Revolutionize Dinner and the World* (Gallery Books, 2018).

174 *Josh Tetrick thought a lot about death*: Chase Purdy first recounted Tetrick's decision to pursue cultivated meat, including Jake's death, in *Billion Dollar Burger: Inside Big Tech's Race for the Future of Food* (Portfolio, 2020).

175 *One hit piece asked*: Helen Holmes, "Is This Embattled Startup the Theranos of Mayonnaise?" *Observer*, February 21, 2019.

175 *Tetrick fired them*: The three fired executives founded Brightseed, which has created an advanced version of Hampton Creek's plant analytics program to search for compounds with medicinal properties. The company has raised more than $100 million, and its partners include ADM and Danone.

176 *the first indistinguishable substitute*: Gene Weingarten, "Gene Weingarten vs. the Meatless Impossible Burger: Will He Taste Victory?" *Washington Post Magazine*, November 16, 2017; Lindsey Hoshaw, "Silicon Valley's Bloody Plant Burger Smells, Tastes and Sizzles Like Meat," *The Salt*, NPR, June 21, 2016; Caitlin Dewey, "Is This the Beginning of the End of Meat?" *Washington Post*, March 17, 2017.

SEVEN: THE FAKE MEAT HYPE CYCLE

178 *The media had crowned it*: Deena Shanker, "The Hottest Thing in Food Is Made of Peas, Soy and Mung Beans," *Bloomberg Businessweek*, August 21, 2019; Mike Brown, "Plant-Based Meat: A Long-Awaited Industry Tipping Point Has Finally Arrived," *Inverse*, July 16, 2019.

179 *"Fake Meat Can Save Us"*: Timothy Egan, "Fake Meat Can Save Us," *New York Times*, June 21, 2019.

181 *analysts were giddy about their future*: Thomas Frank, "Alternative Meat to Become $140 Billion Industry in a Decade, Barclays Predicts," CNBC, May 23, 2019; Damian Carrington, "Most 'Meat' in 2040 Will Not Come from Dead Animals, Says Report," *Guardian*, June 12, 2019; Palash Ghosh, "Will the Beef and Dairy Industries Really Collapse by 2030? Some Are Skeptical," *International Business Times*, November 25, 2019; Rob Leclerc, "Why Technology Could Make Animals Obsolete," AgFunder Network, September 21, 2019.

184 *"This is not just another disgusting tofu burger"*: Eric Bohl, "Taste Test: This Fake Meat Is the Real Deal," Missouri Farm Bureau blog, April 3, 2019, https://mofb.org/taste-test-this-fake-meat-is-the-real-deal/.

190 *it also contained chemicals*: Jacy Reese, *The End of Animal Farming: How Scientists, Entrepreneurs, and Activists Are Building an Animal-Free Food System* (Beacon Press, 2018).

191 *Friedrich had just read an upcoming scientific paper*: David Humbird, "Scale-Up Economics for Cultured Meat," *Biotechnology and Bioengineering*, June 7, 2021. The review was commissioned by Open Philanthropy, GFI's top funder, which is how Friedrich got to see it before its public release.

192 *"Lab-Grown Meat Is Scaling Like the Internet"*: Ron Shiegeta, *Spoon*, October 26, 2020.
193 *The Boston Consulting Group predicted*: Björn Witte et al., *Food for Thought: The Protein Transformation*, Boston Consulting Group and Blue Horizon report, March 2021.
194 *"The transformational outcome did not materialize"*: Katy Askew, "Should We Be Worried About the Outlook for Plant-Based Meat?" *Food Navigator*, February 3, 2023.
195 *The industry was inundated with quasi obituaries*: Matthew Lynn, "The Vege-bubble Turns to Vege-bust," *Telegraph*, June 26, 2022; "Where's the Beef? Here's Why the Fake Meat Fad Sizzled Out," *Washington Post* editorial, May 12, 2023; Michele Simon, "Plant-Based Fail," *Forbes*, February 1, 2023; Deena Shanker, "Beyond Impossible: How Fake Meat Became Just Another Food Fad," *Bloomberg Businessweek*, January 18, 2023.
196 *When U.S. regulators finally approved*: Michael Grunwald, "Afraid of High-Tech Food? Get over It," *Canary Media*, January 17, 2023.
204 *Meati was shutting down*: Elaine Watson, "Meati Plans 'Gut-Wrenching' Mass Layoffs amid 'Bank-Induced Crisis,'" *AgFunder*, March 7, 2025.
206 *Valeti was in a funk*: Matt Reynolds and Joe Fassler, "Insiders Reveal Major Problems at Lab-Grown Meat Startup Upside Foods," *Wired*, September 15, 2023.

EIGHT: THE SOIL FANTASY

212 *They've lost more than 100 billion tons*: Jonathan Sanderman et al., "Soil Carbon Debt of 12,000 Years of Human Land Use," *PNAS* 114, no. 36 (August 21, 2017).
213 *The* Times *review hailed*: Natalia Winkleman, "'Kiss the Ground' Review: Regenerating Hope for the Climate," *New York Times*, September 22, 2020.
213 *The IPCC has estimated that soil carbon sequestration*: Pete Smith et al., "Agriculture," in *Climate Change 2007: Mitigation*, Contribution of Working Group III to the Fourth Assessment Report of the Intergovernmental Panel on Climate Change, eds. Bert Metz et al. (New York: Cambridge University Press, 2007). "Regenerative Agriculture: Healthy Soil Best Bet for Carbon Storage," European Academies Science Advisory Council, May 4, 2022.
214 *"What do climate advocates have in common"*: Robert Paarlberg, "President Biden, Please Don't Get into Carbon Farming," *Wired*, January 22, 2021.
214 *The media have promoted it, too*: Alexandra Topping, "'This Way of Farming Is Really Sexy': The Rise of Regenerative Agriculture," *Guardian*, August 14, 2023; Moises Velasquez-Manoff, "Can Dirt Save the Earth?" *New York Times Magazine*, April 18, 2018. Julia Westbrook, "This New Grain Is Good for the Environment and Good for You," *Eating Well*, April 11, 2019; Dan Charles, "Can This Breakfast Cereal Help Save the Planet?" NPR, April 13, 2019.
214 *The headlines often contrast*: Ola Chrobak, "A Carbon-Neutral Burger? It's Not Impossible," *Popular Science*, October 7, 2019; David McKenzie, "Put Down That Veggie Burger. These Farmers Say Their Cows Can Solve the Climate Crisis," CNN, July 7, 2020; Kim Serverson, "How Will We Eat in 2023? Here Are 10 Predictions," *New York Times*, December 27, 2022; Jennifer Stojkovich, "Five New Trends to Spot in the Future of Food," *Rolling Stone*, April 10, 2023.
215 *regenerative grain called Kernza*: Julia Westbrook, "This New Grain Is Good for the Environment and Good for You," *Eating Well*, April 11, 2019; Dan Charles, "Can This Breakfast Cereal Help Save the Planet?" NPR, April 13, 2019.
216 *"Oh, look! That's beautiful!"*: Some of my reporting on the TomKat Ranch first appeared in *Politico Magazine*. "Tom Steyer Thinks His Ranch Can Save the Planet," October 11, 2019.
218 *Point Blue scientists published a study*: Jacob Weverka et al., "Exploring Plant and Soil Microbial Communities as Indicators of Soil Organic Carbon in a California Rangeland," *Soil Biology and Biochemistry* 178 (March 2023).
218 *USDA has allowed Tyson*: Georgina Gustin, "Environmentalists Are Having a Cow over Tyson Foods' 'Climate Friendly' Beef," *Mother Jones*, May 9, 2024
221 *She pointed me to a gung-ho Rodale report*: Jeff Moyer et al., *Regenerative Agriculture and the Soil Carbon Solution*, Rodale Institute, September 2020.
223 *a 2023 exposé in* Science: Gabriel Popkin, "Shaky Ground," *Science*, July 27, 2023.
224 *Searchinger spent weeks trying to track down*: Tim Searchinger and Janet Ranganathan, "Further

Explanation on the Potential Contribution of Carbon Sequestration on Working Agricultural Lands to Climate Mitigation," World Resources Institute, Technical Perspective, August 24, 2020.

225 *Even if regenerative practices did increase soil carbon*: An example of a study emphasizing how yield drags from practices like cover cropping can have land-use impacts that hurt the climate: David B. Lobell and Nelson B. Villoria, "Reduced Benefits of Climate-Smart Agricultural Policies from Land-Use Spillovers," *Nature Sustainability* 6 (May 18, 2023).

225 *The General Mills–funded study*: "Carbon Footprint Evaluation of Regenerative Grazing at White Oak Pastures," Quantis, February 25, 2019.

225 *A lengthy* Washington Post *puff piece*: Sarah Kaplan, "A Recipe for Fighting Climate Change and Feeding the World," *Washington Post*, October 12, 2021.

226 *The yield penalty for organic row crops*: An example of a pro-organic study that acknowledges a 30 to 40 percent yield penalty: Nancy Pfund and Lucas Strom, "Got Organic? Welcome to the Age of 21st Century Organic Farming," DBL Partners, November 2018, https://www.dbl.vc/resource/got-organic-welcome-to-the-age-of-21st-century-organic-farming/. USDA's Economic Research Service found closer to a 25 percent yield penalty: William D. McBride, "Despite Profit Potential, Organic Field Crop Acreage Remains Low," *Amber Waves*, November 2, 2015. A global meta-analysis of all kinds of organic farms estimated an 18.4 percent overall penalty: Vera De la Cruz et al., "Yield Gap Between Organic and Conventional Farming Systems Across Climate Types and Sub-types: A Meta-analysis," *Agricultural Systems* 211 (October 2023).

226 *Kernza is the first perennial grain crop*: Kernza crackers are delicious, and Kernza pasta tastes like whole-wheat pasta. I don't drink, but my *Climavores* podcast cohost Tamar Haspel liked Kernza beer, too. The only problem is the yields, which is a really big problem. We did an episode about it: Michael Grunwald and Tamar Haspel, hosts, *The Latitude*, podcast, "Climavores: Is Karnza the Climate-Friendly Answer to Wheat?" Latitude Media, January 24, 2023, https://podcasts.apple.com/us/podcast/climavores-is-kernza-the-climate-friendly-answer-to-wheat/id1623272960?i=1000596380478.

226 *In an article in* Nature: Tim Searchinger et al., "Assessing the Efficiency of Changes in Land Use for Mitigating Climate Change," *Nature* 564 (December 13, 2008). See also "Organic Food Worse for the Climate," Chalmers University of Technology press release, December 13, 2008.

227 *That bait-and-switch sounds*: Janet Ranganathan et al., "Regenerative Agriculture: Good for Soil Health, but Limited Potential to Mitigate Climate Change," World Resources Institute Insights, May 12, 2020; Keith Faustian et al., "Climate Mitigation Potential of Regenerative Agriculture Is Significant!" June 2020 response, https://searchinger.princeton.edu/sites/g/files/toruqf4701/files/tsearchi/files/paustian_et_al._response_to_wri_soil_carbon_blog_.pdf. Searchinger and Rangathanan, "Further Explanation."

227 *studies that lumped together*: A classic example of this is D. A. Bossio et al., "The Role of Soil Carbon in Natural Climate Solutions," *Nature Sustainability* 3 (March 2020). It finds large soil carbon benefits from nonfarming activities like "avoided forest conversion," "reforestation," "peatland restoration," and "coastal wetland restoration."

227 *They were reminiscent of Cleveland Cavaliers scrub*: The full Lopez tweet: "I'm going to get it out of the way and point out that Donovan Mitchell and Robin Lopez combined for 72 tonight," January 1, 2023.

229 *one thing the Australians made clear in obscure journals*: Clive A. Kirkby et al., "Stable Soil Organic Matter: A Comparison of C:N:P:S Rations in Australian and Other World Soils," *Geoderma* 163, no 3–4 (July 2011); Clive A. Kirkby et al., "Carbon-Nutrient Stoichiometry to Increase Soil Carbon Sequestration," *Soil Biology and Biochemistry* 60 (May 2013).

229 *Grist wrote them up*: Nathanael Johnson, "The Secret to Richer, Carbon-Capturing Soil? Treat Your Microbes Well," *Grist*, June 3, 2014.

230 *He later watched nitrogen from chicken manure*: Fertilizer was the main source of pollution in the Everglades, too, but the problem there was phosphorus, not nitrogen, an awful problem but less of a climate problem.

231 *while leguminous cover crops*: For example, Janzen et al. (2022) found that soil carbon sequestration on working croplands had the theoretical capacity to offset up to 1.5 percent of human

emissions—way less than claims of 100 percent, but not nothing. H. Henry Janzen et al., "Photosynthetic Limits on Carbon Sequestration in Croplands," *Geoderma* 416 (June 15, 2022). A literature review by the Food Climate Research Network concluded "the contribution of grazing ruminants to soil carbon sequestration is small, time-limited and substantially outweighed by the greenhouse gas emissions they generate," but still found that small contribution could conceivably offset up to 11 percent of livestock emissions. Tara Garnett et al., "Grazed and Confused?" Food and Climate Research Network, University of Oxford, 2017.

232 *"grasslands unplowed, wetlands undrained, and forests uncut"*: Janzen, et al., "Photosynthetic Limits."
232 *Searchinger always assigned his classes*: Michael Pollan, *The Omnivore's Dilemma: A Natural History of Four Meals* (Penguin, 2006); Aldo Leopold, *Sand County Almanac* (Oxford University Press, 1966).
233 *In the 19th century*: Thomas Hager, *The Alchemy of Air: A Jewish Genius, a Doomed Tycoon, and the Scientific Discovery That Fed the World but Fueled the Rise of Hitler* (Broadway Books, 2008).
233 *"without synthetic fertilizer, billions of people"*: Pollan, *Omnivore's Dilemma*, 43.
235 *Born half a century after Cocking*: Some of my reporting on Pivot Bio first appeared in *Canary Media*: "Chemical Fertilizer Is a Climate Disaster. Can High-Tech Biology Fix It?" August 29, 2023.
236 *The Netherlands was paralyzed*: Paul Tullis, "Nitrogen Wars: The Dutch Farmers' Revolt That Turned a Nation Upside Down," *Guardian*, November 16, 2023.
237 *Proven's agricultural success was not yet scientifically proven*: Laura Thompson et al., "Farmer Focus: Nebraska Growers Put Pivot Bio to the Test via On-Farm Research Studies," *Cropwatch*, University of Nebraska–Lincoln, April 7, 2022.

NINE: MORE BEEF, LESS LAND

239 *Fazenda Tropical, a corn, soy, and beef plantation*: Some of my Brazil reporting first appeared in *Canary Media*: "Cattle Are a Huge Climate Threat. Smarter Ranching Can Help," April 23, 2023.
241 *The Brazilian agriculture community is still frosted*: Shari Friedman, "Farms Here, Forests There," David Gardner & Associates for the National Farmers Union and Avoided Deforestation Partners, May 2010.
243 *the ambiguities of calculating beef yields*: Fazenda Tropical uses some of its cropland to feed its cattle, so its stocking rate on its pastures isn't an exact proxy for productivity. But the TomKat Ranch brings in outside hay, so its cattle use extra land, too. Even if it's impossible to calculate from Fazenda Tropical's sky-high stocking rates exactly how much more productive it is than the typical ranch, it's clearly way more productive.
246 *He had just read*: Nate Silver, *The Signal and the Noise: Why So Many Predictions Fail—but Some Don't* (Penguin Press, 2012).
247 *a purple-flowering clover called* Desmodium: Ana Lucia Ferreira, "Little-Known Leguminous Plant Can Increase Beef Production by 60%," Embrapa Agrobiology, October 25, 2022.
247 *Steve Gabel knows beef*: Some of my reporting on Gabel first appeared in: Michael Grunwald, "Sorry, but This Is the Future of Food," *New York Times*, December 13, 2024.
249 *that top 1 percent feeds half of U.S. cattle*: "Cattle on Feed," USDA National Agricultural Statistic Services report, February 23, 2024.
250 *Feedlot-finished beef have lower emissions*: Daniel Blaustein-Rejto et al., "Carbon Opportunity Cost Increases Carbon Footprint Advantage of Grain-Finished Beef," *PLOS ONE* (December 13, 2023).
251 *front-page hit piece about his industry ties*: Hiroko Tabuchi, "He's an Outspoken Defender of Meat. Industry Funds His Research, Files Show," *New York Times*, October 31, 2022.
253 *And global institutions are often just as critical*: "Ten Things You Should Know About Industrial Farming," U.N. Environment Programme, https://www.unep.org/news-and-stories/story/10-things-you-should-know-about-industrial-farming; "Breaking Away from Industrial Food and Farming Systems," Food and Agriculture Organization of the U.N., https://www.fao.org/family-farming/detail/en/c/1443099/.

253 *Three-quarters of America's hog farms*: Holly Cook and Lee Schulz, *The United States Pork Industry 2021: Current Structure and Economic Importance*, National Pork Producers Council, July 2022.

TEN: MORE CROPS, LESS LAND

263 *Many of the 19th-century pioneers*: Allan G. Bogue, *From Prairie to Corn Belt: Farming on the Illinois and Iowa Prairies in the Nineteenth Century* (University of Chicago Press, 1963); Samuel J. Imlay and Eric D. Carter, "Drainage on the Grand Prairie: The Birth of a Hydraulic Society on the Midwestern Frontier," *Journal of Historical Geography* 38, no. 2 (2011).

264 *Since the Green Revolution began*: 2022 Census of Agriculture, U.S. Department of Agriculture, February 2024.

266 *groundwater depletion is expected to fallow*: Ellen Hanak et al., *Managing Water and Farmland Transitions in the San Joaquin Valley*, Public Policy Institute of California, September 2023.

266 *Searchinger and his team calculated*: Patrice Dumas et al., *Pathways to a Sustainable Food Future in Sub-Saharan Africa*, World Resources Institute, 2024.

268 *The industrial ruins of downtown Newark*: Some of my vertical farming reporting also appeared in my *Canary* column: "Why Vertical Farming Just Doesn't Work," June 28, 2023.

270 *"Investors went wild for the utopian promise"*: Donavyn Coffey, "Indoor Agriculture Faces a Reckoning," Ambrook Research, November 12, 2023.

274 *Consumers everywhere fear genetically modified organisms*: Brian Kennedy and Cary Lynne Thigpen, "Many Publics Around World Doubt Safety of Genetically Modified Foods," Pew Research Center, November 11, 2020; Erin Brown, "Jimmy Kimmel Asks Anti-GMO People What GMOs Are—and, Hilariously, They Have No Idea," *Business Insider*, October 10, 2014.

277 *His team began by using a supercomputer*: Elizabeth Kolbert, "Creating a Better Leaf," *New Yorker*, December 6, 2021.

279 *It looked like an ordinary orchard tree*: Some of my reporting about pongamia first appeared in my *Canary Media* column: "This Super-Tree Could Help Feed the World and Fight Climate Change," June 27, 2022.

ELEVEN: HOW TO SAVE THE WORLD

289 *a Danish Parliament hearing in early 2020*: A Twitter thread I did comparing this hearing to U.S. hearings went bizarrely viral, with 50,000 likes and retweets. https://twitter.com/mikegrunwald/status/1225529425187528709?s=51&t=fdU4HRs4gPdORTnkU3doWQ. Incidentally, the hearing was simultaneously translated into English, even though Searchinger and I seemed to be the only non-Danish speakers in the room.

291 *His team had calculated*: Stefan Wirsenius et al., "Comparing the Life Cycle Greenhouse Gas Emissions of Dairy and Pork Systems Across Countries Using Land-Use Carbon Opportunity Costs," World Resources Institute, 2020.

291 *Fewer pigs and cows in Denmark*: Timothy Searchinger et al., "A Pathway to Carbon Neutral Agriculture in Denmark," World Resources Institute, 2021.

297 *hundreds of rural Republican counties are restricting*: Elizabeth Weise and Suhail Bhat, "Across America, Clean Energy Plants Are Being Banned Faster Than They're Being Built," *USA Today*, February 4, 2024.

297 *its record-breaking clean-energy investments*: One little-discussed provision of the IRA provided $10 billion to help rural utilities transition away from coal, so many farmers who don't install their own renewables will still get power from a greener grid.

299 *a report shredding its Miracle Yield assumptions*: Steven Berry et al., "Evaluating the Economic Basis for GTAP and Its Use for Modeling Biofuel Land Use," Yale Tobin Center for Economic Policy, February 20, 2024

299 *Biden was all in for biofuels*: "Remarks by President Biden on Lowering Energy Costs for Working Families," April 12, 2022, www.whitehouse.gov/briefing-room/speeches-remarks/2022/04/12/.

300 *The International Energy Agency's net-zero plans*: "Bioenergy with Carbon Capture and

NOTES

Storage," IEA, https://www.iea.org/energy-system/carbon-capture-utilisation-and-storage/bioenergy-with-carbon-capture-and-storage.

300 *Drax has proposed to retrofit its Yorkshire plant*: Jillian Ambrose, "Drax Gets Go-Ahead for Carbon Capture Project at Estimated 40bn Cost to Billpayers," *Guardian*, January 16, 2024.

301 *In 2021, he abruptly banned fertilizers*: Kenny Torella, "Sri Lanka's Organic Farming Disaster, Explained," *Vox*, July 15, 2022. "Inorganic Fertilizer Ban Could Harm Production with Major Implications," *Daily FT*, June 16, 2021.

301 *has called for* $4.3 trillion: "Cultivating Change: A Collaborative Philanthropic Initiative to Accelerate and Scale Agroecology and Regenerative Approaches," Global Alliance for the Future of Food, May 2024.

301 *The E.U. has set a goal*: European Commission, "Action Plan for the Development of Organic Production in the EU," April 19, 2021, agriculture.ec.europa.eu/farming/organic-farming/organic-action-plan_en.

302 *It's a taxpayer rip-off*: Allowing anti-government extremists who refuse to pay grazing fees out of alleged principle to trespass on federal lands is an even dumber policy, but the feds have consistently avoided confrontations with ranching scofflaws since its armed standoff with the Bundy family in 2014.

302 *USDA also mandates that public schools serve cow milk*: Kenny Torella, "Big Milk Has Taken Over American Schools," *Vox*, May 31, 2024.

302 *I first heard Searchinger denounce subsidized crop insurance*: Yes, he persuaded me to write about it. Michael Grunwald, "Congress Tries to Reseed Federal Crop Insurance," *Washington Post*, November 7, 1999.

303 *Even USDA's research arm found*: Ruben Lubowski et al., "Agricultural Policy Affects Land Use and the Environment," *Amber Waves*, USDA Economic Research Service, September 1, 2006.

304 *I wrote a story*: Michael Grunwald, "Ron DeSantis Is Attacking the Greatest Food Innovation Since the Corndog," *Heatmap News*, February 22, 2024.

306 *a recent study pegged the cost*: Uris Lantz Baldos, "Impacts of US Public R&D Investments on Agricultural Productivity and GHG Emissions," *Journal of Agricultural and Applied Economics* 55, no. 3 (August 2023).

306 *reviewing the manure management literature*: Swati Hedge et al., "Analysis and Recommendations to Mitigate Greenhouse Gas Emissions from U.S. Manure Management," World Resource Institute draft discussion paper, April 2024.

307 *Over the last two decades*: Kelly P. Nelson and Keith Fuglie, "Investment in U.S. Public Agricultural Research and Development Has Fallen by a Third over Past Two Decades, Lags Major Trade Competitors," *Amber Waves*, USDA Economic Research Service, June 6, 2022; Alex Smith, "The Inflation Reduction Act Neglects Agriculture Research. That's a Mistake," Breakthrough Institute, August 2, 2022; Emily Bass and Alex Smith, "Public Financing for Agricultural Decarbonization and Abundance," Breakthrough Institute, May 2024. The ag research community requested $40 billion in the IRA and, depending on what you count as ag research, got between $300 million and $575 million.

308 *he coauthored a paper*: G. V. Subbarao and Timothy D. Searchinger, "A 'More Ammonium Solution' to Mitigate Nitrogen Pollution and Boost Crop Yields," *PNAS* 118, no. 22 (May 26, 2021).

308 *what Bill Gates calls "the green premium"*: Bill Gates, *How to Avoid a Climate Disaster: The Solutions We Have and the Breakthroughs We Need* (Knopf, 2021).

308 *governments pumped in another $1.5 billion*: Good Food Institute, *2023 State of Global Policy: Public Investment in Alternative Proteins to Feed a Growing World*, ClimateWorks Foundation and UK Foreign, Commonwealth & Development Office, "Global Innovation Needs Assessment: Protein Diversity," November 1, 2021.

309 *The Boston Consulting Group concluded*: Benjamin Morach et al., *The Untapped Climate Opportunity in Alternative Proteins*, Boston Consulting Group, August 7, 2022.

311 *The World Bank found*: Sutton, *Recipe for a Livable Planet*.

311 *The UN's mechanism for funding forest conservation*: Georgina Gustin, "New Reports Show Forests Need Far More Funding to Help the Climate, and Even Then, They Can't Do It All," *Inside*

352 NOTES

Climate, May 4, 2022; U.N. Food and Agriculture Organization, *The State of the World's Forests 2022*. REDD, awkwardly, stands for "Reducing Emissions from Deforestation and forest Degradation in developing countries." The plus refers to other forest-related activities that can help the climate, like more sustainable forestry.

316 *Lula's return to power*: Evan Marshall, "Deforestation in 2023: Way Down in the Amazon, Way Up in the Cerrado," *Brazilian Report*, January 5, 2024.

317 *Governments provide $600 billion in agricultural support*: Timothy D. Searchinger et al., "Revising Public Agricultural Support to Mitigate Climate Change," World Bank Group, 2020.

EPILOGUE: CAN WE DO IT?

324 *an op-ed calling for a World War II–style mobilization*: Searchinger recruited neuroscience professor Anthony LaMantia, a close friend who was his roommate during law school, and Gordon Douglas, a professor of medicine who had worked with him on food issues, to coauthor the column to bolster its medical credibility. Tim Searchinger et al., "The U.S. Faces Two Disastrous Scenarios. There's a Third Option," *Washington Post*, March 23, 2020.

326 *only 7 percent of news stories about climate change*: "People Don't See Industrial Meat as a Key Cause of Global Warming—Poll," Madre Brava, March 17, 2023.

327 *It's no longer considered cool in environmental circles*: Michael Grunwald, "What Covid Is Exposing About the Climate Movement," *Politico Magazine*, April 21, 2020; Auden Schendler, "Worrying About Your Carbon Footprint Is Exactly What Big Oil Wants You to Do," *New York Times*, March 31, 2021; Michael E. Mann and Jonathan Brockopp, "You Can't Save the Planet by Going Vegan. Corporate Polluters Must Be Held Accountable," *USA Today*, June 3, 2019; Mary Annaise Heglar, "I Work in the Environmental Movement. I Don't Care If You Recycle," *Vox*, June 4, 2019.

328 *My family hasn't wasted any food at home*: I wrote about Mill for *Heatmap News*. "Close the Freezer. The Future of Composting Is Dehydrated," September 29, 2023.

328 *one study of pet food's*: Peter Alexander et al., "The Global Environmental Paw Print of Pet Food," *Global Environmental Change* 65 (November 2020).

331 *Since 1880, the draining and farming of Everglades peatlands*: Thomas W. Dreschel et al., "Peat Soils of the Everglades of Florida," IntechOpen, December 20, 2017.

333 *The Onion nailed the typical climate-civilian response*: "Yeah, Yeah, Nation Gets It, We're Rapidly Approaching End of Critical Window to Avert Climate Collapse or Whatever," *The Onion*, July 6, 2020.

IMAGE CREDITS

1. World Resources Institute
2. Courtesy of Tim Searchinger
3. Courtesy of Tim Searchinger
4. Courtesy of Tim Searchinger
5. Courtesy of Tim Searchinger
6. Courtesy of *Science*
7. Michael Grunwald
8. Michael Grunwald
9. World Resources Institute
10. World Resources Institute
11. World Resources Institute
12. World Resources Institute
13. World Resources Institute
14. Courtesy of Tim Searchinger
15. World Resources Institute
16. Courtesy of Bruce Friedrich
17. Courtesy of Bruce Friedrich
18. Michael Grunwald
19. Michael Grunwald
20. Courtesy of Pat Brown
21. Courtesy of Josh Tetrick
22. Michael Grunwald
23. Michael Grunwald
24. Michael Grunwald
25. Michael Grunwald
26. Courtesy of Aerofarms
27. Courtesy of Naveen Sikka
28. Michael Grunwald

INDEX

adaptive multi-paddock grazing, 217–18
additionality, 77–78, 221, 314–15
ADM (Archer Daniels Midland)
    Alliance for Abundant Food and Energy, 87
    carbon credits, 237
    Climate-Smart Commodities program and, 220
    corn ethanol and, 23, 26–27
    meat alternatives and, 185–87, 197
    regenerative agriculture and, 214
Advancing Solutions for Alternative Proteins (ASAP), 309
AeroFarms, 268–71
AgFunder, 181
agricultural land
    as carbon sinks, 24–25, 49–50
    crop insurance subsidies for, 302–3
    crop rotations, 142
    deforestation for, 5–8, 9, 15, 17, 121, 340n5
    in Denmark, returning to nature, 293
    integrated crop-livestock systems, 231
    irrigation of, 121
    for livestock, 15, 123–25, 132, 136
    productivity of, 8–9, 119–23, 128–29, 134–39
    reforestation of, 313
    restoring drained peatlands, 311
    shifting south into carbon sinks, 312–13
    soil erosion and, 218–19, 265, 276, 280
    See also indirect land-use change; regenerative agriculture; soil carbon sequestration; yields
agriculture industry
    bailouts for, 307
    cap-and-trade, killing, 89
    carbon emissions from, 9 10, 119, 128, 139–40, 227, 271
    crop analytics and, 267–68
    crop residues, uses for, 58, 228–31, 295–96
    CRP and, 42
    GMO crops, 137–38 (*see also* GMOs)
    methane emissions from, 9, 15, 126–27, 140
    politics and, 22–23, 25–27, 41–42, 45–48, 84, 304–5 (*see also* agriculture lobby)
    reducing emissions from, 297
    regenerative agriculture, 141–43
    See also crop sustainability; fertilizers and manure; livestock; regenerative agriculture; subsidies; *individual companies*
agriculture lobby
    Big Sugar and the Everglades, 37
    corn ethanol and, 23, 26, 84, 87
    ILUC change and, 87–90
    LISA program and, 219
    meat alternatives and, 173, 190
    for milk in school lunches, 302
    power over agricultural policies, 305
    regenerative agriculture subsidies and, 214–15
    USDA research budget cuts and, 307
agriculture technological innovations. *See* technological innovations
agrivoltaic projects, 297
agroecology, 213. *See also* regenerative agriculture
agroforestry, 138, 231
Ahlers, Tommy, 290–91
Albright, Curt, 168
Aleph Farms, 172, 188
Alliance for Abundant Food and Energy, 87

INDEX

Aloha, 286
Alpha Foods, 168, 194
alternative eggs. *See* egg substitutes
alternative meat and dairy. *See* meat and dairy alternatives
Amazon rainforest
　biomass power needs and, 107
　deforestation in, 9, 17, 69–71, 77, 243–45, 316, 342n64
American Egg Board, 164
American Forest & Paper Association, 114
American Heart Association, 200
American Society for the Prevention of Cruelty to Animals (ASPCA), 159
anaerobic digesters of manure, 306–7
Anderson, Dave, 162–63
Andreassen, Johan, 259–62
Andrés, José, 196
Andrew, Joe, 279
animal rights
　livestock and animal welfare, 148–51, 153, 165, 303
　meat and dairy alternatives companies and, 159, 167
　vegan and vegetarian diets for, 147–51, 293
Anthropocene, 6–8
Apeel Sciences, 131
AppHarvest, 269–70
AquaBounty, 258
aquaculture, 258–62
Arby's, 179
Archer Daniels Midland. *See* ADM
Arentowicz, Andrew, 201
Argonne National Laboratory, 21–22, 24–25, 54, 63–64
Ark Biotech, 201
Army Corps of Engineers, 24, 38, 43–45, 46, 341n44
Aronoff, Marcia, 59–61
Arquette, Patricia, 213
artificial intelligence (AI), 266–67, 269
Atlantic Sapphire, 258–62
Attenborough, David, 213
Australia
　pasture expansion in, 139
　seaweed in animal feed, 254
　soil carbon sequestration studies in, 228–31
aviation biofuels, 298–300, 316, 332
Azotic Technologies, 234–38

Backman, James, 236
Baig, Sanah, 209
Balk, Josh, 150, 153–55, 165, 174

Ball, Laurence, 32–33
Bankman-Fried, Sam, 166
Banks, Dean, 223–24
Barclays, 181
Barr, William, 148
Bayer, 214, 236
Beck, Danielle, 173
Beck, Glenn, 184
Becker, Edward, 341n35
beef. *See* livestock
Benioff, Marc, 156
Berman, Richard, 190, 208
Berry, Steven, 50–51, 81, 299
Better Meat Company, 188, 201–2, 204
Beyond Eggs (Eat Just), 154
Beyond Meat
　beginning of, 155–56
　Beyond Burger, 184, 190, 194
　Beyond Sausage, 194
　burger development, 161–62
　downturn of meat alternative industry and, 193–94
　GFI investments in, 168
　growth of, 179, 183–84, 192
　hype cycle of meat alternatives and, 189–90
　initial public offering, 178, 183
　market space for meat alternatives, 160
　steak tips, 200
　Tyson investing in, 171–72
Bezos Earth Fund, 145, 210, 306
Biaggi, Mark, 216–18
Biden, Joe
　Inflation Reduction Act and, 11, 220, 297–300, 350n298
　meat alternatives and, 209
　regenerative agriculture and, 215, 220, 224, 301
　renewable energy funding, 297–98
Big Data, 266–67
Bin Alwaleed, Khaled, 171, 185
biodiversity and habitat loss
　agricultural land leading to, 5–6
　biofuel production leading to, 55
　in Brazil, 69–71
　mass extinction and, 5
　reforestation and, 313
　See also deforestation
bioenergy with carbon capture and storage (BECCS), 112, 300
biofuels
　for aviation, 298–300, 316, 332
　carbon emissions from, 64
　cellulosic ethanol, 58, 73, 91, 116–17, 342n64

deforestation for, 27, 52, 55, 66, 69–71, 77–79, 89–90, 92, 117, 342n64
electric cars vs., 116–17
international popularity and increase in production, 92
from palm oil, 52, 55, 92
pongamia trees and, 279–87
soy biodiesel, 64, 298
sugarcane ethanol, 342n64
See also corn ethanol
"Biofuel Song" (Livebroadkast), 81
biological fertilizers, 234
biological nitrification inhibition, 307–8
biomass power, 95–118
   carbon emissions and, 95–99, 100, 110–14
   in Denmark, 289–91, 293
   EPA draft plan for, 105–8
   in E.U., 95–97, 99–100, 108–10, 113–14
   IPCC on, 110–14
   lack of climate action and reduced incentives for, 116
   Massachusetts deciding against use of, 101–4
   Paris 2050 goals and, 118
   threat to global forests, assessment of, 100–101
   waste vs. roundwood used for, 114–15
birth rates, 126
Bittman, Mark, 156
Blue Apron, 183
Bluehouse fish farm, Florida, 258–62
BlueNalu, 172
Boca, 159
Boca Burgers, 152
Boddey, Robert, 247
Bolsonaro, Jair, 242, 316
Booth, Mary, 102–3, 106, 108, 110–11
Borlaug, Norman, 9, 137, 233–34
Boston Consulting Group, 193, 209, 309
Both Burgers, 201
Bovaer, 257–58
Bowery Farming, 269–70
Bowles, Ian, 101–4, 326
Bowman, Thomas, 163, 169
Boys Only Project, 256–58
BP, 58, 67, 214
Brandolini's Law, 79, 92
Branson, Richard, 171
Brazil
   Cerrado, destruction of, 70–71, 239–43, 316
   deforestation in, 17, 69–71, 77, 79, 241, 244, 342n64
   livestock, high yields of, 239–46, 332, 349n243
   pasture expansion in, 138
   regulation of deforestation in, 315–16
BrightFarms, 270
Brightseed, 346n175
Brown, Ethan
   beginning Beyond Meat, 155–56
   Beyond chicken strips, 189
   on downturn of meat alternative industry, 193–94
   on growth of Beyond Meat, 178, 183–84
   market space for alternative meats, 160–62
   on public acceptance of meat alternatives, 175
Brown, Gabe, 213, 215, 231
Brown, Gene, 114–15
Brown, Patrick
   on cultivated meat, 171
   on eradication of animal agriculture, 134, 157–60
   growth of Impossible Foods and, 179, 197–98
   on health profile of meat alternatives, 190
   heme from soy, 162–63
   Motif, legal battle with, 205
   nondairy cheese curds, 168
   ousting from Impossible Foods, 198
   on regenerative agriculture, 222
   response to negative meat alternative ad, 196
Browner, Carol, 38
Bulkin, Bernard, 109–10
Bündchen, Giselle, 213
Bunge, 236
Burger King, 134, 179, 184, 197
Bush, George H. W., 38–39
Bush, George W., 26, 47, 53, 87

Caaporã company, Brazil, 244–45
Calantha (pesticide), 272–73
California
   animal welfare law in, 303
   Low Carbon Fuel Standard, 65, 67–68, 85–87, 90–92
Campbell's, 220
cap-and-trade
   biomass energy and, 97–98
   EDF and, 59–60, 62
   farm groups killing, 89
   McCain sponsoring legislation on, 56
   no-till carbon credits and, 142
Capiaçu elephant grass, 245, 247
Caplow, Ted, 270–71
carbon credits, 215, 218, 220–24, 237, 246. *See also* cap-and-trade

carbon emissions
　from agriculture, 9–10, 119, 128, 139–40, 227, 271
　from aviation, 298–300, 316
　biomass power and, 95–98, 100, 110–14
　climate change and, 1–2
　from corn ethanol, 24–25, 49–50, 62–65
　COVID-19 lockdowns and, 326–27
　from deforestation, 122
　food waste and, 11
　increases in, 326
　from livestock, 15, 124
　negative, 112
　from wildfires, 17
carbon farming. *See* regenerative agriculture; soil carbon sequestration
carbon footprints, 327–29
carbon-opportunity-cost climate accounting, 292
carbon sinks and storage
　cornfields as, 24–25, 49–50
　ecosystem preservation for, 5
　forests and wetlands as, 5–6, 17, 49–50, 53, 97, 103, 114, 129, 331
　global agriculture shifting south into, 312–13
　regenerative agriculture and, 211–13
carbon taxes, 294, 316–17
Carbon Underground, 214
CARES Act (2020), 324
Cargill
　alt-protein division, 197
　Climate-Smart Commodities program and, 220
　meat alternatives, investing in, 171–72, 180, 186
　regenerative agriculture and, 214
carnivore diets, 193
Carter, John, 70
Cash, David, 102, 104
cassava processing, 131
cattle. *See* livestock
Cavanagh, Amanda, 278
cellulosic ethanol, 58, 73, 91, 116–17, 342n64
Center for Consumer Freedom, 179, 190, 196
Cerrado, Brazil, 70–71, 239–43, 316
CGIAR, 267
Chandna, Alka, 149
Chang, David, 175
Charles III (king of England), 11
Chesapeake Bay, manure pollution in, 42–43
Chevron, 286
chicken. *See* livestock

child mortality rates, 126
Chiles, Lawton, 37–38, 331
China
　increased meat consumption in, 132
　pork production in, 253
Chipotle, 190
Chu, Steven, 88
Churchill, Winston, 169
Clara Foods, 161
Clark, Wesley, 87
clean energy
　increase in use of, 3–4, 296–97
　Inflation Reduction Act and, 220, 297–300, 350n298
　Paris climate goals and, 296
　solar power and, 3, 82, 297, 309
　subsidies for, 309
　wind power, 3, 297, 309
"The Clean Energy Myth" (Grunwald), 14
Clean Meat (Shapiro), 173
climate accounting, 292
climate activists
　biofuels and, 57–59, 65–66, 68, 83
　biomass power plants and, 102–3
　on industrial agriculture, 253
　See also individual organizations
climate anxiety and fatalism, 3, 17, 333
climate change
　agricultural contributions to, 9
　corn ethanol and, 25–27
　decreased yields resulting from, 121, 265–66
　denial of, 3, 220, 242
　effects of, 2, 16–17
　lack of action on, 116
　preindustrial farming and deforestation, 7–8
　public awareness of, 56
　science on, 1–2
　as test, 332–33
　See also climate solutions; greenhouse gases
Climate Robotics, 215
Climate-Smart Commodities program, 220
climate solutions, 289–322
　agricultural reform in Denmark, 289–94
　culture-war politics vs., 303–5
　deforestation regulation, 314–16
　for electricity sector, 296–97
　for fertilizers and manure, 306–8
　financing, 316–18
　Greenhouse Gas Protocol and, 318–22
　individual vs. collective responsibility, 327–30

INDEX

Inflation Reduction Act and, 11, 297–300
investments in R&D, 305–9
land-shifting problem and, 312–14
meat and dairy alternatives, investments in, 308–9
misguided, 301–3
proven techniques, 310–11
reducing beef consumption, 295
taxes on livestock emissions, 293–94
Climax Foods, 200–201
Clinton, Bill, 152
Clinton, Hillary, 57
coal power plants
  beef consumption vs., 15, 305
  corn ethanol and, 90
  Inflation Reduction Act and, 297, 350n298
  shutting down, 3
  wood-burning power plants vs., 95, 100, 104–5, 112, 289
Cocking, Ted, 233–34
Cohn, Gary, 268
Collins, Susan, 113
Columbus grazing method, 217, 243
Common Ground (film), 238
Compassion Over Killing, 150
Conagra, 159
concentrated animal feeding operations (CAFOs), 247–50
Consensus Building Institute, 320
Conservation Reserve Program (CRP), 42, 58
construction industry, 303
Continental Grain, 236
controlled environment agriculture, 268–71
Cooney, Nick, 165
Coons, Chris, 33
corn ethanol, 49–74
  additionality and, 77–78
  animal feed from distillers grains, 342n64
  aviation fuel tax credits for, 298–300
  carbon emissions from, 24–25, 49–50, 62–65
  climate activists and, 65–66
  criticism of Searchinger's findings on, 76–79
  deforestation and, 27, 52, 66, 69–71, 77–79, 89–90
  EDF and, 59–62
  E.U. mandate on, 80–84
  gasoline vs., 21–25, 87–88
  increased yields for, 80–82, 85–86
  ILUC and, 53–55, 62–64, 66, 68–71, 75–76, 79, 85–91
  opportunity costs of land and, 50–53, 65–66, 91, 96–97

as placeholder for better biofuels, 58–59
politics and, 22–23, 25–27, 47–48, 55, 56–59, 68, 85–89
scientists and, 66–68
See also indirect land-use change
Corn Refiners Association, 57
Cornucopia (program), 308
Corzine, Jon, 26
cover crops
  benefits of, 231–32
  as cash crops, 276
  livestock and, 240
  politics and, 215
  soil carbon sequestration and, 142
  for soil erosion, 265
  subsidies for, 220, 299, 310
  yields and, 301, 348n225
Covey, Kris, 222
COVID-19 pandemic, 192–93, 324–25
Cowspiracy (documentary), 160
Cracker Barrel (restaurants), 194
Cramer, Jim, 192
Creating a Sustainable Food Future (Searchinger)
  on deforestation, 340n5
  on emissions and pollution, 139–41
  on food, emissions, and land gaps, 128–30
  on food waste, 130–32
  inspiration for, 124–25
  launch of, 143–44
  on meat alternatives, 132–34
  on regenerative agriculture, 141–43
  research for and writing of, 15, 125–28
  on rice fields and methane emissions, 310
  solutions in, 129–30
  on yield growth, 134–39
Creutzig, Felix, 111–12
Crews, Tim, 226
CRISPR gene editing
  to control gender of animal offspring, 256–58
  for cultivated meat, 196, 207, 208
  for livestock heat tolerance and virus immunity, 256
  for oilseed crop yields, 276
  Searchinger on, 138
crop analytics, 267–68
crop residues
  biofuel from, 58
  for livestock feed, 295–96
  returning carbon to soil, 228–31
crop rotations, 142

crop sustainability, 263–88
　GMOs for, 274–76 (see also GMOs)
　increasing yields for, 263–66
　photosynthesis efficiency, 276–78
　pongamia trees and, 279–87
　precision agriculture and, 266–68
　RNA technology and biopesticides, 272–74
　vertical farms, 268–71
cultivated meat
　ban in Florida, 303–4
　costs of, 182–83
　CRISPR gene editing for, 196, 207, 208
　early development of, 169–76
　innovations and cost reductions in, 206–10
　JBS investments in, 197
　served in Singapore, 192
culture-war politics, 297, 303–5

Dalai Lama, 238
Danone, 187, 214
Daschle, Tom, 26
Datar, Isha, 161, 171
deforestation
　agriculture driving, 5–8, 9, 15, 17, 121, 340n5
　for biofuels production, 27, 52, 55, 66, 69–71, 77–79, 89–90, 92, 117, 342n64
　in Brazil vs. U.S., 241
　carbon emissions from, 122
　construction industry and, 303
　efforts to reduce, 314–16
　for livestock, 123–24, 136, 243–45
　for preindustrial farms, 7–8
　See also biomass power
Delucchi, Mark, 54–55, 72, 78, 111, 341–42n54
Democratic Party
　biomass power, 101–4
　cap-and-trade, 89
　climate action, 56
　COVID-19 pandemic and, 324–25
　Inflation Reduction Act, 11, 220, 297–300, 350n298
　regenerative agriculture, 214–16
　See also individual presidents
Denmark
　agricultural reforms in, 293–94
　biomass energy in, 289–90, 293
　high-yield agriculture in, 291–93
　manure, converting to biogas, 306–7
Department of Agriculture, U.S. See USDA
Department of Energy, U.S., 307
Dern, Laura, 238
DeSantis, Ron, 303–4
Desmodium flowering clover, 247

DeSnayer Dairy, 251–52
dietary changes, 132–34, 151, 328. *See also* meat and dairy alternatives; vegan and vegetarian diets
Diet for a Small Planet (Lappé), 123
Dirtier Than Coal? (Greenpeace U.K. report), 109
Dirt to Soil (Brown), 213
Dodson, Greg, 197
Dogwood Alliance, 114
double-cropping, 138
Doudna, Jennifer, 256
Douglas, Gordon, 352n324
Douglas, Marjory Stoneman, 332–33
Drax Group, 100, 109, 300
Dubai climate summit (2023), 11, 306
Dunkin', 179, 194

Earth Day, 327, 329
Eat Just
　beginning of, 154–57
　cultivated meat, 175–76, 192, 196, 207, 208
　Just Egg, 164–65, 182, 196–97, 209
　Just Mayo, 163–64, 174–75
　name change to, 175–76
EAT-Lancet report, 178–79
Eclipse Foods, 191
education of women and girls, 126
Edwards, Ron, 281–82
"Effective Advocacy: Stealing from the Corporate Playbook" (Friedrich), 149
egg substitutes
　development of, 154–57, 163–65, 174–75
　from fermentation, 161, 182
　growth of, 209
　Just Egg, 164–65, 182, 196–97, 209
　Just Mayo, 163–64, 174–75
Eilish, Billie, 11
Elder, Max, 195
electricity sector, 296–97. *See also* biomass power; clean energy; coal power plants
electric vehicles
　airplanes, 316
　biofuels vs., 116–17
　increase in use of, 3
　to reduce petroleum use, 58
　subsidies for, 309
　tractors, 297
Elizondo, Arturo, 161, 182
Embrapa (Brazilian research agency), 241, 243, 245, 247
Energy Policy, on biomass power, 108–9

INDEX

Environmental Defense Fund (EDF), 36–37, 41, 59–62, 142
Environmental Programme (United Nations), 253
Environmental Protection Agency (EPA)
   on biomass power, 105–8, 113
   Calantha approval, 273–74
   on cellulosic ethanol, 91
   corn ethanol and land-use change models, 85, 87–90
   threats to, 41
Enviva (biomass fuel company), 106, 107, 114–15, 332
Estenoz, Shannon, 38
ethanol
   cellulosic, 58, 73, 91, 116–17, 342n64
   sugarcane, 342n64
   switchgrass, 58, 73, 91, 342n64
   See also corn ethanol
European Academy of Sciences, 322
European Union (E.U.)
   aviation biofuels and, 332
   bans on imports from deforested land, 317
   biofuels mandate in, 80, 83–84
   biomass power in, 95–97, 99–100, 108–10, 113–14
   Great Recession and food crisis, 82–83
   organic farming in, 301
   on regenerative agriculture, 142
   See also individual countries
Everglades National Park, Florida, 37–39, 44–45, 330–31, 348n230
EVERY Company, 161, 180, 182

Faaij, André, 111
Faber, Scott, 40, 236
factory feedlots, 247–50
fake meat. *See* meat and dairy alternatives
Fallon, Jimmy, 196
farmed fish, 138, 258–62
Farm Foundation, 75
farmland. *See* agricultural land
Farm Security and Rural Investment Act (2002), 45–48
Farrell, Alex, 66–69, 72, 74, 79, 326
Farrell, Mark, 68, 74
Fazenda Tropical plantation, Brazil, 239–44, 349n243
"Feeding a Hot and Hungry Planet" speech (Searchinger), 125
fermentation
   for cheese making, 160
   dairy-free ice cream, 181–82
   egg substitutes from, 161, 182
   meat alternatives from, 162–63, 201–4, 308
   mycoprotein from, 186–88, 202–4
   precision, 160–63, 308
fertilizers and manure
   biological nitrification inhibition, 307–8
   carbon sequestration and, 228–30
   converting to biogas, 306–7
   greenhouse gasses released by, 9, 12, 231, 306–7
   Gulf of Mexico dead zone and, 230, 265
   self-fertilizing crops, 234–38
   water pollution from, 140, 218, 230, 233–34, 348n230
   yield increases and, 120–23, 142
50/50 Foods, 201
Fikes, Liz, 203
fires. *See* wildfires
fish alternatives, 167–68, 172
fish and seafood, 138, 258–62
flexitarian diets, 151
floods
   farmland and, 265
   of Mississippi River in 1993, 40
   Missouri River, flood-control project for, 44, 46
Florida
   ban on cultivated meat, 303–4
   Everglades National Park, 37–39, 44–45, 330–31, 348n230
   fish farming in, 258–62
Folkenflik, David, 42
Food and Agriculture Organization (FAO, United Nations), 124, 128, 141, 213, 251
Food and Drug Administration (FDA), 258
Food Climate Research Network, 349n231
food crises
   diversion of land for biomass fuel plantations, 115
   diversion of land for corn ethanol production, 71, 81, 85
   grain for livestock vs. humans, 123, 166
   Great Recession and, 82
   population growth and, 119, 126, 127
food technological innovations. *See* technological innovations
food waste, 11, 130–32, 143, 311, 328
Fora Foods, 168, 194
forest land
   agroforestry and, 138, 231
   as carbon sinks, 5–6, 17, 49–50, 53, 97, 103, 114
   pine plantations, 114–15
   See also deforestation

forestry interests, 105–6, 113–14, 303
fossil fuels, 22, 24–25, 64, 328. *See also* carbon emissions; coal power plants
Four Per Thousand carbon farming campaign, 213, 222–23, 231
France
  biofuels mandate for the E.U. and, 84
  diesel subsidies protests in, 304
  fossil fuel taxes in, 294
Francis (pope), 11
Friedrich, Bruce
  background, 145–48
  cultivated meat and, 169–73, 210
  on downturn of meat alternative industry, 193
  on funding for meat and dairy alternatives, 309
  GFI and, 165–69 (*see also* Good Food Institute)
  on growth of meat and dairy alternatives, 181
  Impossible Burger and, 176–77
  on industry in-fighting, 205
  on meat and dairy alternatives, 152–53, 156, 159, 189
  on meat industry selling meat alternatives, 180
  New Crop Capital fund and, 167–69, 171
  PETA, working at, 149–51
  secret memo on meat and dairy alternatives, 189–92, 346n191
  TED Talk on meat alternatives, 166
Fuller, Buckminster, 201

Gabel, Steve, 247–49
The Game Changers (documentary), 178
Gardein, 159
Gartner Hype Cycle, 199
gasoline, 22, 64, 328
Gates, Bill
  climate action, 11
  on cost of cleaner new technologies, 308
  cultivated meat, investing in, 171
  meat alternatives, investing in, 156, 158, 162, 187
  self-fertilizing crops, investing in, 235
Gates Foundation, 145, 277
Geistlinger, Tim, 161–62
gene editing. *See* CRISPR gene editing
General Mills, 192, 194, 213, 215, 225
genetically modified organisms. *See* GMOs
Gingko BioWorks, 191
Gingrich, Newt, 56

glaciers, disappearance of, 2
Glass Walls (film), 151
Global Alliance for the Future of Food, 301
Global Methane Hub, 306
Global Report on Food Crises (2024), 16–17
Global Warming Solutions Act (California, 2006), 56
Global Warming Solutions Act (Massachusetts, 2008), 101
Glover, Juleanna, 273
glyphosate, 137–38
GMOs (genetically modified organisms)
  bans on, 302
  innovations through, 274–76
  in meat alternatives, 184
  pest-resistant corn, 265
  public perception of, 162
  Roundup-ready crops, 137–38
  yields of, 138, 275–78
  See also CRISPR gene editing
Golden Rice, 275
Goldman Sachs, 279
Goodall, Jane, 238
Good Catch, 167–68
Good Food Institute (GFI)
  ASAP, 309
  Bezos Earth Fund and, 210
  cultivated meat and, 173, 208
  development and focus of, 165–69
  funding scientific grants, 191
  growth of, 180
  Open Philanthropy and, 346n191
  San Fransico conference in 2019, 178, 185–87
  San Fransico conference in 2023, 193
  supporting meat alternative startups, 189
Gore, Al, 40, 43, 56, 57, 213, 221–22
Gore, Kasia, 196
grass-fed livestock, 216–18, 221, 250, 301
Grassley, Charles, 88–89, 300
Great Green Wall of the Sahel, 231
Great Recession (2008), 82–84, 151
Greene, Nathanael, 58, 65–66, 72–74, 326
greenhouse gases
  from fertilizers and manure, 9, 12, 231, 306–7
  individual vs. collective responsibility for, 327–30
  Supreme Court on, 56
  yields and, 226–27
  See also carbon emissions; climate change; methane emissions

INDEX

Greenhouse gases, Regulated Emissions, and Energy use in Technologies (GREET), 24, 62, 64, 78, 298–99
Greenhouse Gas Protocol, 318–22
greenhouses, 270
GreenLight Biosciences, 272–73
Green New Deal, 132
Greenpeace, 109, 292–93
Green Revolution
  beginning of, 9
  Brazil and, 241
  continuation of, 137
  fertilizers and, 233–34
  sub-Saharan Africa and, 266
  yield increases and, 120–21, 128–29, 264
greenwashing, 220, 222, 231, 321
Grist (environmental publication), 79
Gro Intelligence, 267–68
Gross, Robert, 32
groundwater depletion, 265–66
Growth Energy (biofuels group), 87, 88
guano, 233
Guimarães, Roberto, Jr., 243
Gulf of Mexico dead zone, 230, 265
Gulf Stream, changes in, 2

Haber-Bosch fertilizer, 233
habitat loss. *See* biodiversity and habitat loss
Hahn, Alan, 186, 195, 203
Hall, Kevin, 200
Hampton Creek, 154–57, 163–65, 174–76. *See also* Eat Just
Hanson, Craig, 125, 143
Hardee's, 179, 185, 194
Harkin, Tom, 89
Harper, Sara Hessenflow, 59–62, 142, 213
Harrelson, Woody, 211–12
Harris, Will, 213, 215, 225, 231
Haspel, Tamar, 348n226
Hastert, Dennis, 26
Hayes, David, 224, 237–38
Heimlich, Ralph, 51–52, 64, 68
Hemami protein, 192, 204–5
Hertel, Thomas, 85–87, 90
Hodson, Paul, 80–81, 82–84
Hoefner, Ferd, 219
Holm, Thue, 260–61
Hooray Foods, 189, 194
How to Be Great at Doing Good (Cooney), 165
Huggins, Tyler, 203–4

Hurricane Katrina (2005), 45
hydroponics, 142–43

IBM, 215
Impossible Foods
  attempts to eradicate animal agriculture, 134, 157–60
  beginning of, 133–34
  growth of, 179, 184–85, 192
  hype cycle of meat alternatives and, 190, 197–99
  Impossible Burger, 134, 162–63, 175–77
  Impossible Sausage, 194
  Impossible Whopper, 179, 184
  Motif, legal battle with, 204–5
  plant-based vs. meat burgers, 176–77
An Inconvenient Truth (documentary), 56
India
  beef from, 250
  Bt cotton grown in, 138
  increased meat consumption in, 132
  pongamia trees in, 279–87
Indigenous peoples, 7–8
Indigo Ag, 215, 222–24, 226
indirect land-use change (ILUC)
  biomass power and pine plantations, 115
  cellulosic ethanol production and, 342n64
  corn ethanol mandates, provisions for, 83–85
  corn ethanol production and, 53–55, 62–64, 66, 68–71, 75–76, 79, 85–91
  EPA conclusions on, 87–90
  increasing yields to combat, 80–82, 85–86
  lobbying to discard, 87–90
  regenerative agriculture and, 225, 227
individual vs. collective responsibility, 327–30
indoor farming, 268–71
Industrial Revolution, 2
Infarm (vertical farm), 269–70
Inflation Reduction Act (IRA, 2022), 11, 220, 297–300, 350n298
insecticides, 137
insect protein foods, 310
integrated crop-livestock systems, 231, 240
Intergovernmental Panel on Climate Change (IPCC, United Nations)
  on biofuels, 49
  on biomass power and carbon emissions, 95–96, 98, 100, 110–14
  on climate tipping points, 2
  on deforestation, 5, 122
  on potential croplands, 52–53
  on regenerative agriculture, 213, 223–24
  warming predictions, 4

International Energy Agency, 300
Iowa State University, 62
irrigated farmland, 121

Janzen, H. Henry, 348–49n231
Jatropha biodiesel, 92, 117
JBS, 180, 186, 192, 197, 220
Jenkins, Jennifer, 106, 115
John Deere, 264–65, 297
Jonas, Thomas, 186–88
JPMorgan Chase, 215
Just (food company). *See* Eat Just

Kamayura tribe, Brazil, 69
Kammen, Dan, 67
Kanter, David, 237–38
Kardashian, Kim, 194
Kebreab, Ermias, 254–55, 305
Kellogg, John, 152
Kellogg's, 159
Kennedy, John F., 296
Kennedy, Robert F., Jr., 213, 215, 301–2, 326
Kernza grain, 215, 225–26, 348n226
Kerr, Chris, 167–69, 172
KFC, 149, 179
Khademhosseini, Ali, 201
Khosla, Vinod
  on biomass energy, 103
  on cellulosic ethanol, 58, 116–17
  meat alternatives, investing in, 156, 158, 179
  Searchinger's response to criticism of, 79
Kimmel, Jimmy, 275
Kirkby, Clive, 228–30
Kirkegaard, John, 228–31
Kiss the Ground (documentary), 211–13
Kite Hill, 160, 168
Kleiner Perkins, 156
Kodjak, Drew, 72
Koppel, Daniel, 267
Kotok (chief of Kamayura tribe), 69
Kraft, 159
Kurzweil, Allen, 32
Kyoto Protocol (1997), 2, 26, 98–99

lab-grown meat. *See* cultivated meat
Lal, Rattan, 226, 230
LaMantia, Anthony, 352n324
Lamont, Ewan, 224
land. *See* agricultural land; forest land
Land Institute, 215, 221, 225–26
Lappé, Frances Moore, 123, 194
Laranja, Luis Fernando, 245–46
Lashof, Dan, 65

Lee, Catie, 274
Lee, Charles, 37, 331
Lee, Jane, 197
Leonard, Michael, 205
Leopold, Aldo, 232–33
Leucaena shrubs, 129, 247, 310
Liebreich, Michael, 12
life-cycle analyses
  of biofuels, 51, 55, 60, 71
  development of, 24
  of meat alternatives, 179
  of regenerative beef, 215
A Life on Our Planet (film), 213
LightLife, 159
Li Ka-shing, 156, 158, 179
Lipman, David, 185
Livebroadkast, 81
livestock
  adaptive multi-paddock grazing, 217–18
  agricultural land used for, 15, 123, 132, 136
  animal welfare and, 148–51, 153, 165, 303
  carbon emissions from, 15, 124
  carbon pricing for, 329
  Columbus grazing method, 217, 243
  crop residue feed for, 295–96
  deforestation for, 123–24, 136, 243–45
  diversion of corn feed for corn ethanol, 61, 342n64
  diversion of land for biomass fuel plantations, 115
  farmers' subsidies, 302
  grass-fed, 216–18, 221, 250, 301
  increasing productivity of, 134–35, 291–92
  insect proteins for, 310
  integrated crop-livestock systems, 231, 240
  manure from. *See* fertilizers and manure
  meatfluencers promoting consumption of, 193
  methane emissions from, 9, 15, 132–33, 140, 231, 250, 304–7
  for pet food, 328
  on public land, 302, 351n302
  reducing consumption of, 132–34, 146, 328
    (*see also* meat and dairy alternatives)
  regenerative agriculture and, 216–18, 225
  silvopasture systems for, 129, 245, 310
  sustainability of (*see* livestock sustainability)
  taxes on emissions from, 293–94
  water pollution from, 42–43
Livestock's Long Shadow (FAO report), 124, 251
livestock sustainability, 239–62
  advocacy for industrial farming practices, 250–54

INDEX

CAFOs for, 247–50
fish farming, 258–62
high yields in Brazil, 239–46, 349n243
technology for, 254–57
Loam (seed coating company), 215
Locks, Charlton, 241, 246–47
logging, 303. *See also* deforestation
Long, Steve, 108, 277–78
Lorenzen, Tyler, 186, 192, 195
Low Carbon Fuel Standard (California), 65, 67–68, 85–87, 90–92
Low-Input Sustainable Agriculture (LISA) program, 219
Lula da Silva, Luiz Inácio, 242, 315–16

MacKay, David, 109–10
Maggi, Blario, 70, 241
Magnum Feedyard, Colorado, 247–50
Malthus, Thomas, 122
Manchin, Joe, 297
Mann, Charles, 121
Mann, Michael, 327
manure. *See* fertilizers and manure
Maple Leaf Foods, 159, 194
March, Joshua, 208–9
Marcus, Jerome, 35
Markey, Ed, 89
Mars (company), 214
Massachusetts, biomass power used in, 101–4
mass extinction, 5
Matthew 25, 147
Matzner, Franz, 66, 68
McCain, John, 56, 57
McCartney, Paul, 150–51
McDonald's, 150, 192, 193–94, 220
McGuinness, Peter, 198–99
McKinsey, 285
McPlant burgers, 192–94
meat and dairy alternatives, 145–77
    cultivated meat, 169–73 (*see also* cultivated meat)
    development of, 153–59, 162–63
    GFI and, 165–67
    government investment in, 308–9
    issues with companies, 161–65, 175–76
    market space for, 159–61, 167–69
    New Crop Capital fund for, 167–69, 171
    problem solving with, 17–18, 133–34
    public opinion of, 152–53, 176–77
    reducing livestock consumption through, 145–46
    See also egg substitutes; *individual companies*
meat and dairy alternatives hype cycle, 178–210
    continued innovation in industry, 200–204

costs, consumer response to, 200
cultivated meat, 206–10 (*see also* cultivated meat)
downturn of alternative meat companies, 192–97
Friedrich's concerns over, 189–92
growth of alternative meat companies, 178–88
health profile of, 179–80, 190, 200
Impossible Foods and, 197–99
industry in-fighting, 204–5
meat industry selling alternative meats, 180–81
mycoprotein, 202–4
meat-eating diets. *See* cultivated meat; livestock
meatfluencers, 193
Meati, 203–4
Meat the Future (film), 192
Meet Your Meat (film), 150–51
Melendez, Mary Jane, 215
Mencken, H. L., 25, 30
Menker, Sara, 267–68
Mercy for Animals, 150, 165
methane emissions
    from animals, 9, 15, 132–33, 140, 250, 304–5
    from manure, 231, 306–7
    from rice fields, 126–27, 140, 310
    seaweed in animal feed to reduce, 254, 257–58
Micol, Laurent, 244–45
Mill Industries, 328
Mimikakis, John, 45–46
Miracle Yield theory, 85–86, 121–22, 299
Miranda, Renata Bueno, 316
misinformation, 79, 194
Mission Barns, 207
Mississippi River flood (1993), 40
Missouri River flood-control project, 44, 46
Mitloehner, Frank, 250–51
Mitsubishi Corporation, 286
Miyoko's Creamery, 160, 168, 194
molecular farming, 275
Molino, Steve, 160
Momoa, Jason, 238
Monsanto, 87, 88, 137–38
Moomaw, William, 53, 103, 110–11, 321
Moore, Michael, 150
Moringa trees, 247, 310
MorningStar, 159
Mosa Meat, 172, 208
Moskovitz, Dustin, 166
Motif FoodWorks, 191, 192, 201, 204
Mraz, Jason, 213
Musk, Elon, 11, 184

Musk, Kimbal, 171
Muth, Chuck, 183
Muufri, 160
mycelia, 201–3
mycoprotein, 186–88, 202–4
MycoTechnology, 186, 195, 203
Myers, Morey, 36

National Cattlemen's Beef Association, 173
National Corn Growers Association, 61
National Pork Producers Council, 61
National Sustainable Agriculture Coalition, 219
Native Americans, 7–8
natural buffer strips, 227–28
Natural Resources Defense Council (NRDC), 57–59, 65–66, 72
Natural Systems Agriculture, 226. *See also* regenerative agriculture
Nature Conservancy, 226
Nature's Fynd, 186–88, 192, 203
negative emissions, 112
Nestlé, 81, 159, 193, 214
Neutral (milk company), 218–19
New Age Eats, 193, 196
New Crop Capital fund, 167–69, 171–72, 180, 194, 275
New Harvest (research institute), 161
New Wave Foods, 168, 181, 194
New Zealand, 294
Nielsen, Ebsen, 291–92
nitrate pollution, 140
nitrification inhibition, 307–8
The Nitrogen Dilemma (film), 238
nitrogen fertilizer. *See* fertilizers and manure
nitrogen-fixing bacteria, 234
Nobell Foods, 168, 275
North Carolina, agricultural wastewater systems in, 140
Nossa Senhora ranch, Brazil, 244–45
NotCo, 192
no-till
　Biden initiative for, 301
　carbon credits for, 142, 223–24
　cover crops for, 265
　soil carbon sequestration and, 223
　soil erosion and, 219, 265
　subsidies for, 220, 299, 310
　yields and, 301
Novo Nordisk Foundation, 308

Oatly, 192, 194
Obama, Barack
　on biofuels, 57, 88

　on biomass power, 113
　cap-and-trade bill, 89, 97–98
Ocean Hugger Foods, 189, 194
O'Conner, Bill, 46
Oge, Margo, 88–89
O'Hare, Michael, 52, 86
Omeat, 201
The Omnivore's Dilemma (Pollan), 232, 233
OpenAg Initiative, 142–43
Open Philanthropy, 346n191
Oppenheimer, Michael, 97, 108
opportunity costs of land
　agricultural expansion and, 313
　carbon-opportunity-cost climate accounting, 292, 321–22
　corn ethanol and, 50–53, 65–66, 91, 96–97
　defined, 50
　meat consumption and, 123–24, 329
　regenerative agriculture and, 225–28, 301
　trees and, 103, 108, 114, 116
organic farming, 142, 227, 257, 301
Oxfam, 81

Packard Foundation, 301
palm oil production, 52, 55, 92
Pandya, Ryan, 182
Paris Agreement (2015)
　biomass power and, 118
　electricity sector and, 296
　food-related emissions, reducing, 10, 128
　goals of, 2
　reforestation and, 5
　regenerative agriculture, 213
　rice and methane emissions, 127
Patrick, Deval, 101–2
Paul, Chris, 183
peatlands, 311, 331
pellet industry. *See* biomass power
Pelosi, Nancy, 56, 325
People for the Ethical Treatment of Animals (PETA), 149–51
Peoples, Oliver, 276
Peppou, George, 186
PepsiCo, 192, 220
Perdue, 85–86, 90–91, 181, 188
Perfect Day (dairy alternative company)
　dairy-free cream cheese, 192
　dairy-free ice cream, 181–82
　downturn of meat alternative industry and, 194
　growth of, 180, 209
　market space for dairy alternatives, 160–61
permafrost, melting of, 2

Perry, David, 223
Perry, Katy, 179
Perschel, Robert, 103
personal food computers, 142–43
pesticides, 142, 219, 265, 272–74, 301
PETA (People for the Ethical Treatment of Animals), 149–51
Peterson, Collin, 88–89
pet food, 328
phosphorus fertilizers, 348n230
photosynthesis, 276–78
pigs. *See* livestock
pine plantations, 114–15
Pingree, Chellie, 215–16
Pivot Bio, 235–37, 274
plant-based diets. *See* meat and dairy alternatives; vegan and vegetarian diets
Plant Based Food Association, 166, 171
plastic pollution, 303
Platform for Big Data in Agriculture, 267
Platt, Roger, 31–32
Plenty (vertical farm), 269–70
Plevin, Richard, 67, 111, 343n89
Plowshares movement, 148
Podesta, John, 299
Point Blue Conservation Science, 217–18
politics
  agricultural interests in, 45–46, 84, 304–5
  aviation biofuel and, 298–300
  climate consciousness in, 56
  corn ethanol and, 22–23, 25–27, 41–42, 47–48, 55, 56–59, 68, 85–89
  cultivated meat and, 303–4
  forestry interests in, 105, 113
  regenerative agriculture and, 214–16, 219–21
  wetlands restoration and, 41–43
  See also Democratic Party; Republican Party; *individual presidents*
Pollan, Michael, 12, 57, 213–14, 232, 233
pongamia trees, 279–88, 310
population growth
  agricultural history of humans and, 7
  food crises and, 119, 126, 127
  increased demand on agricultural land, 6
  land as finite resource and, 16
  Malthusian predictions for, 122
  research on, 126
pork. *See* livestock
Portman, Natalie, 269
Porzig, Elizabeth, 218
Post, Mark, 169, 172, 208

Pottenger, Jay, 35
Powlson, David, 227
precision agriculture, 137, 266–68
precision fermentation, 160–63, 308
Project Drawdown, 328, 331
Project Jake, 175
Prospera Technologies, 267
Proven (biofertilizer), 235–37
Pruitt, Acott, 113
Purdue land-use model, 85–86, 90, 298–99
Purdy, Chase, 346n175
Puris, 186, 192, 195
Putin, Vladimir, 234, 237

Quint, Yossi, 201
Quorn, 152, 187

Rajapaksa, Gotabaya, 301
rangeland. *See* agricultural land
Raskin, Jamie, 324–25
recycled cooking oil, 64, 117
REDD+ forest conservation funding (UN), 311, 314–15
Redefine Meat, 180
ReFED, 310–11
reforestation, 313–15
regenerative agriculture, 211–38
  benefits of, 212–13, 231–33
  carbon credits and, 220–24
  carbon storage and, 211–14
  fertilizer leading to carbon sequestration, 228–30
  fertilizer problems, 233–34
  integrated crop-livestock systems, 240–41
  livestock grazing and, 216–19, 225, 239–46, 349n243
  natural buffer strips, benefits of, 227–28
  politics and, 214–16, 219–21
  problems with, 141–43
  promoters of, 213–14
  Searchinger's criticisms of, 224–28
  self-fertilizing crops, 234–38
  yields from, 141–42, 226–27, 301–2
  See also cover crops; no-till; soil carbon sequestration
regenivore, 214
renewable energy. *See* clean energy
Renewable Fuels Standard, 25, 56, 65–66, 299
Reopen America Act (2020, proposed), 325
reproductive health services, 126

Republican Party
  biofuels and, 88, 300
  biomass energy, 113
  cap-and-trade, 56, 89
  COVID-19 pandemic and, 325
  cultivated meat, 303–4
  culture-war politics, 297
  Green New Deal, 132
  opposition to climate action, 304–5
  regenerative farming, 301–2
  wetlands conservation and, 39
  See also individual presidents
Rhiza (mycoprotein), 202
Rice, Dirk, 264–66
rice fields
  ban on fertilizers and reduced yields, 301
  GMO Golden Rice, 275 (*see also* GMOs)
  methane emissions from, 126–27, 140, 310
  rice straw as animal feed, 296
RIPE (Realizing Increased Photosynthetic Efficiency) program, 277–78
RNA cellular messaging technology, 272–74
Roberts, David, 12
Rockefeller Foundation, 214, 301
Rodale Institute, 213, 220, 221, 223
Rogan, Joe, 213, 304
Rogers, James, 131
Ronald, Pam, 275
Ronnen, Tal, 152–53, 160, 168
Rosenberg, David, 268–71
Roundup-ready crops, 137–38
Rubisco, 277–78
Ruddiman, William, 7
Runkle, Milo, 150, 165

Sadhguru, 11
Sahel region, 231
Sandalow, David, 48
A Sand County Almanac (Leopold), 232
Sarkozy, Nicolas, 84
Savory, Allan, 217–18
Savory Institute, 213
Schinner, Miyoko, 160, 194
school lunches, 302
Schrag, Daniel, 106–8
Schroeder, Doug, 266
Schwarzenegger, Arnold, 57, 64, 87
SCiFi Foods, 196, 207, 208–9
Searchinger, Brian, 28, 31, 33
Searchinger, Gene, 30–31, 97
Searchinger, Marian, 30–31
Searchinger, Tim, 21–49
  background and education, 27–34

  climate fight success of, 330–34
  on climate financing, 316–17
  COVID-19 pandemic and testing research, 323–25, 352n324
  on deforestation regulation, 314–16
  EDF and, 41
  Greenhouse Gas Protocol and, 318–22
  on land-shifting problem, 312–14
  law career, 34–47, 341n35
  on meat consumption, 13–16
  Szymanek, marriage and family with, 34, 39, 41, 47
  wetlands, advocating for, 37–40, 41–43, 44–45, 330–31
Searchinger, Tim, biomass research
  on biomass power and carbon emissions loophole, 96–99
  biomass power publications, 99, 103, 108–9
  Danish Parliament hearing, testifying at, 289–91
  EPA draft plan and, 105–8
  E.U., warnings to, 108–10
  logging, climate impacts of, 303
  Massachusetts, convincing to reduce subsidies, 101–4
  Paris 2050 goals and, 118
  threat to global forests, assessment of, 100–101, 103–4
  on waste vs. roundwood used for, 114–15
Searchinger, Tim, corn ethanol research
  academic responses to corn ethanol research, 71–74
  on Amazon rainforest deforestation, 69–71
  on biofuels and ILUC, 53–55
  biofuels policymaking and, 81–85
  biofuels research, questioning, 21–22, 24–26, 47–48, 49–53
  on California's biofuels mandate, 91–92
  climate activists, biofuels meeting with, 65–66
  confronting academics on corn ethanol research, 65–68, 75–77
  corn ethanol land-use emissions modeling, 62–65
  EDF, corn ethanol meeting with, 60–62
  Farm Foundation workshop speech, 75–76
  increased corn yields and reduced land-use change, 80–82, 85–86
  Science paper on biofuels, 75–79
Searchinger, Tim, food and agriculture research
  on benefits of regenerative agriculture, 231–33
  on Biden climate action, 298–300

INDEX

biological nitrification inhibition, 307–8
career shift to, 92–94, 120
crop residue livestock feed, 295–96
Denmark, high-yield agriculture in, 291–92
on land-use emissions of sub-Saharan Africa, 266
livestock efficiency, 239–46, 250, 291–92
meat and dairy alternatives, 145–46, 185, 210
methane studies, 305–7
nitrogen fertilizer and carbon sequestration, 230–31
on regenerative agriculture, 219, 222–28, 230–31
self-fertilizing crops, 234–38
yield increases and, 120–23
See also Creating a Sustainable Food Future
seaweed animal feed, 254, 257–58
Shapiro, Paul, 150, 173, 188, 201–4
Sheehan, Meg, 102, 104
Shell Oil, 214
shipping industry, 316–17
Shopify, 215
Sikka, Naveen, 278–87
silvopasture systems, 129, 245, 310
Simon, Michele, 166, 171, 195
Sinclair, Upton, 76
Singleton, Mark, 272
single-use plastic, 303
Smithfield, 180
Snoop Dogg, 183
Snyder, Rod, 89
social media "meatfluencers," 193
soil carbon sequestration
　cover crops and, 231, 348–49n231
　crop residues and, 228–31
　Four Per Thousand campaign, 213
　livestock grazing and, 218
　measuring, 221–22
　no-till and, 142, 223
　opportunity costs of land, 225–28
　pongamia trees and, 280
　protecting nature for, 232
　Rodale report on, 221
　subsidies for, 59, 221, 228
　See also regenerative agriculture
soil erosion, 218–19, 265, 276, 280
Soil Health Institute, 220
Soil Inventory Project, 222, 224
solar power, 3, 82, 297, 309
solutions. *See* climate solutions
Solyndra, 309
Southeast Asia, 52, 55, 92
soy
　biodiesel from, 64, 298

genetically modified, 275
　in meat alternatives, 184, 190
　public perception of, 162
Soylent, 160
Sperling, Dan, 85, 87, 90
Sri Lanka, 301
Starbucks, 197, 332
Steer, Andrew, 143
Stewart, Martha, 269
Steyer, Tom, 216–18, 232
Stop Spewing Carbon, 102
Stroer, Rachel, 221
Subbarao, Guntur, 307–8
sub-Saharan Africa, 126, 250, 266
subsidies
　for biomass power, 101–4, 109
　for clean energy, 309
　climate solutions through, 317
　for cover crops and no-till practices, 220, 299, 310
　for crop insurance, 302–3
　for livestock farmers, 302
　for methane-reducing rice growing techniques, 310
　for regenerative agriculture, 59, 214–15, 221, 228
Subway, 179
sugarcane ethanol, 342n64
sugarcane farms, 37
Sustainable Agriculture Research and Education (SARE), 219
sustainable intensification, 9
Sweet Earth, 159
switchgrass ethanol, 58, 73, 91, 342n64
Szymanek, Brigitte, 34, 39, 41, 47

Taylor, Kat, 216–18
technological innovations
　Apeel food coating, 131
　climate change solutions, 17–18
　electric tractors, 297
　in food industry, 11 (*see also* egg substitutes; meat and dairy alternatives)
　for livestock efficiency, 254–57
　meat alternatives, 133–34
　nitrogen-fixing bacteria, 234
　personal food computers, 142–43
　precision agriculture, 137, 266–68
　tractor innovations, 140, 264–66
　vertical farms, 268–71
　yield increases and, 120–23
　See also meat and dairy alternatives
Temme, Karsten, 235–36

Terraton Initiative (Indigo Ag), 223–24
Terviva, 280–87
Tetrick, Josh
  cultivated meat, 181–83, 192, 208
  egg substitute company of, 153–57, 174–76, 182, 196–97
  Just Mayo lawsuit and, 163–64
  market space for meat alternatives, 160
  on popularity of meat alternatives, 193
Thank You for Smoking (film), 190
Thiel, Peter, 156, 158
30x30 goal (UN), 312
Thunberg, Greta, 3–4, 11
Timberlake, Justin, 269
Tolles, Chris, 222
TomKat Ranch, California, 216–18, 222, 225, 349n243
tractors, 264–66, 297
trash collection, 329
Trillion Trees, 315
triple-cropping, 241
TruBeef Organic, 218–19
Trump, Donald
  climate action, opposition to, 113, 297, 326
  coal, promotion of, 296
  COVID-19 pandemic and, 324–25
  farmworkers and, 284
  increased fossil fuel use and, 1, 2
  trade wars of, 307
Tufts University, 53, 308
Turmes, Claude, 83–84
Tyson, 171–72, 180–81, 218–19, 220

ultra-processed foods, 194–95, 200
Unilever, 246–47
United Kingdom, 100, 109, 131
United Nations
  Champions of the Earth award, 179
  Development Programme, 143–44
  Environment Programme, 143
  REDD+ forest conservation funding, 311, 314–15
  on regenerative agriculture, 219
  30x30 goal, 312
  See also Food and Agriculture Organization; Intergovernmental Panel on Climate Change
University of California, Berkeley, 49, 51–52, 341–42n54
University of Illinois, 276, 308
University of Wisconsin, 272
Upside Foods, 170–71, 192, 196, 206

USDA (United States Department of Agriculture)
  corn ethanol and, 22
  on crop insurance subsidies, 303
  declines in research budget, 307
  mandating milk in schools, 302
  meat and dairy alternatives, investments in, 308–9
  regenerative agriculture and, 220, 224
  rice fields, methane emissions reductions and, 310

Valeti, Uma, 169–71, 173, 192, 206–7
Valmont, 267
Vance, J. D., 269, 304
Van Den Broek, Karen, 240, 242–43
Van Den Broek, Mario, 239–40, 242–43
Van Eenennaam, Alison, 255–58
Van Puijenbroek, Paul, 252–53
Vavrus, Steve, 8
vegan and vegetarian diets
  in hospitals, 332
  impact of, 132–34, 328
  impracticality of, 18, 132, 243
  individual responsibility and, 327, 328
  lack of interest in meat and dairy alternatives, 194
  to protect animals, 147–51, 293
  See also meat and dairy alternatives
veggie burgers, 152
Verra, 223
vertical farms, 268–71
Vilsack, Tom, 88, 215, 220, 224, 298–300

Wahlberg, Mark, 238
Waite, Richard, 125, 319
Walton Foundation, 301
Wang, Michael, 24, 54, 63–64, 75–78
wastewater systems, 140
water shortages and pollution
  agricultural practices contributing to, 6, 9, 42–43
  climate change and droughts, 265–66
  corn ethanol production and, 23
  in Everglades National Park, 37–38
  from fertilizers, 140, 218, 230, 233–34, 348n230
  nitrates, 140
  rice fields and methane emissions, 310
Waxman, Henry, 89
weather changes, 2, 121. *See also* floods
Weingarten, Gene, 149, 152, 176–77, 184
Welch, Jack, 171

INDEX

wetlands
    as carbon sinks, 129, 331
    rewetting farmed peatlands, 311
    Searchinger advocating for, 37–39, 41–43, 44–45, 330–31
Whitaker, Jim, 310
Who Killed the Electric Car? (documentary), 58
Whole Foods, 163
Wilcove, David, 39–40, 117
wildfires, 17, 333
wildlife
    biodiversity and habitat loss, 5–6, 55, 69–71, 313
    mass extinction and, 5
    See also deforestation
Wildtype, 207
Willey, Zach, 60
Williams, Serena, 179
wind power, 3, 297, 309
wood-burning power plants. *See* biomass power
Woods Hole Oceanographic Institution, 63
World Bank, 143, 213, 311
World Business Council for Sustainable Development, 322
World Economic Forum, 303
World Resources Institute (WRI)
    agricultural modeling of, 125–26, 128
    carbon-opportunity-cost climate accounting of, 292
    Creating a Sustainable Food Future and, 125
    decarbonizing Denmark's agriculture, 291
    food waste research, 131
    Greenhouse Gas Protocol and, 319–22
    on livestock and land use, 132
    methane studies, 306

Yard Stick (soil carbon measurement), 222, 224
Yellen, Janet, 298–99
Yield10 Bioscience, 276
yields
    climate change and decreases in, 121, 265–66
    corn ethanol and increases in, 80–81
    crop efficiency and, 263–66
    deforestation and, 313–14
    in Denmark, 291–93
    fertilizer use and, 140, 233–38, 301, 307–8
    of fish farms, 261
    gene editing and, 276
    GMOs and, 138, 275–78 (*see also* GMOs)
    greenhouse gas emissions and, 226–27
    Green Revolution and, 9, 120–21, 128–29, 264
    from Kernza, 226, 348n226
    livestock efficiency and, 134–39, 239–50, 254, 349n243
    Miracle Yield theory, 85–86, 121–22, 299
    of organic farms, 141–42, 227
    pongamia trees and, 279–88, 310
    protecting natural land by increasing, 313–14
    regenerative agriculture and, 225–27, 301–2
    rice and methane emissions, 127, 310
    self-fertilizing seed and, 237–38
    of soy vs. corn, 64
    yield elasticity, 80–82, 85–86

Zambia, 141–42
Zarur, Andrey, 273–74

ABOUT THE AUTHOR

MICHAEL GRUNWALD is the bestselling author of two widely acclaimed books, *The Swamp* and *The New New Deal*. He's a former staff writer for *The Washington Post*, *Time*, and *Politico* and winner of the George Polk Award for national reporting, the Worth Bingham Prize for investigative reporting, and many other journalism prizes. He lives in Miami.